International Symposium on Exotic Nuclear Structure From Nucleons

(ENSFN 2012)

Journal of Physics: Conference Series Volume 445

Tokyo, Japan
10-12 October 2012

ISBN: 978-1-62748-677-4
ISSN: 1742-6588

Printed from e-media with permission by:

Curran Associates, Inc.
57 Morehouse Lane
Red Hook, NY 12571

Some format issues inherent in the e-media version may also appear in this print version.

Printed by Curran Associates, Inc. (2013)

Institute of Physics
Dirac House, Temple Back
Bristol BS1 6BE UK

Phone: 44 1 17 929 7481
Fax: 44 1 17 920 0979

techtracking@iop.org

Additional copies of this publication are available from:

Curran Associates, Inc.
57 Morehouse Lane
Red Hook, NY 12571 USA
Phone: 845-758-0400
Fax: 845-758-2634
Email: curran@proceedings.com
Web: www.proceedings.com

Preface

The International Symposium on "Exotic Nuclear Structure From Nucleons (ENSFN2012)" was held at the Koshiba Hall, the University of Tokyo, Japan, from October 10th to 12th, 2012. This symposium was supported by RIKEN Nishina Center (RNC) and the Center for Nuclear Study (CNS), University of Tokyo.

This symposium was devoted to discussing recent achievement and perspectives in the structure of exotic nuclei from the viewpoint of the nuclear force.

The following subjects were covered in this symposium from both theoretical and experimental sides:

- Evolution of shell structure and collectivity in exotic nuclei
- Ab-initio theory and its application to exotic nuclei
- Advancement in large-scale nuclear-structure calculations
- Effective Hamiltonian and energy density functional
- Spin-isospin responses
- New aspects of two- and three-body forces
- Impact on nuclear astrophysics

Emphasis was placed on the development of large-scale nuclear-structure calculations and the new experimental information on exotic nuclei.

Around 80 participants attended this symposium and we enjoyed 37 excellent invited talks and 9 selected oral presentations. A special talk was presented to celebrate the 60th birthday of professor Takaharu Otsuka, who has made invaluable contribution to the progress in the fields covered in this symposium.

The organizing committee consisted of T. Abe (Tokyo), M. Honma (Aizu; chair), N. Itagaki (YITP, Kyoto), T. Mizusaki (Senshu), T. Nakatsukasa (RIKEN), H. Sakurai (Tokyo/RIKEN), N. Shimizu (CNS, Tokyo; scientific secretary), S. Shimoura (CNS, Tokyo), Y. Utsuno (JAEA/CNS, Tokyo; scientific secretary).

Finally, we would like to thank all the speakers and the participants as well as the other organizers for their contributions which made the symposium very successful.

Michio Honma, Yutaka Utsuno and Noritaka Shimizu
Editors, Tokyo, April 2012

International Symposium on
Exotic Nuclear Structure From Nucleons
ENSFN
2012
October 10-12, 2012
Koshiba Hall, University of Tokyo
Topics
Evolution of shell structure and collectivity in exotic nuclei
Ab-initio theory and its application to exotic nuclei
Advancement in large-scale nuclear-structure calculations
Effective Hamiltonian and energy density functional
Spin-isospin responses
New aspects of two- and three-body forces
Impact on nuclear astrophysics
Organizing Committee
Takashi Abe (Tokyo)
Michio Honma (Aizu; chair)
Naoyuki Itagaki (YITP, Kyoto)
Takahiro Mizusaki (Senshu)
Takashi Nakatsukasa (RIKEN)
Hiroyoshi Sakurai (Tokyo/RIKEN)
Noritaka Shimizu (CNS, Tokyo; scientific secretary)
Susumu Shimoura (CNS, Tokyo)
Yutaka Utsuno (JAEA; scientific secretary)
Website
https://sites.google.com/a/cns.s.u-tokyo.ac.jp/ensfn/
email: ensfn@cns.s.u-tokyo.ac.jp
Sponsored by
RIKEN Nishina Center,
Center for Nuclear Study, University of Tokyo
RIKEN NISHINA CENTER
CNS

Journal of Physics: Conference Series **445** (2013) 011001 doi:10.1088/1742-6596/445/1/011001

ENSFN2012 Scientific Program

October 10

9:00-9:30 Registration

Chairperson: W. Nazarewicz (Univ. Tennessee/ORNL)

9:30-9:40 M. Honma (Univ. Aizu)
Opening address

9:40-10:05 M. Hjorth-Jensen (Univ. Oslo)
Evolution of shell structure in neutron-rich isotopes and the stability of nuclear matter

10:05-10:30 I. Tanihata (RCNP, Osaka. Univ.)
An evidence of the tensor force in nuclear configuration

10:30-10:55 K. Sekiguchi (Tohoku Univ.)
Experimental approach to three nucleon forces via few nucleon systems

Break (10:55-11:25)

Chairperson: N. Itagaki (YITP, Kyoto Univ.)

11:25-11:50 B. R. Barrett (Univ. Arizona)
Microscopic calculations of nuclear structure beyond the 0p-shell

11:50-12:15 K. Takayanagi (Sophia Univ.)
Effective Hamiltonian in non-degenerate model space

12:15-12:40 N. Shimizu (CNS, Univ. Tokyo)
History and future perspectives of the Monte Carlo shell model method

12:40-12:55 L. Liu (Peking Univ.)
No-core Monte Carlo shell-model calculation for ^{10}Be and ^{12}Be low-lying spectra

Lunch (12:55-14:10)

Chairperson: H. Sakurai (Univ. Tokyo/RNC, RIKEN)

14:10-14:35 A. Gade (MSU)
Nuclear spectroscopy with fast exotic beams

14:35-15:00 G. Neyens (KU Leuven)
Ground state moments of neutron-rich K and Al isotope as probes for changes in shell structure

15:00-15:25 Y. Utsuno (JAEA)
Towards unified description of the shell evolution

Break (15:25-15:55)

Chairperson: P. Van Isacker (GANIL)

15:55-16:20 N. Aoi (RCNP, Osaka Univ.)
Study of shell evolution with in-beam gamma-ray spectroscopy

16:20-16:45 T. Duguet (CEA Saclay)
How shall we talk about the shell structure?

Reception (17:30-19:30) at "GOTEN", Sanjo Conference Hall B1

October 11

Chairperson: R. V. F. Janssens (ANL)

9:00-9:25 A. Schwenk (TU Darmstadt)
Three-nucleon forces and nuclei at the extremes

9:25-9:50 B. A. Brown (MSU)
Pairing and shell gaps

9:50-10:15 H. Nakada (Chiba Univ.)
Mean-field and RPA approaches to stable and unstable nuclei with semi-realistic NN interactions

10:15-10:30 D. Steppenbeck (CNS, U. Tokyo)
Investigating the strength of the N=34 subshell closure in ^{54}Ca

Break (10:30-11:00)

Chairperson: B. A. Brown (MSU)

11:00-11:25 P. Van Isacker (GANIL)
Shell-model derivation of the shears mechanism

11:25-11:50 N. Tajima (Fukui Univ.)
Angular correlation between proton and neutron rotors

11:50-12:15 K. Nomura (Univ. Koeln)
Interacting boson model with energy density functionals

12:15-12:30 S. Zhu (ANL)
First β-decay study with CARIBU and Gammasphere: $^{142,144}Cs \rightarrow {}^{142,144}Ba$

Lunch (12:30-13:30)

Chairperson: T. Nakatsukasa (RNC, RIKEN)

13:30-13:55 W. Nazarewicz (Univ. Tennessee/ORNL)
Microscopic description of the fission process

13:55-14:20 J. Meng (Peking Univ.)
Exotic nuclear structure and origin of the heavy nuclei with covariant density functional theory

14:20-14:45 Y. Iwata (GSI)
Time-dependent density functional calculations for fusion and fission

14:45-15:10 T. Uesaka (RNC, RIKEN)
Spin-isospin responses studies with RI-beams

Break (15:10-15:30)

Chairperson: J. Meng (Peking Univ.)

15:30-15:55 S. Shimoura (CNS, Univ. Tokyo)
Physics opportunities using double-charge exchange reactions

15:55-16:20 M. Matsuo (Niigata Univ.)
Continuum dipole response near the threshold and the direct neutron capture cross section at astrophysical energies

16:20-16:35 M.-K. Cheoun (Soongsil Univ.)
High-lying Gamow-Teller excited states in the deformed nuclei, ^{76}Ge, ^{82}Se *and* $N = 20$ *nuclei in the island of inversion by the deformed quasi-particle RPA*

16:35-16:50 S. Ebata (CNS, Univ. Tokyo)
Systematic investigation of E1 strength for the isotopes from $Z = 28$ *to 50*

Break (16:50-17:10)

Chairperson: T. Motobayashi (RNC, RIKEN)

17:10-17:35 A. B. Balantekin (Univ. Wisconsin-Madison)
Neutrinos and rare isotopes

17:35-18:00 G. J. Mathews (Univ. Notre Dame)
Updated nuclear equation of state for core-collapse supernovae and neutron stars: Effects of 3-body forces, QCD, and magnetic fields

18:00-18:25 T. Kajino (NAO)
Nuclear weak interactions, supernova nucleosynthesis and neutrino oscillation

18:25-18:40 M. Famiano (W Mich. Univ.)
Nuclear structure inputs to the production of heavy nuclei in accretion-induced black hole formation

Journal of Physics: Conference Series **445** (2013) 011001 doi:10.1088/1742-6596/445/1/011001

October 12

Chairperson: A. Gade (MSU)

9:00-9:25	R. V. F. Janssens (ANL) *Changes in shell structure in A ~ 60 neutron-rich nuclei as a function of spin*
9:25-9:50	S. M. Lenzi (Univ. Padova) *Deformation and shape coexistence in neutron-rich nuclei at N=40*
9:50-10:05	K. Yoshida (Niigata Univ.) *Skyrme energy-density functional approach to collective modes of excitation in exotic nuclei*
10:05-10:20	Y. Tsunoda (Univ. Tokyo) *Study of medium-mass nuclei by large-scale shell model calculations*

Break (10:20-10:50)

Chairperson: Y. Akaishi (KEK)

10:50-11:15	D. J. Dean (ORNL) *Pairing reentrance phenomenon in heated rotating nuclei in the shell-model Monte Carlo approach*
11:15-11:40	N. Pietralla (TU Darmstadt) *Exotic ideas on mixed proton-neutron symmetry and their manifestation in heavy nuclei*
11:40-12:05	S. Pittel (Univ. Delaware) *Spherical to deformed shape transitions in the nucleon pair shell model*
12:05-12:30	N. Yoshinaga (Saitama Univ.) *Band structure of neutron rich Se and Ge isotopes*

Conference Photo & Lunch (12:30-13:40)

Chairperson: H. Horiuchi (RCNP, Osaka Univ.)

13:40-14:05	T. Nakamura (Tokyo Inst. Tech.) *Deformed halo nuclei probed by breakup reactions*
14:05-14:30	R. Kanungo (St. Mary's Univ.) *Soft dipole resonance and nucleon distribution in halo nuclei*
14:30-14:55	P. Maris (Iowa St. Univ.) *Emergence of rotational bands in Be isotopes from ab initio calculations*

Break (14:55-15:25)

Chairperson: S. Shimoura (CNS, Univ. Tokyo)

15:25-15:50	J. Maruhn (Univ. Frankfurt) *Exotic cluster structures in the mean-field theory*
15:50-16:15	Y. Kanada-En'yo (Kyoto Univ.) *Structures of ground and excited states in C isotopes*
16:15-16:30	T. Yoshida (CNS, Univ. Tokyo) *Density profiles in the intrinsic frame of light nuclei obtained from Monte Carlo shell model calculation*
16:30-16:40	T. Otsuka (Univ. Tokyo) *Closing*

Experimental approach to three nucleon forces via few nucleon systems

K Sekiguchi

Aoba 6-3, Aramaki, Aoba-ku, Sendai, 980-8578, JAPAN

E-mail: kimiko@lambda.phys.tohoku.ac.jp

Abstract. Recent progress on three nucleon force study with three nucleon scattering at at intermediate energies ($E \gtrsim 100$ MeV/nucleon) are presented, especially focusing on the experimental work on deuteron–proton elastic scattering at RIKEN.

1. Introduction

A hot topic of present day few-nucleon system studies is to explore the properties of three-nucleon forces (3NFs) that appear when more than two nucleons ($A \geq 3$) interact. 3NFs arise naturally in the standard meson exchange picture in which the main ingredient is considered to be a 2π–exchange between three nucleons along with the Δ–isobar excitation, initially proposed by Fujita and Miyazawa a half century ago [1]. Further augmentations have led to the Tucson–Melbourne (TM) [2], the Urbana IX 3NF [3], *etc.* New impetus to study 3NFs has come from chiral effective field theory (χEFT) descriptions of nuclear interactions. In that framework consistent two-, three-, and many-nucleon forces are derived on the same footing [4, 5]. The first non-zero contribution to 3NFs appears in χEFT at the next-to-next-to-leading order (N^{2}LO) of the chiral expansion. That explains why 3NFs are relatively small compared to NN forces (2NFs) and why their effects are easily masked. Therefore, it is, in general, hard to find evidence for them.

The first evidence for a 3NF was found in the three-nucleon bound states, ^{3}H and ^{3}He [6, 7]. The binding energies of these nuclei are not reproduced by exact solutions of three-nucleon Faddeev equations employing modern NN forces only, *i.e.* AV18 [8], CD Bonn [9], Nijmegen I, II and 93 [10]. The underbinding of ^{3}H and ^{3}He can be explained by adding a 3NF, mostly based on 2π-exchange, acting between three nucleons [6, 7, 11]. The importance of 3NFs has been further supported by the binding energies of light mass nuclei and by the empirical saturation point of symmetric nuclear matter. *Ab initio* microscopic calculations of light mass nuclei, such as Green's Function Monte Carlo [12] and no-core shell model calculations [13], highlight the necessity of including 3NFs to explain the binding energies and low-lying levels of these nuclei. As for the density of symmetric nuclear matter, it has been reported that all NN potentials provide saturation at too high a density, and a short-range repulsive 3NF is one possibility to shift the theoretical results to the empirical point [14].

Three nucleon (3N) reactions have been studied for a long time as one of the most promising tools to explore the properties of 3NFs, because this system provides a rich set of energy dependent spin observables and differential cross sections. At lower energies ($E/A \leq 20$ MeV),

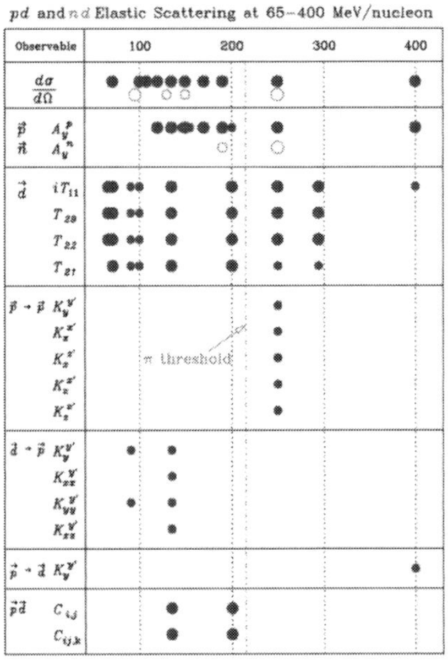

Figure 1. Compilations of recent experiments of *pd* and *nd* elastic scattering at 70–400 MeV/nucleon. Solid circles denote *pd* experiments and open circles denote *nd* experiments. The measurements with large circles cover the wide angular range, while those with small circles cover the limited angular range.

very high precision measurements were carried out in proton–deuteron (*pd*) and neutron–deuteron (*nd*) scattering, and in breakup reactions. However, theoretically predicted 3NF effects are rather small and a generally good description for nucleon–deuteron (*Nd*) elastic scattering data is obtained by exact solutions of 3N Faddeev equations employing only NN forces [15, 16].

Study of the 3NF has changed since the end of 1990's. The following advances have made it possible to explore the 3NF effects contained in 3N scattering. **(i)** Generation of the so-called realistic NN forces (*e.g.*, AV18 [8], CDBonn [9], Nijmegen I, II and 93 [10]) which reproduce a rich set of experimental NN data for laboratory energy up to 350 MeV with an accuracy of $\chi^2 \sim 1$. **(ii)** Achievement of rigorous numerical Faddeev calculations based on the realistic NN potentials below the π-threshold energy (the incident nucleon energy $E/A \leq 215$ MeV) [15] [1]. **(iii)** Development of experimental techniques to obtain precision data for 3N scattering

In the last decade the experimental studies of intermediate-energy *pd* and *nd* elastic scattering have been extensively performed by groups at RIKEN, KVI, RCNP, and IUCF providing precision data for cross sections and a variety of spin observables [17, 18, 19, 20, 21, 22, 23, 24, 25, 26]. This is partly due to the fact that the first indication of 3NF was pointed out [27, 28] in the elastic channel. Compilations of recent experiments for *pd* and *nd* elastic scattering at intermediate energies are shown in Fig.1. It should be noted that the experimental study of *dp* scattering have been recently extended at the new facility of RIKEN RI beam factory (RIBF) [29, 30] where polarized deuteron beams are available up to $\sim$ 400 MeV/nucleon.

In Section 2 the experiments performed at RIKEN RIBF is described briefly. The experimental results for elastic *dp* scattering are compared with the theoretical predictions in

[1] As for the χEFT potential calculations have been performed up to the NNLO for the 3N scattering system, providing reasonable agreement with experimental data at a laboratory energy of 100 MeV/nucleon [5]. Since here we discuss on the experimental results at $E \gtrsim 100$ MeV/nucleon the calculations based on the χEFT potential are not shown.

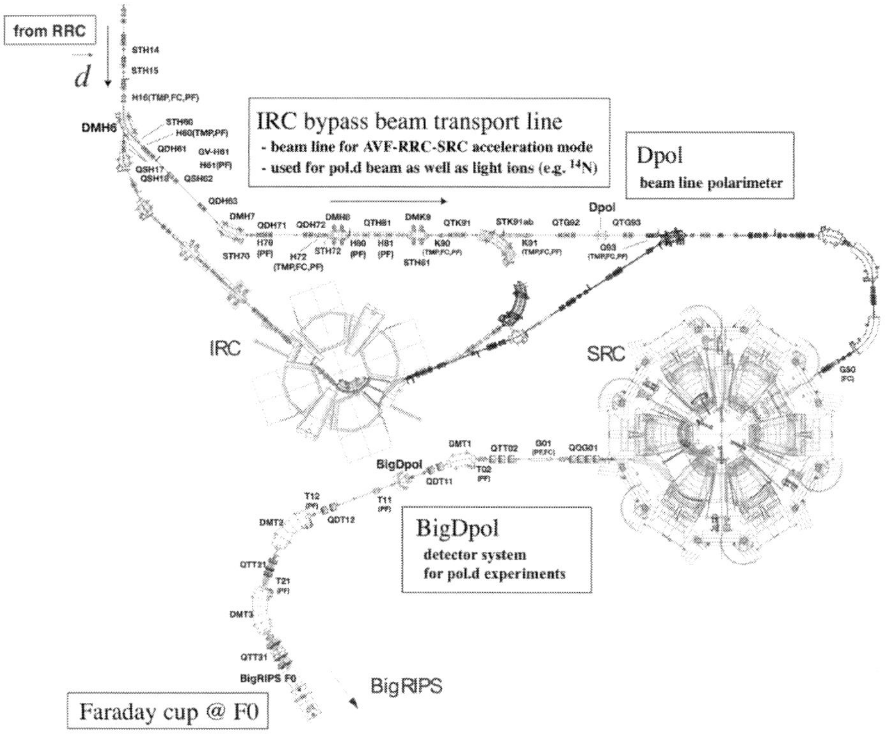

Figure 2. Schematic view of the experimental setup for *dp* elastic scattering with polarized deuteron beams at RIKEN RIBF.

Section 3. Summary follows in Section 4.

The recent achievements of study of 3NFs in intermediate–energy *Nd* scattering are discussed.

2. Experiments with polarized deuteron beams at RIBF

The schematic view of the experimental setup is shown in Fig. 2. At RIBF the vector and tensor polarized deuteron beams were accelerated at first by the injector cyclotrons AVF and RRC up to 90 (100) MeV/nucleon, and then up to 250 (294) MeV/ nucleon by the new superconducting cyclotron SRC. The polarization axis of the deuteron beam was rotated with a Wien filter system prior to acceleration of the beams. Single turn extraction of the beam was required and achieved all for the three cyclotrons, AVF, RRC and SRC, in order to maintain the polarization amplitudes during acceleration. In the measurement typical values of the beam polarizations were 80% of the theoretical maximum values which were monitored continuously with a beam line polarimeter Dpol prior to acceleration by the SRC using the reaction of elastic *dp* scattering at 90 (100) MeV/nucleon.

The measurement for elastic *dp* scattering was performed with the detector system BigDpol which was installed at the extraction beam line of the SRC. A polyethylene (CH_2) target with a thickness of 330 mg/cm^2 was used as a hydrogen target. In the BigDpol four pairs of plastic scintillators coupled with photo-multiplier tubes were mounted in two independent planes, 90° apart in azimuthal angle and operated in kinematical coincidence of elastic *dp* scattering. The measured angles in the center of mass system are $\theta_{\rm c.m.}$ = 40°–162°. In the experiment the deuteron beams were stopped in the Faraday cup which was installed at the focal plane F0 of the BigRIPS spectrometer.

3. Experimental Results and Theoretical Predictions

In Fig. 3 some representative experimental results for *pd* and *nd* elastic scattering are compared with the Faddeev calculations with and w/o 3NFs. The red (blue) bands are the calculations with (without) TM′99 3NF [31], which is a version of the Tucson-Melbourne 3NF more consistent with chiral symmetry [32, 33], based on the modern NN potentials, *i.e.* CD Bonn, AV18, Nijmegen I and II. The solid lines are the calculations based on the AV18 potential with including the Urbana IX 3NF.

For the cross section specific features are seen depending on the scattering angles in the center of mass system $\theta_{\rm c.m.}$. **(i)** At forward angles $\theta_{\rm c.m.} \lesssim 80^\circ$, where the direct processes by the NN interactions are dominant, the theoretical calculations based on the various NN potentials are well converged and the predicted 3NF effects are very small. The experimental data are well described by the calculations except for the very forward angles. This discrepancy comes from that fact that the calculations shown in the figure do not take into account the Coulomb interactions between protons [34]. **(ii)** At middle angles $\theta_{\rm c.m.} \sim 80^\circ$–$140^\circ$, where the cross sections take minimum, the clear discrepancies between the data and the calculations based on the NN potentials are found. They become larger as an incident energy increases. The discrepancies are explained by taking into account the 2π exchange type 3NF models (TM′99, and Urbana IX). **(iii)** At backward angles $\theta_{\rm c.m.} \gtrsim 140^\circ$, where the exchange processes by the NN interactions are dominant, the differences begin to appear between the experimental data and the calculations even including the 3NF potentials with increasing an incident energy. Since this feature are clearly seen at higher energies the relativistic effects have been estimated by using the Lorentz boosted NN potentials with the TM′99 [35, 36, 37]. However the relativistic effects have turned out to be small and only slightly alter the cross sections (see Fig. 4).

As for the polarization observables the energy dependence of the predicted 3NF effects and the difference between the theory and the data is not always similar to that of the cross section. The deuteron vector analyzing power iT_{11} has features similar to those of the cross section. Meanwhile the tensor analyzing power T_{22} reveals different energy dependence from that of iT_{11}. Starting from $\sim$ 100 MeV/nucleon large 3NF effects are predicted. At 135 MeV/nucleon and below adding 3NFs worsens the description of data in a large angular region. It is contrary to what happens at the highest energies above 250 MeV/nucleon, where large 3NF effects are supported by the T_{22} data. The relativistic effects are estimated to be small also for these polarization observables for *Nd* elastic scattering (see Fig. 4).

The results obtained for *Nd* elastic scattering indicate that some significant components are missing in the calculations, especially in the regions of higher momentum transfer.

4. Summary

3NFs are now accepted as key elements in understanding various nuclear phenomena, such as the binding of light mass nuclei and the empirical saturation point of nuclear matter density. The *Nd* scattering data provide rich sources with which to explore the properties of 3NFs such as momentum and spin dependence. In this talk the recent achievements of study of 3NFs in intermediate–energy *Nd* scattering are discussed.

In the last decade extensive experimental study of *pd* and *nd* elastic scattering at intermediate energies ($E \gtrsim 100$ MeV) were performed at several facilities. As for the *pd*/*nd* elastic scattering the energy and angular dependent results of the cross sections as well as the polarization observables show that the experimental data are generally described by the calculations with the 2π-exchange 3NFs at the angles $\theta_{\rm c.m.} \lesssim 140^\circ$. However the serious discrepancies between the data and the calculations appear at the backward angles and increase with the incident energy. The discrepancies are not remedied even by including the 2π-exchange 3NFs nor taking into account the relativity. This feature indicates that some significant components are missing in the calculations in the higher momentum transfer region.

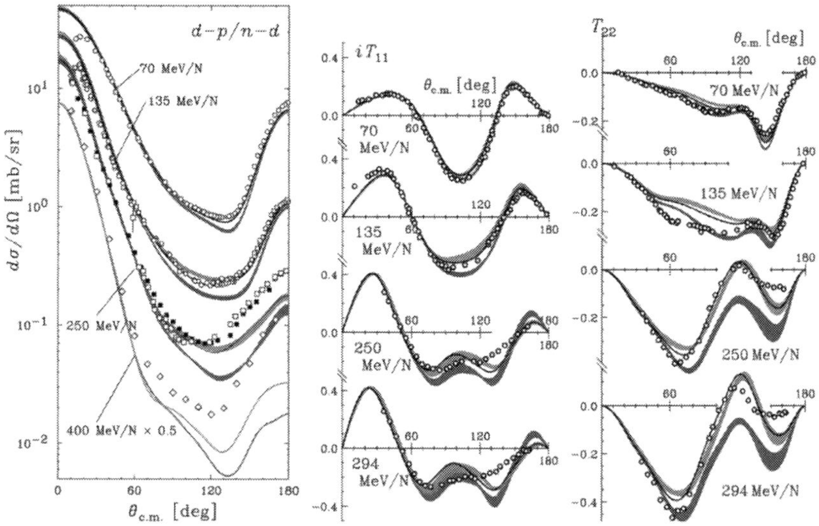

Figure 3. Differential cross sections and deuteron analyzing powers iT_{11}, T_{22} for elastic Nd scattering at 70–400 MeV/nucleon (MeV/N). The red (blue) bands are the calculations with (w/o) TM99 3NF based on the modern NN potentials, namely CD Bonn, AV18, Nijmegen I and II. The solid lines are the calculations with including Urbana IX 3NF based on AV18 potential. For the cross sections the open circles are the data in Refs. [17, 18, 20]. The solid squares and open circles are the *pd* [21] and *nd* [25] data at 250 MeV/nucleon, respectively. The open diamonds show the data at 400 MeV/nucleon [26]. For the deuteron analyzing powers the data at 70 and 135 MeV/nucleon are from Refs. [17, 18, 19]. The data at 250 and 294 MeV/nucleon are taken at the RIBF [29, 30].

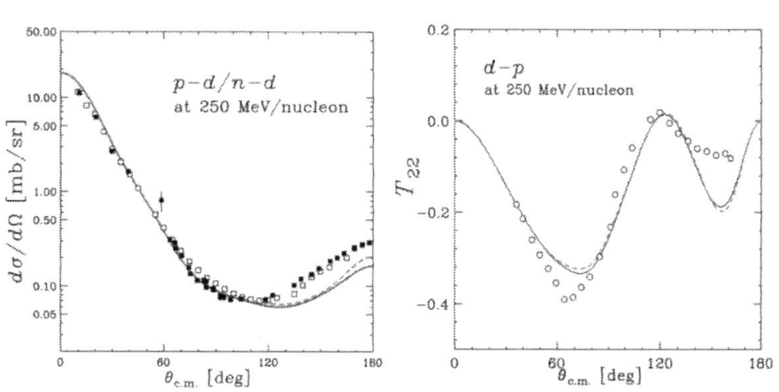

Figure 4. Differential cross section and the tensor analyzing power T_{22} for Nd elastic scattering at 250 MeV/nucleon. Faddeev calculations based on the CD Bonn potential with the TM′99 3NF are shown with the blue solid lines. The calculations based on the Lorentz boosted NN potential with the 3NF are shown with the red dashed lines. For description of symbols see Fig.3.

As the next step of 3NF study in the few nucleon scattering it would be interesting to see how well the theoretical approaches, *e.g.* inclusion of 3NFs other than 2π–exchange types, and the potentials based on chiral effective field theory, describe these data. Experimentally, it is interesting to measure spin correlation coefficients as well as polarization transfer coefficients for elastic *pd* scattering at higher energies 200–400 MeV/nucleon. Various kinematic configurations of the exclusive *pd* breakup reactions should also be measured in order to study the properties of 3NFs as well as relativistic effects. As a first step from few to many body systems it is interesting to extend the measurements to 4N scattering systems which would provide a valuable source of information on 3NFs.

References

[1] Fujita J and Miyazawa H 1957 *Prog. Theor. Phys.* **17** 360
[2] Coon S A and Glöckle W 1981 *Phys. Rev.* C **23** 1790
[3] Pudliner B S *et al.* 1997 *Phys. Rev.* C **56** 1720
[4] van Kolck U 1994 *Phys. Rev.* C **49** 2932
[5] Epelbaum E, Hammer H W and Meißner U G 2009 *Rev. Mod. Phys.* **81** 1773
[6] Chen C. R. *et al.* 1986 *Phys. Rev.* C **33** 1740
[7] Sasakawa T and Ishikawa S 1986 *Few-Body Syst.* **1** 3
[8] Wiringa R B, Stoks V G J and Schiavilla R 1995 *Phys. Rev.* C **51** 38
[9] Machleidt R 2001 *Phys. Rev.* C **63** 024001
[10] Stoks V G J *et al.* 1994 *Phys. Rev.* C **49** 2950
[11] Nogga A *et al.* 2002 *Phys. Rev.* C **65** 054003
[12] Pieper S C *et al.* 2001 Phys. Rev. C **66** 044310
[13] Navrátil P and Ormand W E 2003 Phys. Rev. C **68** 034305
[14] see for example, Akmal A *et al*, 1998 *Phys. Rev.* C **58** 1804
[15] Glöckle W, Witała H, Hüber D, Kamada H and Golak J 1996 *Phys. Rep.* **274** 107
[16] Kievsky A, Viviani M and Rosati S 2001 *Phys. Rev.* C **64** 024002
[17] Sakai H. *et al.* 2000 *Phys. Rev. Lett.* **84** 5288
[18] Sekiguchi K *et al.* 2002 *Phys. Rev.* C **65** 034003
[19] Sekiguchi K *et al.* 2004 *Phys. Rev.* C **70** 014001
[20] Sekiguchi K *et al.* 2005 *Phys. Rev. Lett.* **95** 162301
[21] Hatanaka K *et al.* 2002 *Phys. Rev.* C **66** 044002
[22] Bieber R *et al.* 2000 *Phys. Rev. Lett.* **84** 606; Ermisch K *et al.* 2001 *Phys. Rev. Lett.* **86** 5862; Ermisch K *et al.* 2003 *Phys. Rev.* C **68** 051001; Ermisch K *et al.* 2005 *Phys. Rev.* C **71** 064004; Amir-Ahmadi H R *et al.* 2007 *Phys. Rev.* C **75** 041001; Mardanpour H *et al.* 2007 *Eur. Phys. J.* A **31** 383
[23] Stephenson E J *et al.* 1999 *Phys. Rev.* C **60** 061001; Cadman R V *et al.* 2001 *Phys. Rev. Lett.* **86** 967; Przewoski B v *et al.* 2006 *Phys. Rev.* C **74** 064003
[24] Mermod P *et al.* 2005 *Phys. Rev.* C **72** 061002
[25] Maeda Y *et al.* 2007 *Phys. Rev.* C **76** 014004
[26] Tamii A *et al.* 2007 *AIP Conference Proceedings* **915** 765
[27] 1998 Witała H *et al. Phys. Rev. Lett.* **81** 1183
[28] Nemoto S *et al.* 1998 *Phys. Rev.* C **58** 2599
[29] Sekiguchi K *et al.* 2011 *Phys. Rev.* C **83** 061001
[30] Sekiguchi K *et al.* 2012 *paper in preparation*
[31] Coon S A and Han H K 2001 *Few Body Syst.* **30** 131
[32] Friar J L *et al.* 1999 *Phys. Rev.* C **59** 53.
[33] Hüber D *et al.* 2001 *Few-Body Syst.* **30** 95
[34] Deltuva A *et al.* 2005 *Phys. Rev.* C **71** 054005
[35] Witała H *et al.* 2005 *Phys. Rev.* C **71** 054001
[36] Witała H *et al.* 2008 *Phys. Rev.* C **77** 034004
[37] Witała H *private communications*
[38] Witała H *et al. Phys. Rev.* C **83** 044001

International Symposium on Exotic Nuclear Structure From Nucleons (ENSFN 2012) IOP Publishing
Journal of Physics: Conference Series **445** (2013) 012002 doi:10.1088/1742-6596/445/1/012002

Microscopic calculations of nuclear structure beyond the 0p-shell

B R Barrett

Department of Physics, PO Box 210081, University of Arizona, Tucson, AZ, 85721-0081 USA

E-mail: bbarrett@physics.arizona.edu

Abstract. Significant progress has been made in our microscopic understanding of the properties of light nuclei, starting from the basic nucleon-nucleon and three-nucleon interactions among the particles inside the nucleus. The main challenge now is to extend these successes for light nuclei ($A \leq 16$) to heavier mass nuclei. Here we discuss three recent approaches within the No Core Shell Model for possibly going beyond the $0p$-shell.

1. Introduction

Over the last twenty years significant advances have been made in our ability to perform *ab initio* calculations for describing the properties of atomic nuclei. These advances have come in the form of 1.) a better understanding of how to construct consistent nucleon-nucleon (NN), three-nucleon (NNN) and higher-nucleon interactions based on QCD, using Effective Field Theory (EFT) and Chiral Perturbation Theory (CPT) [1, 2]; 2.) the development of new and/or improved nuclear many-body techniques, such as the Green Function Monte Carlo (GFMC) approach [3], the No Core Shell Model (NCSM) formalism [4], the Coupled Cluster (CC) method [5], *etc.*, for performing the many-nucleon calculations; and 3.) tremendous advances in computer hardware and software, which allow for much larger and more complicated numerical calculations to be performed. Although important progress has been made in our understanding of the interactions among nucleons based on QCD, an analytical form for these interactions has not yet been obtained, so that the NN, NNN and higher-N interactions, both phenomenologial and QCD-based, that are currently available for use in numerical calculations, still contain parameters, which must be fitted to experimental data or results based on such data, *e.g.*, the NN phase shifts, *etc.* The QCD based interactions do have the advantage of consistently showing the evolution of higher-nucleon forces, but have continuing problems regarding regulators, ultraviolet cutoffs, power counting, and renormalization group invariance. These are all issues beyond the scope and subject of this presentation, which deals with the many-body techniques that use such interactions as input. In particular, this talk deals with the question of how the NCSM can be modified, extended and/or improved, so as to handle heavier-mass nuclei beyond the $0p$-shell.

2. Understanding nuclei beyond the $0p$-shell

One of the major problems facing most nuclear many-body methods for calculating the properties of atomic nuclei microscopically is the rapid growth of the model space, in which the calculations must be performed so as to obtain converged results, with the increasing number of nucleons in the nucleus, A. So far, this has restricted most techniques to calculations for nuclei within the $0p$-shell.

A great deal of time and effort is presently being devoted to this problem. Some of these new approaches are 1.) the *ab initio* Shell Model with a Core [6, 7], 2.) the Importance Truncation method [8], 3.) the NCSM in an EFT Framework [9, 10, 11], 4.) the Monte Carlo-NCSM [12], 5.) the In-Medium Similarity Renormalization Group (SRG) [13], the Symmetry-adapted NCSM [SP(3,R)] [14], and a number of other techniques. Here we will limit our discussion to topics 1.) to 3.), because of our involvement in these three efforts.

3. Extending the NCSM to Heavier-Mass Nuclei

3.1. The ab initio Shell Model with a Core

The basic idea of this approach is to use the usual NCSM method to calculate microscopically the input for standard Shell Model calculations, *i.e.*, the energy of the inert core, the single-particle (s.p.) energies, and the two-nucleon effective-interaction matrix elements within the major shell of interest [6].

3.1.1. Formalism The formalism for the *ab initio* Shell-Model-with-a-Core approach can be outlined as follows:

(i) Perform a NCSM calculation in a sufficiently large model space for a nucleus with two nucleons outside of a doubly magic closed shell, such as, ^{6}Li or ^{6}He, so as to obtain converged results for the ground state (g.s.) and first few excited states.

(ii) Project the six-nucleon wave functions and energies, obtained in the above NCSM calculation, into the $0\hbar\Omega$ model space, *i.e.*, by using an unitary transformation. This produces what we call the six-body cluster results, that is, six-body matrix elements in the restricted model space, which contain all the six-body correlations from the full (or large) model space. Because we have restricted the energy in the smaller model space to be $0\hbar\Omega$, the four nucleons in the $0s$-shell do not have enough energy to be excited into the $0p$-shell and, thus, essentially form a *frozen* or *inert* ^{4}He core. The remaining two-body matrix elements of the two nucleons in the $0p$-shell now contain all the correlation information of the six-nucleon system.

(iii) Separate these two-body matrix elements into a ^{4}He-core contribution, s.p. energies for the $0p_{1/2}$ and $0p_{3/2}$ configurations and residual two-nucleon effective interactions in the $0p$-shell. Using the core and s.p. energies along with the residual two-nucleon matrix elements in a standard Shell Model calculation for two-particles in the $0p$-shell will reproduce the same results as those for the same configurations in the full NCSM calculation, by construction.

(iv) Calculate, in a similar manner, the seven-body cluster for $A = 7$ nuclei, which will yield the 3N matrix elements.

(v) The A-dependence of these terms can be found by performing six-body-cluster calculations for all $0p$-shell nuclei with $A > 7$.

3.1.2. Results for 0p-shell nuclei As stated above, the $0p$-shell results for $A = 6$ nuclei are exact by construction. The results for $A = 7$ nuclei, using only the six-body-cluster terms, are close to the exact NCSM determinations, suggesting that the three-nucleon contributions are small. As Fig. 1 indicates, these three-nucleon contributions are also small for nuclei with $A > 7$, at least when the CD-Bonn NN potential is used [15]. Other physical quantities can also be computed, besides the binding energies and excitation spectra, such as EM moments and transition strengths [7].

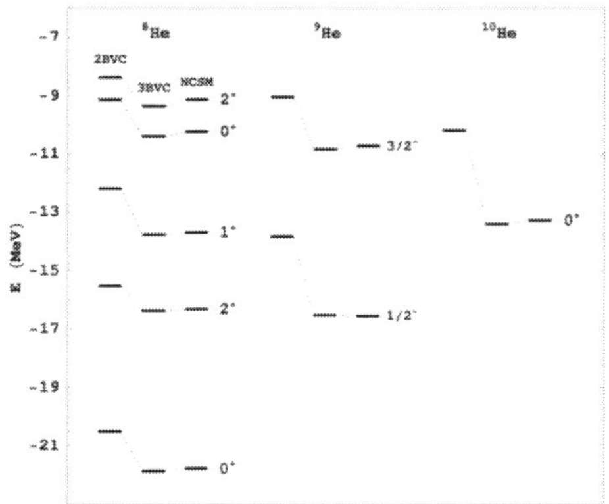

Figure 1. Comparison of spectra for ^{8}He, ^{9}He and ^{10}He from SSM calculations using up to two-body components (2BVC) and up to three-body components (3BVC) with exact NCSM results, all calculated for $N_{max} = 6$ and $\hbar\Omega =$ 20 MeV using the CD-Bonn NN potential [15].

3.2. The Importance Truncated NCSM

The basic idea of the Importance Truncation (IT) approach is to decrease the size of the model space required in a given NCSM calculation by eliminating configurations based on an *importance criterium*. This approach was pioneered in the NCSM by Roth and Navrátil [16] and is referred to as the IT-NCSM. This initial formulation used a multiparticle-multihole (*e.g.*, $2p - 2h$, *etc.*,) approach to generate the basis states coming from the next larger N_{max} space. However, this approach was criticized for possibly having size-extensive issues [17, 18]. Thus, an improved version of the IT-NCSM approach was developed [8, 19], which includes all basis states in the truncation space, but in a sequential manner, as discussed below. The most complete description of this IT-NCSM method is given in [19].

The IT-NCSM procedure is based on multi-configutation perturbation theory (PT), as originally conceived in quantum chemistry [20, 21, 22]. The basic idea centers around a size parameter κ that determines, which many body basis states are kept in a certain N_{max} model space. One wants to calculate the first-order PT contributions to the total wave function, coming from components in the next larger model space, *i.e.*, $N_{max} \Rightarrow N_{max} + 2 \Rightarrow N_{max} + 4 \Rightarrow$ *etc.*, namely,

$$|\psi^{(1)}_{N_{max}+2,\mathrm{IT}}\rangle = \sum_{\nu \in N_{max}+2} \frac{\langle\phi_\nu|W|\Psi_{\mathrm{ref},N_{max}}\rangle}{\epsilon_\nu - \epsilon_{\mathrm{ref,sp}}} |\phi_\nu\rangle \tag{1}$$

In Eq. (1) $|\psi^{(1)}_{N_{max}+2,\mathrm{IT}}\rangle$ denotes the approximate wave function of the full space $N_{max} + 2$ wave function; $|\phi_\nu\rangle$, the $N_{max} + 2$ many-body basis states; $|\Psi_{\mathrm{ref},N_{max}}\rangle$, the previously calculated reference state in the N_{max} space; and W, the perturbation operator. The two terms in the denominator refer to the single-particle energies of the states ϕ_ν and $\Psi_{ref,N_{max}}$, respectively. The energy $\epsilon_{ref,sp}$ is always taken to be the lowest unperturbed energy configuration of the nucleus. For example, in the case of ^{6}Li, this would correspond to taking $\epsilon_{ref,sp} = 2\hbar\Omega$, because there are two valence nucleons in the $N = 1$ shell. We neglect the zero-point motion of the HO, *i.e.*, $(3/2)\hbar\Omega$, since we only require the difference in energy of the s.p. energies. Furthermore, for ^{6}Li $\epsilon_\nu = (6 + 2)\hbar\Omega$ for the basis states in $N_{max} = 4$. It is convenient to define the W as $W = H - H_0$, where H is the initial Hamiltonian and H_0 is that part of H, which connects only many-body basis states that lie in the initial basis space, *i.e.*, from 0 up to N_{max}. Thus, H_0 does not connect basis states from the reference space to the $N_{max} + 2$ space. Consequently, we can rewrite Eq. (1) in the form

$$|\psi^{(1)}_{N_{max}+2,\mathrm{IT}}\rangle = \sum_{\nu \in N_{max}+2} \frac{\langle \phi_\nu | H | \Psi_{\mathrm{ref},N_{max}} \rangle}{\epsilon_\nu - \epsilon_{\mathrm{ref,sp}}} |\phi_\nu\rangle, \tag{2}$$

so that the *importance measure*, κ_ν, of a basis state (ϕ_ν) can be defined as

$$\kappa_\nu = \frac{|\langle \phi_\nu | H | \Psi_{\mathrm{ref},N_{max}} \rangle|}{\epsilon_\nu - \epsilon_{\mathrm{ref,sp}}}. \tag{3}$$

The importance measure κ_ν can now be used to set a threshold limit, on which basis states are included in the truncated $N_{max}+2$ space. If the threshold value for κ_ν is taken to be a few 10^{-5}, say $2*10^{-5}$, then only basis states (ϕ_ν) in the $N_{max}+2$ space with $\kappa_\nu \geq 2*10^{-5}$ are kept. Basis states (ϕ_ν) with κ_ν lower than this threshold will be discarded, starting the truncation of the $N_{max}+2$ space. Once the truncated $N_{max}+2$ basis has been formed, it can be used to diagonalize the Hamiltonian H in this truncated space. This yields a new g.s. wave function, $|\Psi^{(N_{max}+2)}_{\mathrm{ref}}\rangle$, which is then used as the reference state for evaluating all the basis states in the $N_{max}+4$ basis space, using the same method as described above, and so on.

This sequential method has the advantage that it automatically generates all basis states in the evaluated N_{max} space, so it only needs to be done once per basis space. Furthermore, one can generate a sequence of g.s. energies for each truncated N_{max} basis space, which can be easily extrapolated to obtain the g.s. energy at $N_{max} = \infty$.

In practice, the calculations must be done for several values of κ_ν, because it is not possible, by definition, to do the calculations for $\kappa_\nu = 0$. This would be equivalent to performing the NCSM calculation in the full $N_{max}+2$ space. If we could do this, then there would be no need to truncate the $N_{max}+2$ space. Thus, we need results for several values of κ_ν, which will guide us in constructing extrapolation procedures for going to the limit $\kappa_\nu \to 0$.

Our research group is currently investigating refinements of this method, for which preliminary results in the $0p$-shell are yielding encouraging results. One of these refinements is to include one or more excited states as additional reference states. Doing this, we find a much improved rate of convergence in the IT-NCSM process, as shown in Fig. 2 [8].

3.3. The NCSM in an EFT framework

For many years, the standard procedure in nuclear many-body theory has been to start with a *given* free-space NN potential (and NNN potential) and then to use this potential to construct an appropriate effective interaction in a model space for calculating nuclear properties. However, it is also possible to use the philosophy of Effective Field Theory (EFT), *i.e.*, a separation of the relevant physics based on different physical scales, *e.g.*, high-momentum physics versus low-momentum physics, to construct effective interactions in model spaces without going through the intermediate step of first constructing a free-space NN potential (and NNN potential) [9, 10, 11]. Not only would such a procedure be more efficient, but it would also have the possibility of yielding effective interactions that would converge more rapidly in the different many-body techniques, thereby allowing calculations to be performed for heavier-mass nuclei.

3.3.1. Formalism Effective Field Theory (EFT) [1, 2] is based on a separation of scales, such as the nucleon mass or the pion mass, where the *details* of the physics at energies greater than the separation energy (representing the short-distance physics) are irrelevant for understanding the physics below the separation energy (representing the long-distance physics). In EFT the low-energy degrees of freedom are explicitly included, while the high-energy degrees of freedom are integrated out.

One then constructs a potential, in this case in the NCSM model space, which is consistent with the symmetries of the QCD Lagrangian and is expanded as a Taylor series in the momenta or as a contact gradient expansion in coordinate space. Each term in this potential expansion contains a constant. These constants are known as the low-energy constants (LECs). For now, the LECs are determined by fitting to experimental data.

In the case of the NCSM the calculations are performed in an harmonic-oscillator (HO) space for two-nucleons up to some maximum energy, $E_{max} = (N_{max} + \frac{3}{2})\hbar\omega$. Thus, the calculations are carried out in a truncated model space, so the interactions in this truncated space represent the *renormalized effective interactions*. In standard NCSM calculations one needs to repeat the calculations for increasing N_{max} to check that the results are converging in an appropriate manner. In EFT one introduces a cutoff Λ, known as the Ultraviolet (UV) cutoff, which depends on $N_{max}\hbar\omega$. To obtain a correct EFT result, one must achieve Renormalization Group Invariance (RGI). This means that the LECs, which are redetermined as Λ increases, become independent of Λ. In other words, for a high enough Λ, the LECs have absorbed the high-energy physics and the calculated physical observables are approximately independent of the cutoff Λ. The Taylor series expansion of the potential provides a power counting or hierarchy between the different contributions, which yield results that are improvable order by order, *i.e.*, leading order (LO) > next-to-leading order (NLO) > next-to-next-to-leading order (NNLO) > *etc.*

3.3.2. Results Our first investigation was for an EFT without pions, *i.e.*, a separation-scale energy of $\approx$ 100 MeV, which is known as the pionless EFT. In this case the Hamiltonian in the NCSM model space contains three potential terms consistent with the QCD symmetries, namely the NN potentials in the triple S and single S channels and a NNN potential in the $S = \frac{1}{2}$ channel. One then uses this Hamiltonian to compute the three LECs associated, respectively, with the three potential terms above. That is, the values of the LECs are adjusted until they reproduce, in our case, the binding energies of the deuteron, the triton and the α particle. Once these constants are determined, the Hamiltonian in the NCSM model space is completely known and can be used to compute other nuclear properties, such as the energy of the first-excited 0^+ state of the α particle or the binding energy of ^{6}Li. The determination of the Hamiltonian must be done for increasing values of Λ to achieve RGI. Because the calculations are performed using HO wave functions, *i.e.*, in an HO potential well, one must also take the limit as $\omega \to 0$ (the Infrared (IR) cutoff), since the nucleus is a self-bound system. Thus, the LECs are functions of both Λ and ω.

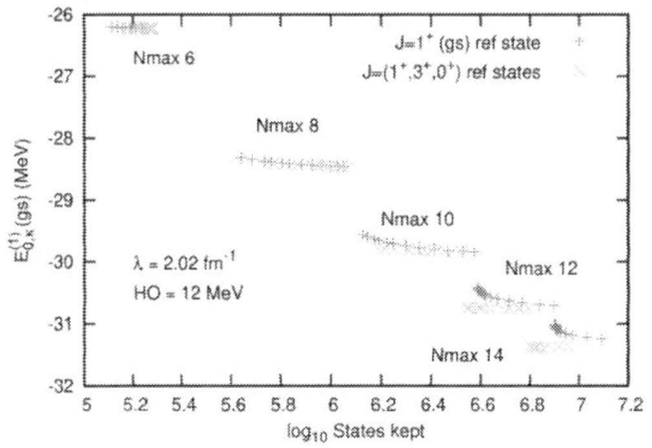

Figure 2. The g.s. energy at $\hbar\Omega$ = 12 MeV vs the number of basis states kept. The (+) signs are when only the g.s. is used as a reference states, whereas the (×) signs are the behavior, when the lowest three states are used.

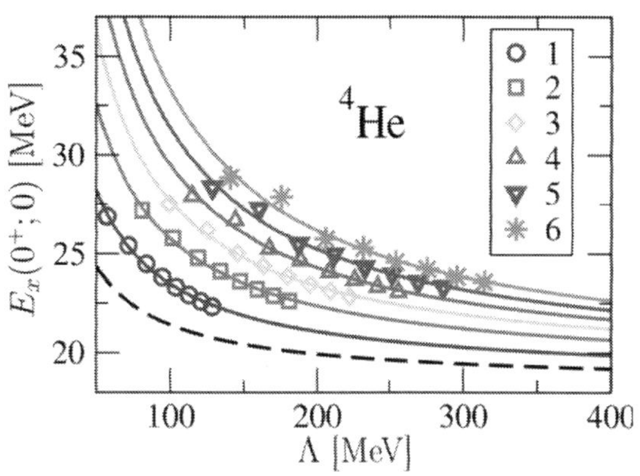

Figure 3. Energy of the first-excited $(0^+, 0)$ state in ^{4}He as a function of the UV cutoff Λ for different values of the frequency $\hbar\omega$ in MeV. The dashed curve marks the limit $\hbar\omega \to 0$.

Figure 3 shows our results for the energy of the first-excited 0^+ state in the α particle, for increasing Λ and decreasing ω, which yields $E(J^\pi, T, {}^4He) = E(0_2^+, 0)_{theor} = 18.8$ MeV, in the limits $\Lambda \to \infty$ and $\omega \to 0$, compared with $E(0_2^+, 0, {}^4He)_{expt} = 20.21$ MeV. These agree within 10%, which is remarkable agreement for a LO calculation and indicates that the description of this level is insensitive to details of the short-range physics.

Following the same procedure, we also evaluated for the first time in the pionless EFT the binding energy of ^{6}Li. For large values of the UV cutoff Λ and small values of the IR cutoff ω, we estimate the binding energy to be about 23 MeV, to be compared with the experimental result of 31.99 MeV. While not as precise as the first-excited-0^+-state energy in ^{4}He, this agreement between the theoretical and experimental values for the binding energy of ^{6}Li to within 30% is consistent with the expected errors in LO in the pionless EFT.

Applications to two and three nucleons in a trap are made in Ref. [11]. Work is currently in progress for extending our formalism to the pionfull EFT, *i.e.,* the EFT with pions.

Acknowledgments

B.R.B. wishes to acknowledge his collaborators on these projects, *i.e.,* M. C. Birse, M. K. G. Kruse, A. F. Lisetskiy, J. Rotureau, I. Stetcu, U. van Kolck, and J. P. Vary and partial support of this research by NSF grant number PHY-0854912; and to thank M.K.G.K. for assistance with some of the figures.

References

[1] van Kolck U 1999 *Prog. Part. Nucl. Phys.* **43** 337
[2] Bedaque P F and van Kolck U 2002 *Annu. Rev. Nucl. Part. Sci.* **52** 339; references therein
[3] Marcucci L E, Pervin M, Pieper S C, Schiavilla R and Wiringa R B 2008 *Phys. Rev.* C **78** 065501; references therein
[4] Barrett B R, Navrátil P and Vary J P 2013 *Prog. Part. Nucl. Phys.* **69** 131; references therein
[5] Hagen G, Papenbrock T, Dean D J and Hjorth-Jensen M 2010 *Phys. Rev.* C **82** 034330; references therein
[6] Lisetskiy A F, Barrett B R, Kruse M K G, Navrátil P, Stetcu I and Vary J P 2008 *Phys. Rev.* C **78** 044302
[7] Lisetskiy A F, Kruse M K G, Barrett B R, Navrátil P, Stetcu I and Vary J P 2009 *Phys. Rev.* C **80** 024315
[8] Kruse M K G, Jurgenson E D, Navrátil P, Barrett B R and Ormand W E 2013 *Phys. Rev.* C **87** 044301
[9] Stetcu I, Barrett B R and van Kolck U 2007 *Phys. Lett.* B **653** 358
[10] Stetcu I, Rotureau J, Barrett B R and van Kolck U 2010 *J. Phys.* G **37** 064033
[11] Rotureau J, Stetcu I, Barrett B R and van Kolck U 2012 *Phys. Rev.* C **85** 034003
[12] Abe T, Maris P, Otsuka T, Shimizu N, Utsuno Y and Vary J P 2012 *Phys. Rev.* C **86** 054301
[13] Hergert H, Bogner S K, Binder S, Calci A, Langhammer J, Roth R and Schwenk A 2013 *Phys. Rev.* C **87** 034307
[14] Dytrych T, Sviratcheva K D, Bahri C, Draayer J P and Vary J P 2007 *Phys. Rev.* C **76** 014315
[15] Machleidt R, Sammarruca F and Song Y 1996 *Phys. Rev.* C **53** R1483
[16] Roth R and Navratil P 2007 *Phys. Rev. Lett.* **99** 092501
[17] D. J. Dean, G. Hagen, M. Hjorth-Jensen, T. Papenbrock and A. Schwenk 2008 *Phys. Rev. Lett.* **101** 119201
[18] Roth R and Navátil P 2008 *Phys. Rev. Lett.* **101** 119202
[19] Roth R 2009 *Phys. Rev.* C **79** 064324
[20] Sherrill C D and Schaefer III H F 1999 *Advances in Quantum Chemistry* vol 34 (Academic Press) pp 143-269
[21] Rolik Z, Szabados Á and Surján P R 2003 *J. Chem. Phys.* **119** 1922
[22] Surján P R, Rolik Z, Szabados Á and Köhalmi D 2004 *Annalen der Physik* **13** 223

Effective Hamiltonian in non-degenerate model space

K Takayanagi[1], N Tsunoda[2], M Hjorth-Jensen[3,4] and T Otsuka[2]

[1] Department of Physics, Sophia University, 7-1 Kioi-cho, Tokyo 102-8554, Japan
[2] Department of Physics, University of Tokyo, Hongo, Tokyo 113-0033, Japan
[3] Department of Physics and Center of Mathematics for Applications, University of Oslo, N-0316 Oslo, Norway
[4] National Superconducting Cyclotron Laboratory, Michigan State University, East Lansing, Michigan 48824, USA

E-mail: `k-takaya@sophia.ac.jp`

Abstract. We present a novel method to calculate a microscopic effective interaction in many-body systems. This method applies not only to degenerate model spaces but also to non-degenerate model spaces, and therefore allows us to construct microscopically the effective interaction in multi-major-shell model spaces. As an application of this novel method, we examine the effective interaction between valence neutrons of ^{18}O in the *sd*-shell (model space of single-major-shell), and in the sdf_7p_3-shell (model space of multi-major-shells). We find that the microscopic cross-shell interactions are much more attractive than have been expected by *ad hoc* methods. Our approach presents an essential step to a microscopic description of nuclei in multi-major-shells, which is crucial, e.g., to describe many of the neutron-rich nuclei.

1. Introduction

The shell model has been playing an important role in nuclear physics. In order to describe nuclear properties in a satisfactory fashion, however, we still need to improve the theory of the effective interaction, the only input of the nuclear shell model. Phenomenological effective interactions, being obtained by tuning two-body matrix elements to reproduce observed experimental data, have described many properties of nuclei successfully [1, 2, 3, 4, 5]. Microscopic effective interactions, on the other hand, have been calculated in the Kuo-Krenciglowa (KK) [6] or the Lee-Suzuki (LS) methods [7] in degenerate model spaces only; these methods cannot be applied, e.g., to a calculation of ^{18}O if we take the non-degenerate model (valence) space composed of *sd* and *pf*-shells.

It has now become clear that many unstable nuclei require two or more major-shells for their description, e.g., physics in the island of inversion cannot be explained with the single-major-shell degrees of freedom alone [3]. To date, investigations of these nuclei have been performed using phenomenological effective interactions. It is, therefore, highly desirable to establish a theory to construct the microscopic effective interaction in the multi-major-shell model spaces starting from realistic nucleon-nucleon interactions.

Recently, the KK and the LS methods have been extended so that they apply not only to the degenerate, but also to the non-degenerate model spaces [8, 9]. In this situation, we naturally adopt the extended KK (EKK) method to construct the microscopic effective interaction in multi-major-shell model spaces. We shall see that our theory is a natural extension of the well known folded diagram theory [6, 10].

In Sec.2, we briefly explain the formal theory of effective interaction and the EKK method. Then in Sec.3, we apply the EKK method to ^{18}O, and present discussions using test calculations of the effective interaction in a single major shell (*sd*-shell) and in two major shells (sdf_7p_3-shell). We shall see clearly that our theory has established a way to calculate the microscopic effective interaction in multi-major-shell model spaces. In Sec.5 we give a brief conclusion.

2. Formal theory of effective interaction

We consider a quantum system that is described by $H = H_0 + V$ in a Hilbert space of dimension D, where H_0 is the unperturbed Hamiltonian and V is the perturbation:

$$H|\Psi_\lambda\rangle = E_\lambda|\Psi_\lambda\rangle, \quad \lambda = 1, \cdots, D. \tag{1}$$

Because the dimension D is very large in many cases, we divide the Hilbert space of dimension D into a model space (P-space) of dimension d to describe the system and its complement (Q-space). Let us denote the projectors onto the P and Q-spaces as P and Q, respectively. We then define the effective Hamiltonian, $H_{\rm eff} = PH_0P + V_{\rm eff}$, in the P-space that describes P-space projections of selected d eigenstates of the full Hamiltonian H:

$$H_{\rm eff}\ P|\Psi_i\rangle = E_i\ P|\Psi_i\rangle, \quad i = 1, \cdots, d. \tag{2}$$

We now explain how to calculate $H_{\rm eff}$.

2.1. Extended Kuo-Krenciglowa (KK) method

The effective Hamiltonian $H_{\rm eff}$ can be obtained most simply by the similarity transformation of the Hamiltonian, $\mathcal{H} = e^{-\omega}He^{\omega}$. Here we determine $\omega = Q\omega P$ by the following decoupling equation [7, 11]:

$$0 = Q\mathcal{H}P = QVP - \omega PHP + QHQ\omega - \omega PVQ\omega, \tag{3}$$

which decouples the P-space part from the rest in the transformed Schrödinger equation, and gives the desired effective Hamiltonian $H_{\rm eff}$ and the effective interaction $V_{\rm eff}$ as follows.

$$H_{\rm eff} = P\mathcal{H}P = PHP + PVQ\omega, \quad V_{\rm eff} = PVP + PVQ\omega. \tag{4}$$

The EKK method is devised to solve Eqs. (3) and (4) without imposing the assumption of the degenerate model space [8, 9], and gives the following iterative scheme for $H_{\rm eff}$:

$$H_{\rm eff}^{(n)} = H_{\rm BH}(E) + \sum_{k=1}^{\infty} \hat{Q}_k(E)(H_{\rm eff}^{(n-1)} - E)^k, \tag{5}$$

where $H_{\rm eff}^{(n)}$ stands for $H_{\rm eff}$ at the nth step of iteration, and $\hat{Q}(E)$ and $\hat{Q}_k(E)$ are the $\hat{Q}$-box and its derivatives given by

$$\hat{Q}(E) = PVP + PVQ\frac{1}{E - QHQ}QVP, \quad \hat{Q}_k(E) = \frac{1}{k!}\frac{{\rm d}^k\hat{Q}(E)}{{\rm d}E^k}. \tag{6}$$

In Eq. (5), we have introduced an arbitrary parameter E that shifts the origin of the Hamiltonian. In the right hand side of Eq. (5), the first term, $H_{\rm BH}(E) = PH_0P + \hat{Q}(E)$, is the Bloch-Horowitz Hamiltonian, and is the input of the iteration, $H_{\rm eff}^{(1)}$. In case of convergence, $H_{\rm eff}^{(n)} \to H_{\rm eff}$, Eq. (5) gives $H_{\rm eff}$ of Eq. (4) that satisfies

$$H_{\rm eff} = H_{\rm BH}(E) + \frac{{\rm d}\hat{Q}(E)}{{\rm d}E}(H_{\rm eff} - E) + \frac{1}{2!}\frac{{\rm d}^2\hat{Q}(E)}{{\rm d}E^2}(H_{\rm eff} - E)^2 + \cdots. \tag{7}$$

Note that Eq. (7) is interpreted as a Taylor series expansion of $H_{\rm eff}$ around $H_{\rm BH}(E)$, and changing E corresponds to shifting the origin of the expansion. This explains that the left hand side of Eq.(7) is independent of E, while the right hand side is seemingly dependent on E.

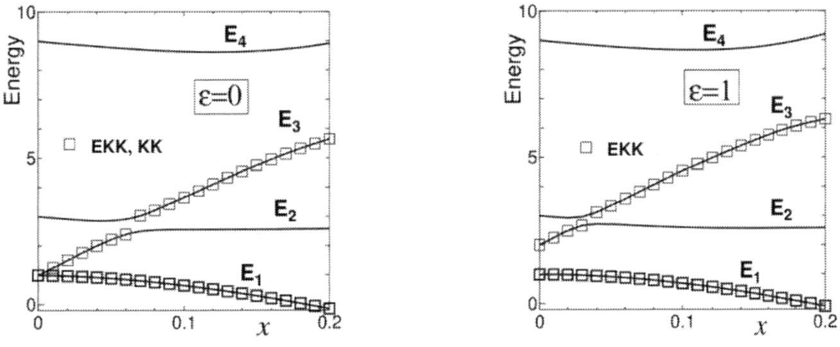

Figure 1. Open squares show eigenenergies of H_{eff} obtained by iterative methods (KK,EKK) in the degenerate (left, $\varepsilon = 0$) and the non-degenerate (right, $\varepsilon = 1$) P-spaces. Solid curves show exact eigenenergies of the Hamiltonian H as functions of the strength parameter x.

2.2. Toy model calculation

Here we carry out the EKK iterative scheme (5) using a toy model. We take a four-dimensional ($D = 4$) Hilbert space that is spanned by the basis vectors $\{|\lambda\rangle,\ \lambda = 1, 2, 3, 4\}$, and a two-dimensional ($d = 2$) P-space $\{|1\rangle, |2\rangle\}$. Our Hamiltonian in the above basis is given by

$$H = H_0 + V = \begin{pmatrix} 1 & & & \\ & 1+\varepsilon & & \\ & & 3 & \\ & & & 9 \end{pmatrix} + x \begin{pmatrix} 0 & 5 & -5 & 5 \\ 5 & 25 & 5 & -8 \\ -5 & 5 & -5 & 1 \\ 5 & -8 & 1 & -5 \end{pmatrix}, \tag{8}$$

where a nonzero value of ε removes the degeneracy of the P-space, and the perturbation V is proportional to a strength parameter x. By diagonalizing $H = H_0 + V$, we obtain four eigenstates which we denote as $\{|\Psi_\lambda\rangle,\ \lambda = 1, 2, 3, 4\}$ with eigenenergies $E_1 < E_2 < E_3 < E_4$.

Fig.1 shows the numerical results for the degenerate (left, $\varepsilon = 0$) and the non-degenerate (right, $\varepsilon = 1$) P-spaces. We see that that the EKK method can reproduce excellently two ($= d$) eigenstates both in the degenerate and non-degenerate P-spaces [8, 9], while the standard KK method is available in the degenerate P-space only. Fig.1 demonstrates very well that the EKK method can give microscopic H_{eff} (and V_{eff}) in non-degenerate P-spaces, which had long been beyond the scope of former theories. We are now ready to set out for microscopic shell model calculations in multi-major-shell model spaces.

3. Application to ^{18}O

Here we apply the above EKK theory of the effective interaction to ^{18}O, a simple nucleus composed of two neutrons on top of the ^{16}O core. As the P-space for ^{18}O, we take the sd-shell (degenerate case), and the sdf_7p_3-shell (non-degenerate case) composed of the sd-shell and $0f_{7/2}$ and $1p_{3/2}$ orbits. Detailed description of the theory and comprehensive study of other nuclei will be published elsewhere [12].

3.1. Many-body theory of effective interaction

In the many-body system, H_{eff} of Eq. (7) has a well defined expression in terms of the folded diagrams using the valence-linked $\hat{Q}$-box as in the standard KK theory [6, 10], because the energy derivative of the $\hat{Q}$-box represents the folding procedure. In order to calculate H_{eff}, all that we have to know is the valence-linked $\hat{Q}$-box and its energy derivatives at the arbitrarily specified energy E [12].

3.2. scheme of calculation

Our input interaction V is the low-momentum interaction $V_{\text{low}k}$ with a sharp cutoff $\Lambda = 2.0\ \text{fm}^{-1}$, which is derived from the $\chi\text{N}^3\text{LO}$ interaction [13]. Our total Hilbert space is composed of the harmonic oscillator basis states in the lowest seven major shells. The single particle energies in the shell model diagonalization are taken from the USD interaction [1, 14].

Let us note here that the new parameter E in our EKK theory provides us with a useful test of calculation [12]; Eq. (7) shows that H_{eff} does not depend on the value of E, provided that the $\hat{Q}$-box in Eq. (7) is given exactly. Any approximation may spoil the E-independence of H_{eff}, and therefore of physical quantities described by H_{eff}. This in turn implies that E-independence of physical quantities is a signal of convergence. In this work, we have calculated the $\hat{Q}$-box diagrams up to the third order in powers of V. In Sec.3.3. we shall see that our numerical results for Eq. (7) are almost E-independent, suggesting that H_{eff} is convergent if $\hat{Q}(E)$ in Eq. (7) is calculated up to the third order in V. Note that the above useful test of convergence is possible only for the EKK method; there is no such criterion in the original KK method.

3.3. Numerical results

Here we present numerical results for ^{18}O in the degenerate *sd*-shell and in the non-degenerate sdf_7p_3-shell; we show (i) two-body matrix elements of the effective interaction V_{eff}, and (ii) several energy levels obtained by the shell model calculation of ^{18}O, paying special attention on their E-independence.

3.3.1. degenerate sd-shell Fig.2 shows our numerical results in the degenerate *sd*-shell model space, where both of the standard KK and the EKK methods are available. To see the E-independence of the EKK results for V_{eff}, we have varied the parameter E in the range of $-5 \leq E \leq 2$ MeV in Fig. 2. Here the origin of energy is so fixed that the EKK method with $E = 0$ coincides with the KK method.

Let us start with the upper panel, where we present the monopole part of V_{eff} in the *sd*-shell. The dotted lines (which make a shaded band in the figure) are calculated by dropping the folded diagram contributions, i.e., by replacing V_{eff} simply by $\hat{Q}(E)$. We can see clearly that $\hat{Q}(E)$ is strongly dependent on E. Now we take into account all the folded diagram contributions in the right hand side of Eq. (7). The results are shown by solid lines for $E = -5, 0, 2$ MeV, which almost coincide with each other. This shows clearly that the folded diagrams cancel the E-dependence of $\hat{Q}(E)$ to yield the E-independent H_{eff} (and V_{eff}) in Eq. (7).

The lower panels display several energy levels of ^{18}O with respect to ^{16}O calculated by the shell model with our V_{eff}. Let us look at the results without the folded diagram contributions (left panel). They are obviously E-dependent, and are decreasing functions of E, which is explained naturally by the E-dependence of $\hat{Q}(E)$. On the other hand, the calculation with the folded diagrams (right panel) shows clearly that the results are almost independent of the parameter E, as they should.

The above observation shows that the evaluation of the $\hat{Q}$-box up to third order in V is sufficient to establish the E-independence of the right hand side of Eq. (7), and therefore of V_{eff}. This also implies that the $\hat{Q}$-box through third order gives convergent results for V_{eff}, and that we do not need to move on to the fourth order calculation of the $\hat{Q}$-box.

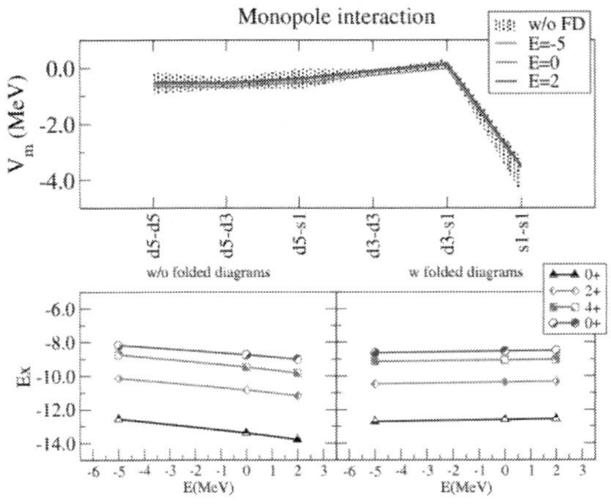

Figure 2. E-dependence of the EKK results in the sd-shell (degenerate model space). upper panel: the monopole part of V_{eff}, ($-5 \leq E \leq 2$ MeV). lower panels: level energies of ^{18}O without (left) and with (right) folded diagram contributions. See the text.

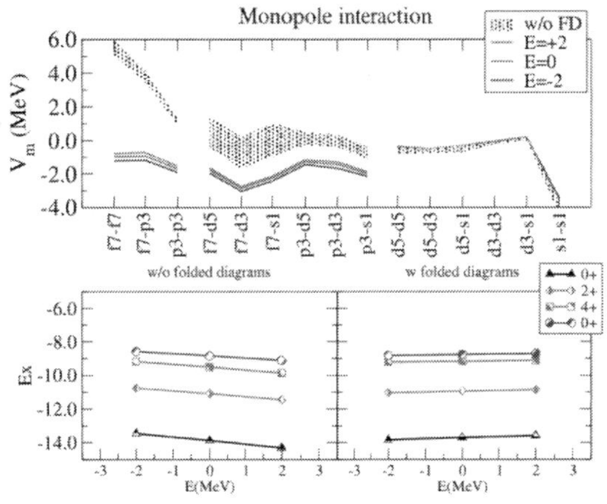

Figure 3. E-dependence of the EKK results in the sdf_7p_3-shell (non-degenerate model space). The range of E is $-2 \leq E \leq 2$ MeV. Other notation is the same as for Fig. 2

3.3.2. non-degenerate sdf_7p_3-shell Here we examine ^{18}O in the non-degenerate P-space, sdf_7p_3-shell, composed of the sd-shell and $0f_{7/2}$ and $1p_{3/2}$ orbits. In this non-degenerate model space, the standard KK method does not work, and the EKK method only is available. To date, therefore, there have been only phenomenological interactions in this model space [15].

Fig.3 shows the numerical results of the EKK method in the sdf_7p_3-shell, in the same way as Fig. 2 in the degenerate sd-shell. Let us first look into the monopole part of V_{eff}. Here we have cross-shell interactions (f_7-d_5,..,p_3-s_1), in addition to the interactions within the sd-shell. We see that the cross-shell interaction is strongly dependent on E without the folded diagram contributions. This can be explained by noticing that the corresponding $\hat{Q}(E)$ has intermediate states with small energy denominators. The figure clearly shows that the folded diagram contributions cancel the above strong E-dependence of $\hat{Q}(E)$, to make the resultant V_{eff} almost independent of E. This observation confirms that the EKK method, implemented by the $\hat{Q}$-box through third order, is robust not only in the degenerate sd-shell, but also in the non-degenerate sdf_7p_3-shell.

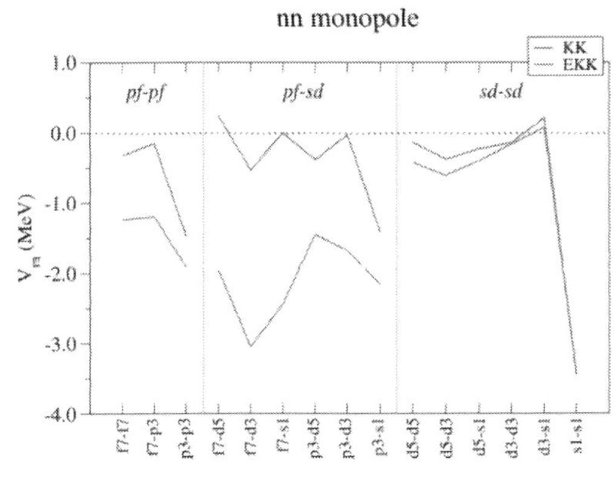

Figure 4. The monopole part of V_{eff} in the sdf_7p_3 model space, obtained by the EKK method (lower lines, red) and by the *ad hoc* KK method (upper lines, green).

3.3.3. comparison of KK and EKK methods Here we compare the KK and the EKK methods in the non-degenerate sdf_7p_3-shell model space. Given the non-degenerate model space, the standard way has been to use the KK method by shifting the single particle energies by hand to make them degenerate, and by treating the energy shift as a perturbation. The resultant effective interaction is not, however, as meaningful as the EKK effective interaction, since it is an *ad hoc* interaction between orbits with artificial single particle energies. Here we show how different they are.

Fig. 4 shows the monopole part of $V_{\rm eff}$ in the sdf_7p_3-shell. The lower (red) lines show the results of the EKK method, and the upper (green) lines stand for the results of the KK method with the *ad hoc* modification of the unperturbed Hamiltonian. It is really striking that the cross-shell interaction (f_7-d_5,..,p_3-s_1) is much more attractive in the EKK method than in the *ad hoc* KK method. This discrepancy clearly comes from the difference in the energy denominator; excitations from the artificially degenerate sdf_7p_3-shell have larger excitation energies than from the non-degenerate sdf_7p_3-shell, to make their contributions smaller in $\hat{Q}(E)$. We expect that the strong cross-shell attraction would have a significant effect, e.g., on the shell evolution, which is one of the hottest topics in nuclear physics.

4. Conclusion

We have presented a novel many-body theory (EKK) to calculate the effective interaction $V_{\rm eff}$ for the shell model which is applicable not only to degenerate (single-major-shell) model spaces, but also to non-degenerate (multi-major-shell) model spaces. On the basis of our numerical results, we have clarified the followings.

First, we have shown that the EKK method works beautifully in many-body systems. In the degenerate model space, it gives the same results as the former methods. In the non-degenerate model space, which is beyond the range of the former theories, our EKK method nicely works.

Second, we have shown that our results for $V_{\rm eff}$ are almost E-independent, as they should, if we calculate the $\hat{Q}$-box up to third order in the perturbation theory. This in turn suggests strongly that the $\hat{Q}$-box through third order is sufficient to give convergent results for $V_{\rm eff}$.

Third, we have shown that the difference is significant between the EKK method and the KK method with an *ad hoc* modification of the unperturbed Hamiltonian, especially for cross-shell interactions. This should have a large impact, e.g., on the investigation of neutron-rich nuclei where it is essential to incorporate two-major-shell degrees of freedom.

At the end, we stress that our method has established a robust way to calculate the effective interaction in non-degenerate model spaces, or in multi-major-shells, in a microscopic way. We believe that this is an indispensable step to make the nuclear shell model a reliable theory based on a microscopic effective interaction derived from realistic nucleon-nucleon interactions.

References

[1] Brown B A, Richter W A, Julies R E and Wildenthal B H 1988 *Annals of Physics* **182** 191
[2] Honma M, Otsuka T, Brown B A and Mizusaki T 2002 *Phys. Rev.* C **65** 061301
[3] Utsuno Y, Otsuka T, Glasmacher T, Mizusaki T and Honma M 2004 *Phys. Rev.* C **70** 044307
[4] Poves A and Zuker A 1981 *Physics Reports* **70** 235
[5] Poves A, Sánchez-Solano J, Caurier E and Nowacki F 2001 *Nuclear Physics* A **694** 157
[6] Krenciglowa E M and Kuo T T S 1974 *Nuclear Physics* A **235** 171
[7] Suzuki K and Lee S Y 1980 *Prog. Theor. Phys.* **64** 2091
[8] Takayanagi K 2011 *Nuclear Physics* A **852** 61
[9] Takayanagi K 2011 *Nuclear Physics* A **864** 91
[10] Kuo T T S and Osnes E 1990 *Lecture Notes in Physics* vol 364 (Berlin: Springer)
[11] Takayanagi K 2013 *Nuclear Physics* A **899** 107
[12] Tsunoda N, Takayanagi K, Hjorth-Jensen M and Otsuka T *in preparation*
[13] Epelbaum E, Hammer H -W and Meissner U -G 2009 *Rev. Mod. Phys.* **81** 1773
[14] Brown B A and Wildenthal B H 1988 *Annu. Rev. Nucl. Part. Sci.* **38** 29
[15] Utsuno Y, Otsuka T, Mizusaki T and Honma M 1999 *Phys. Rev.* C **60** 054315

Journal of Physics: Conference Series **445** (2013) 012004 doi:10.1088/1742-6596/445/1/012004

History and future perspectives of the Monte Carlo shell model -from Alphleet to K computer-

Noritaka Shimizu[1], Takaharu Otsuka[1,2,3], Yutaka Utsuno[4], Takahiro Mizusaki[5], Michio Honma[6] and Takashi Abe[2]

[1]Center for Nuclear Study, University of Tokyo, Tokyo 113-0033, Japan
[2]Department of Physics, University of Tokyo, Tokyo 113-0033, Japan
[3]National Superconducting Cyclotron Laboratory, Michigan State University, East Lansing, Michigan, USA
[4]Advanced Science Research Center, Japan Atomic Energy Agency, Tokai, Ibaraki 319-1195, Japan
[5]Institute of Natural Sciences, Senshu University, Tokyo, 101-8425, Japan
[6]Center for Mathematical Sciences, University of Aizu, Ikki-machi, Aizu-Wakamatsu, Fukushima 965-8580, Japan

E-mail: shimizu@cns.s.u-tokyo.ac.jp

Abstract. We report a history of the developments of the Monte Carlo shell model (MCSM). The MCSM was proposed in order to perform large-scale shell-model calculations which direct diagonalization method cannot reach. Since 1999 PC clusters were introduced for parallel computation of the MCSM. Since 2011 we participated the High Performance Computing Infrastructure Strategic Program and developed a new MCSM code for current massively parallel computers such as K computer. We discuss future perspectives concerning a new framework and parallel computation of the MCSM by incorporating conjugate gradient method and energy-variance extrapolation.

1. Introduction

The nuclear shell model calculation has been very successful in understanding the nuclear structure on the basis of nucleons interacting via the nuclear force. It can provide precisely the properties, excitation schemes and transition probabilities of the low-lying states of a nucleus. The conventional, standard solver for shell-model calculations is the exact diagonalization of Hamiltonian matrix by the Lanczos algorithm in a given model space. However, the dimension of the matrix grows exponentially as the number of valence nucleons increases and surpasses the numerical feasibility limit of the Lanczos method (e.g. 5×10^{15} for ^{68}Ni in $pfg9d5$-shell, discussed in Ref.[1]). In order to perform such large-scale shell model calculations, we have developed the Monte Carlo shell model (MCSM) method. The MCSM was firstly proposed by a part of the authors (M. Honma, T. Mizusaki and T. Otsuka) in 1995 [2], and has been developed by the nuclear theory group of the University of Tokyo. Since the MCSM algorithm is suited for parallel computations, we started the large-scale nuclear structure calculation project with introducing PC clusters since 1999. In the proceedings, we briefly report a history of the MCSM and utilization of PC clusters for parallel computations in view of computational aspect in Sect.2, and summarize the framework of the MCSM in Sect.3, the details of which can be found in Ref.[3].

2. Computational aspects of the MCSM, Alphleet cluster and K computer

In 1999, a research project for large-scale shell-model calculations was launched jointly by the RIKEN Accelerator Research Facility (currently known as RIKEN Nishina center) and the nuclear theory group of the University of Tokyo. Since the MCSM calculation is suitable for parallel computation, we introduced a PC cluster, Alphleet-1, mainly for the MCSM calculations. It consisted of 73 workstations, each of which had 2 Alpha ev6 CPUs and was connected with each other by Myrinet network [4, 5]. A photograph of the system is shown in the left of Fig.1. The total system performance of LINPACK benchmark was 62GFLOPs, which was ranked as the 169th fastest computer in the world at Nov. 1999 [6]. The main achievements of this project were summarized in Ref.[7]. Especially, this project played an essential role to establish the effective interaction of pf shell, GXPF1 [8].

After the completion of the project with Alphleet-1 cluster, a succeeding joint project has started since the year 2001 based on a collaboration agreement between Center for Nuclear Study, the University of Tokyo (CNS) and RIKEN Nishina Center. In this project, PC clusters are provided by KAKENHI grants [9] through the Department of Physics, the University of Tokyo, but infrastructures (e.g. electricity, cooling, etc.) and running costs are supported primarily by CNS and RIKEN Nishina Center. Under this project, we introduced a computer cluster named Alphleet-2, which was comprised of 176 Alpha CPUs from 2001 to 2003. Its photo is shown in the right of Fig.1. Four racks on the right end of the photo were added in 2003 and were comprised of 64 CPUs and 128 GB main memory. Each computation node of the cluster are connected with each other via Myrinet network and has no hard disk drive. The features of these clusters are summarized in Tab.1. This project has been continued at present. Especially in 2012, we introduced a PC server equipped with 40 Intel Xeon cores and 660GB main memory mainly for conventional shell-model calculations with the Lanczos method.

Table 1. Features of Alphleet-1 and Alphleet-2.

	Fiscal Year	Brand name	Memory	Nodes
Alphleet-1	1999	DEC DS-20 (Alpha 2 CPUs)	512MB	73
Alphleet-2	2001	Compaq ES-45 (Alpha 4 CPUs)	16GB	28
	2003	Hewlett-Packard GS-1280 (Alpha 32 CPUs)	64GB	2

However, the number of processing units of the state-of-the-art massively parallel computers surpasses ten thousands, which cannot be caught up by a PC cluster of a local research group. In 2009, we decided to write a new MCSM code from scratch so that the MCSM calculations can be run on recent massively parallel computers. This new code was equipped with many advantages: hybrid parallel with Message Passing Interface library and OpenMP, capability of isospin-breaking interaction, applicability to *no-core* shell model calculations, and a new efficient algorithm for the computation of Slater determinants [10]. This new code played an essential role to develop a new framework of the MCSM utilizing the energy-variance extrapolation [17], which is discussed in Sect.3.2.

Based on these experiences of parallel computations of the MCSM and its achievements, we have participated in activities of High Performance Computing Infrastructure Strategic Programs for Innovative Research (SPIRE) Field 5 "The origin of matter and the universe" since 2011. The SPIRE project aims at an integral understanding of the origin and structure of matter and the Universe utilizing the K computer, which is the world's fastest supercomputer in 2011 [11]. In this project, we are in charge of the elucidation of nuclear properties by ultra large-scale simulations of quantum many-body systems. In 2011, we carried out benchmark tests and performance tuning of the MCSM code at K computer.

3. Theoretical developments of the MCSM method

The major features of the recently developed MCSM framework are the variational procedure to obtain the approximated wave function, which is described in Sect.3.1, and the energy-variance extrapolation, which is described in Sect. 3.2.

3.1. Variational procedure

In the framework of the MCSM, the shell-model wave function is approximated as a linear combination of angular-momentum- and parity-projected Slater determinants,

$$|\Psi_{N_b}\rangle = \sum_{n=1}^{N_b} \sum_{K=-I}^{I} f_{n,K}^{(N_b)} P_{MK}^{I\pi} |\phi_n\rangle, \tag{1}$$

where N_b is the number of the Slater-determinant basis states. The $P_{MK}^{I\pi}$ operator is the angular-momentum and parity projector defined as

$$P_{MK}^{I\pi} = \frac{1+\pi\Pi}{2} \frac{2I+1}{8\pi^2} \int d\Omega \, {D_{MK}^{I}}^{*}(\Omega) e^{i\alpha J_z} e^{i\beta J_y} e^{i\gamma J_z}, \tag{2}$$

where $\Omega \equiv (\alpha, \beta, \gamma)$ are the Euler angles and $D_{MK}^{I}(\Omega)$ denotes Wigner's D-function. Π stands for the parity transformation. Each $|\phi_n\rangle$ is a deformed Slater determinant defined as

$$|\phi\rangle = \prod_{k=1}^{N_f} \left(\sum_{l=1}^{N_{sp}} D_{lk} c_l^\dagger \right) |-\rangle, \tag{3}$$

which is parametrized by the complex $N_{sp} \times N_f$ matrix D. N_f and N_{sp} are the numbers of fermions and single-particle states, respectively. The $|-\rangle$ denotes an inert core in the conventional shell-model calculations or the vacuum in no-core shell-model calculations.

The coefficients $f_{n,K}^{(N_b)}$ are determined by the diagonalization of the Hamiltonian matrix in the subspace spanned by the projected Slater determinants, $P_{MK}^{I\pi}|\phi_n\rangle$ with $-I \leq K \leq I$ and $1 \leq n \leq N_b$. This diagonalization also determines the energy, $E_{N_b} \equiv \langle\Psi_{N_b}|H|\Psi_{N_b}\rangle$, as a function of N_b. In the early stage of the development of the MCSM, the wave function is a linear combination of Slater determinants without angular-momentum projection, or only with z-component of the angular-momentum projection [12]. We increase N_b until E_{N_b} converges enough, or the extrapolated energy converges.

In the original MCSM calculation, the Slater determinant basis, $|\phi_n\rangle$, are selected from many candidates generated stochastically utilizing the auxiliary-field Monte Carlo technique, the detail of which is summarized in Ref. [7]. Recently, we have introduced an additional procedure, named sequential conjugate gradient (SCG) process, to obtain better basis states [13]. In this procedure we determine $D^{(n)}$ variationally to minimize $E_{N_b=n}$ utilizing the conjugate gradient method.

Here, we demonstrate how the original MCSM and the additional SCG process work with ^{56}Ni in the pf shell as an example. The effective interaction is taken as FPD6 interaction [14]. Its M-scheme dimension is 1,084,455,228, which is tractable by recent shell model codes with the Lanczos method [15]. At the time of the Alphleet-1 project, it was far beyond the limitation of the Lanczos method, and the MCSM enabled us to discuss the shape co-existence of ^{56}Ni. Since ^{56}Ni has been studied further by shell model calculations [16], it should be a good example for a benchmark test.

Figure 1. Left: Alphleet-1. Right: Alphleet-2.

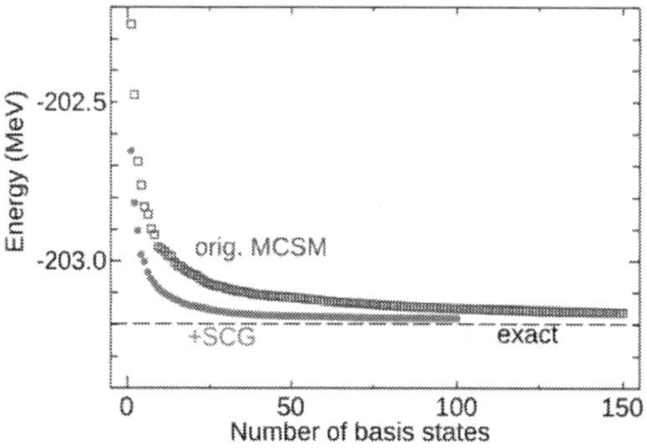

Figure 2. Energy convergence for the ground state of ^{56}Ni. against number of basis states. The blue open squares and red filled circles denote the result of the original MCSM and that of the SCG method, respectively.

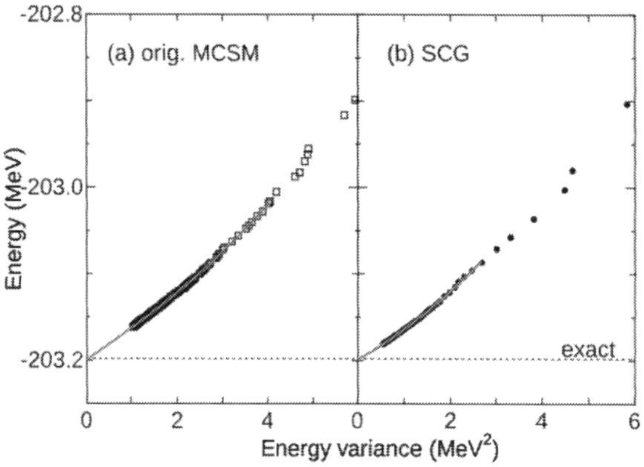

Figure 3. Energy and energy variance of the wave functions provided by (a) the original MCSM, and (b) SCG method for ^{56}Ni. The red lines are fitted for the points. The dashed line denotes the exact shell-model energy. See text for details.

Figure 2 shows that the energy expectation values of the MCSM wave function generated by two kinds of methods converge to the exact shell-model energy with increasing a number of basis states, N_b. The energy of the SCG is much better convergence than that of the original MCSM method. While the exact shell-model energy is −203.198MeV, that of the original MCSM is −201.161MeV with 150 basis states. The same energy (−201.161MeV) is reached by the SCG method with only 30 basis states. The energy of the SCG finally reaches −201.180MeV with 100 basis states. Nevertheless, the small gap between the SCG energy and the exact one, 18keV, remains. In order to remove this gap, we introduced the energy-variance extrapolation, which enables us to estimate the exact eigenvalue precisely even in case the Lanczos method cannot be applied [17].

3.2. Energy variance extrapolation method

The energy-variance extrapolation method is based on the fact that the energy variance of the exact eigenstate is zero. The energy variance of the approximated wave function is not exactly zero, but rather small and approaches zero as the approximation is improved. In the framework of the energy-variance extrapolation, we draw the energy $E_{N_b} = \langle \Psi_{N_b}|H|\Psi_{N_b}\rangle$ against the energy variance $\langle \Delta H^2\rangle_{N_b} = \langle \Psi_{N_b}|H^2|\Psi_{N_b}\rangle - E^2_{N_b}$. The variance usually approaches zero as N_b increases, and the point in the energy-variance plot approaches the y-axis. These points are fitted by a first- or second-order polynomial such as

$$E = c_0 + c_1\langle \Delta H^2\rangle + c_2(\langle \Delta H^2\rangle)^2, \tag{4}$$

where these coefficients c_0, c_1, and c_2 are determined by a least square fit. By extrapolating the fitted curve into the y-intercept we obtain the extrapolated energy, namely, c_0.

The variational procedure discussed in Sect. 3.1 provides us with a successive sequence of the wave functions, $|\Psi_{N_b}\rangle$ with $1 \le N_b \le N_m$, where N_m is the maximum of N_b. This sequence is utilized in the energy-variance extrapolation method. In practice we evaluate the energy E_{N_b} and energy variance $\langle \Delta H^2\rangle_{N_b}$ for each N_b.

Figure 3 shows the energy and energy variance of the variational wave functions obtained by the original MCSM and the SCG method in Sect.3.1. As N_b increases the energy-variance point moves smoothly and approaches the y-axis or variance zero. The fitted curves for these points are shown as red solid lines, and also approach the y-axis smoothly. The extrapolated energies, or y-intercepts of the fitted curves, agree quite well with the exact one, which is shown as a dashed line. The minimum variance of the SCG wave function is $\langle \Delta H^2\rangle_{N_b=100} = 0.54\text{MeV}^2$, which is smaller and gives a better approximation than the result of the original MCSM, $\langle \Delta H^2\rangle_{N_b=150} = 1.04\text{MeV}^2$. It means that the SCG method improves the accuracy of the MCSM wave function to a certain extent. Both the extrapolated energies of the original MCSM and the SCG agree with the exact energy quite well within a keV.

4. Summary

We reported the collaboration projects concerning the MCSM over the latest decade, including the introduction of Alphleet-1 and Alphleet-2 clusters and the recent code developments for utilizing state-of-the-art massively parallel computers. As the computational environment evolves, the code of the MCSM has been developed intermittently. In parallel, the framework of the MCSM has been extended such as the introduction of the conjugate gradient method and the energy-variance extrapolation. Currently we have finished the benchmark tests and start production runs on K computer under the SPIRE project. The recent achievements of this activity are seen in Refs.[1, 3, 10, 18, 19, 20].

Acknowledgments

We acknowledge E. Ikezawa, M. Kase, M. Ishihara, H. Sakai, S. Shimoura, A. Taketani, I. Tanihata and Y. Watanabe for their support of Alphleet project and CNS-RIKEN joint project for large-scale nuclear structure calculations. We also acknowledge L. Liu, P. Maris, Y. Tsunoda, J. P. Vary and T. Yoshida for our recent collaborations. This work has been supported by Grants-in-Aids for Specially Promoted Research (13002001), for Scientific Research (23244049, 20244022), for Scientific Research on Innovative Areas (20105003), and for Young Scientists (20740127) from JSPS, the SPIRE Field 5 from MEXT, and the CNS-RIKEN joint project for large-scale nuclear structure calculations. The numerical calculations were performed mainly on FX10 at the University of Tokyo. The Lanczos shell-model calculations were performed by the code MSHELL64 [15].

References

[1] Tsunoda Y *et al.* contribution of this proceedings
[2] Honma M, Mizusaki T and Otsuka T 1995 *Phys. Rev. Lett.* **75** 1284
[3] Shimizu N, Abe T, Tsunoda Y, Utsuno Y, Yoshida T, Mizusaki T, Honma M and Otsuka T 2012 *Prog. Theor. Exp. Phys.* **2012** 01A205
[4] Otsuka T, Shimizu N, Haruyama S, Honma M, Mizusaki T, Taketani A, Utsuno Y and Watanabe Y 2000 *Prog. Theor. Phys. Suppl.* **138** 24
[5] Watanabe Y, Shimizu N, Haruyama S, Honma M, Mizusaki T, Taketani A, Utsuno Y and Otsuka T 2000 *Prog. Theor. Phys. Suppl.* **138** 43
[6] TOP500 Supercomputer Sites, http://www.top500.org/
[7] Otsuka T, Honma M, Mizusaki T, Shimizu N and Utsuno Y 2001 *Prog. Part. Nucl. Phys.* **47** 319
[8] Honma M, Otsuka T, Brown B A and Mizusaki T 2002 *Phys. Rev.* C **65** 061301(R)
[9] Grant-in-Aids for Specially Promoted Research (13002001), for Scientific Research (23244049, 20244022) from Japan Society for the Promotion of Science
[10] Utsuno Y, Shimizu N, Otsuka T and Abe T 2013 *Comp. Phys. Comm.* **184** 102
[11] K computer, http://www.aics.riken.jp/en/
[12] Honma M, Mizusaki T and Otsuka T 1996 *Phys. Rev. Lett.* **77** 3315
[13] Shimizu N, Utsuno Y, Mizusaki T, Honma M, Tsunoda Y and Otsuka T 2012 *Phys. Rev.* C **85** 054301
[14] Richter W A, van der Merwe M G, Julies R E and Brown B A 1991 *Nucl. Phys.* A **523** 325
[15] Mizusaki T, Shimizu N, Utsuno Y and Honma M, code MSHELL64, unpublished
[16] Horoi M, Brown B A, Otsuka T, Honma M and Mizusaki T 2006 *Phys. Rev.* C 061305(R)
[17] Shimizu N, Utsuno Y, Mizusaki T, Otsuka T, Abe T and Honma M 2010 *Phys. Rev.* C **82** 061305(R)
[18] Abe T, Maris P, Otsuka T, Shimizu N, Utsuno U and Vary J P 2012 *Phys. Rev.* C **86** 054301
[19] Liu L, Otsuka T, Shimizu N, Utsuno Y and Roth R contribution of this proceedings
[20] Yoshida T *et al.* contribution of this proceedings

No-Core MCSM calculation for ^{10}Be and ^{12}Be low-lying spectra

Lang Liu[1], Takaharu Otsuka[2,3,4], Noritaka Shimizu[2], Yutaka Utsuno[5] and Robert Roth[6]

[1] School of science, Jiangnan University, No. 1800, Lihu Avenue, Wuxi, 214122, Jiangsu, China
[2] Department of Physics, University of Tokyo, Hongo, Tokyo 113-0033, Japan
[3] Center for Nuclear Study, University of Tokyo, Hongo, Tokyo 113-0033, Japan
[4] National Superconducting Cyclotron Laboratory, Michigan State University, East Lansing, Michigan, 48824, USA
[5] Japan Atomic Energy Agency, Tokai, Ibaraki, 319-1195 Japan
[6] Institut für Kernphysik, Technische Universität Darmstadt, D-64289 Darmstadt, Germany

E-mail: liulang@pku.edu.cn

Abstract. The low-lying excited states of ^{10}Be and ^{12}Be are investigated within a no-core Monte Carlo Shell Model (MCSM) framework employing a realistic potential obtained via the Unitary Correlation Operator Method. The excitation energies of the 2_1^+ and 2_2^+ states and the B(E2; $2_{1,2}^+ \rightarrow 0_{g.s.}^+$) for ^{10}Be in the MCSM with a standard treatment of spurious center-of-mass motion show good agreement with experimental data. Some properties of low-lying states of ^{10}Be are studied in terms of quadrupole moments and E2 transitions. The E2 transition probability of ^{10}C, the mirror nucleus of ^{10}Be, is also presented with a good agreement to experiment. The triaxial deformation of ^{10}Be and ^{10}C is discussed in terms of the B(E2) values.

1. Introduction

In the last decades, many progresses have been made in obtaining an accurate representation of realistic nucleon-nucleon (NN) potentials. One can construct a two-body potential phenomenologically by fitting experimental data on nucleon-nucleon (NN) scattering, as it is done in the Argonne V18 potential [1], the CD-Bonn potential [2] and the Nijmegen potentials [3]. Alternatively, the chiral N^{3}LO potential can be constructed in the framework of chiral effective field theory [4–6]. By using these realistic nuclear interactions, *ab initio* nuclear many-body calculations have been performed. In Green's Function Monte Carlo (GFMC) calculations the exact ground-state wave function is calculated by treating the many-body Green's functions in a Monte Carlo approach [7–9]. The GFMC calculations of light nuclei up to ^{12}C with the Argonne interaction reproduce the experimental nuclear binding energies and radii as well as the spectra. Another *ab initio* approach for nuclei up to A=14 is the No-Core Shell Model (NCSM) [10–12].

The straightforward application of those realistic interactions in nuclear many-body calculations is difficult due to the strong short-range repulsion and tensor correlation. The Unitary Correlation Operator Method (UCOM) is one of the methods to tackle this problem by introducing a unitary transformation [13–15]. In the UCOM approach two unitary transformation operators are defined, a central correlation operator and a tensor correlation operator, which correspond to two most important correlations: the central correlations induced by the strong short-range repulsion and the tensor correlations, respectively. The so-called UCOM potential can be used in various kinds of many-body calculations, such as no-core shell model calculations [16–18].

In the shell model calculations, the direct diagonalization of the Hamiltonian matrix in the full valence-nucleon Hilbert space is difficult, as the dimension of such a space becomes larger and larger when one moves from light nuclei to heavier nuclei. One way to overcome this difficulty, the Quantum Monte Carlo Diagonalization (QMCD) method has been proposed for solving quantum many-body systems with a two-body interaction [19–22]. The QMCD can describe not only the ground state but also excited states, including their energies, wave functions and hence transition matrix elements. Thus, on the basis of the QMCD method, the Monte Carlo Shell Model (MCSM) has been introduced [23] for nuclear shell model calculations [24–27]. An extrapolation method in the Monte Carlo Shell Model has been proposed very recently [28–30]. We shall present, in this paper, for the first time, how such *ab initio* no-core MCSM calculations work. The results to be shown in this paper will play a key role in judging as to whether one should move ahead to more systematic calculations with the revised method or not.

2. Low lying spectra of ^{10}Be and ^{12}Be

The theoretical framework and numerical detail are not presented in this paper. The details of these parts can be found in Ref. [31].

The convergence of low-lying excitation energies as a function of the MCSM dimension has been examined firstly. In the MCSM calculation, the diagonalization is performed in a subspace comprised of 25 to 50 optimally generated basis states. The size (dimension) of this subspace is quite small compared to that of the entire Hilbert space taken in the direct diagonalization in the conventional shell model. This advantage will be even more obvious for heavier nuclei by the fact that the full diagonalization in $e_{\max} = 3$ is hardly feasible with other calculational techniques available presently.

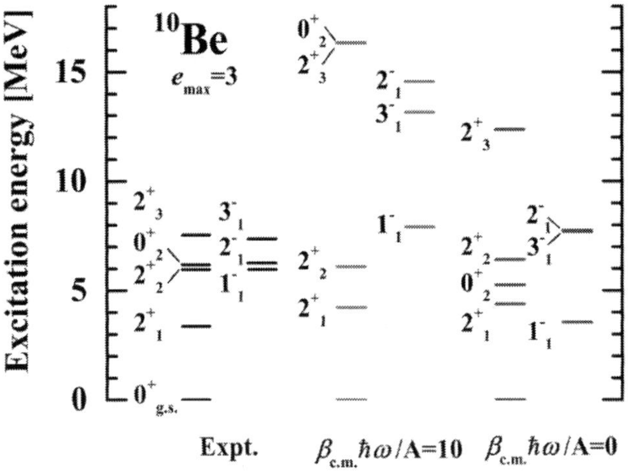

Figure 1. (Color online) Some low-lying spectra of ^{10}Be in the $e_{\max} = 3$ model space. Black bars indicate experimental levels. Red levels are theoretical results obtained with the suppression of spurious center-of-mass motion ($\beta_{c.m.} \cdot \hbar\omega/A = 10$ MeV). Blue levels are obtained without removing the spurious center-of-mass motion.

We now discuss properties of the 0_1^+ and $2_{1,2}^+$ states of ^{10}Be. Figure 1 shows energy levels of these states. Some other low-lying states of ^{10}Be are shown also. While the MCSM results are about 1 MeV and 0.6 MeV higher than the experimental values for 2_1^+ and 2_2^+, respectively, the basic patterns and scale are reproduced well by the MCSM calculation. In particular, the low-lying 2_2^+ level is a characteristic indicator of triaxial deformation, as discussed later.

We now investigate these excited states in terms of the quadrupole moments and E2 transitions. The quadrupole moments of protons and neutrons for the 2_1^+ and 2_2^+ states of ^{10}Be are calculated. One finds that beyond MCSM dimension of 30, those quadrupole moments reach stable values. The nucleus ^{10}Be has a negative quadrupole moment for the 2_1^+ state. In contrast, the 2_2^+ state shows a positive quadrupole moment. These features are also predicted in Ref. [1]. We note that the protons have stronger deformation than neutrons in both states of ^{10}Be, because there are two valence protons and four valence neutrons in the p-shell in major configurations, and the former produce stronger deformation than the latter.

The B(E2) values from the $2_{1,2}^+$ states to the ground state of ^{10}Be are evaluated. Some B(E2) values are calculated also for the mirror nucleus, ^{10}C, in the isospin formalism, as shown in Table 1. This table indicates that MCSM value of B(E2; $2_1^+ \to 0_{g.s.}^+$) appears to be in rather good agreement with the corresponding experimental data [32, 33] for both ^{10}Be and ^{10}C. This is of certain importance because from the viewpoint of the liquid-drop model, B(E2) value is proportional to Z^2, and thereby the value of ^{10}C is expected to be larger than the corresponding one of ^{10}Be, by a factor of $6^2/4^2$ in a naive expectation. While we take only bare charge ($e_p = e$ and $e_n = 0$ with e being the unit charge), we can still produce almost the same values of B(E2; $2_1^+ \to 0_{g.s.}^+$) of ^{10}Be and ^{10}C. This is because although there are two more protons in ^{10}C than in ^{10}Be, they do not necessarily increase quadrupole deformation, partly due to the $0p_{3/2}$ closed-shell formation.

The nuclei ^{10}C and ^{10}Be belong to the same isospin multiplet of T=1. In the notation of Timmer [34], which makes direct use of the isospin formalism, the B(E2) value of ^{10}C should be smaller than that in ^{10}Be, as $T_z = -1$ for ^{10}C and $T_z = 1$ for ^{10}Be.

Assuming that the 0_1^+ and 2_1^+ states of ^{10}Be belong to the same $K = 0$ rotational band, the intrinsic quadrupole moment Q_0 can be evaluated from the B(E2; $0_1^+ \to 2_1^+$) value and the spectroscopic quadrupole moment. Q_0 evaluated by the spectroscopic quadrupole moment is 20.5 e fm^2, which is consistent to the one (21.6 e fm^2) extracted from the B(E2; $2_1^+ \to 0_1^+$) value. This similarity seems to suggest an axially symmetric deformation in the yrast band. On the other hand, the B(E2; $2_2^+ \to 2_1^+$) is sizable, which hints at a notable triaxial deformation of ^{10}Be. B(E2; $2_2^+ \to 0_1^+$)= 0.32 e^2 fm^4 leads us to a triaxial deformation with $\gamma = 11.4°$ in the Davidov-Fillipov model [35]. Thus, the present results are of interest in view of nuclear shapes, although it may be an open question as to whether the classical picture of shapes can make sense in such light nuclei. However, for ^{12}Be, We definitely need a larger model space, and it is not tractable presently.

In summary, for the first time, we have applied the no-core MCSM with realistic UCOM-transformed interactions to the investigation of structure of ^{10}Be and ^{12}Be. We calculate some low-lying states of ^{10}Be and ^{12}Be in an $e_{\rm max}$=3 model space. The results for the 2_1^+ and 2_2^+ states of ^{10}Be show a reasonable agreement with experimental data. Some properties of low-lying states of ^{10}Be are studied in terms of quadrupole moments and E2 transitions.

Table 1. (a) B(E2; $2_1^+ \to 0_{g.s.}^+$), (b) B(E2; $2_2^+ \to 0_{g.s.}^+$) and (c) B(E2; $2_2^+ \to 2_1^+$) values (e^2 fm^4) of ^{10}Be and those values of the mirror nucleus ^{10}C obtained by the MCSM and the experimental data [32, 33].

	^{10}Be			^{10}C		
	(a)	(b)	(c)	(a)	(b)	(c)
Exp.	9.2(3)	0.11(2)		8.8(3)		
MCSM	9.29	0.32	3.28	9.30	2.15	12.81

Acknowledgments
This work has been supported by Grants-in-Aid for Scientific Research ((A)20244022) and for Scientific Research on Innovative Areas (20105003) from JSPS. It has also been supported by the CNS-RIKEN joint project for large-scale nuclear structure calculations. R. Roth acknowledges support from the DFG (SFB 634) and from HIC for FAIR.

References
[1] Wiringa R B, Stoks V G J and Schiavilla R 1995 *Phys. Rev.* C **51** 38–51
[2] Machleidt R 2001 *Phys. Rev.* C **63** 024001
[3] Stoks V G J, Klomp R A M, Terheggen C P F and de Swart J J 1994 *Phys. Rev.* C **49** 2950
[4] Entem D R and Machleidt R 2003 *Phys. Rev.* C **68** 041001(R)
[5] Epelbaum E 2006 *Prog. Part. Nucl. Phys.* **57** 654–741
[6] Machleidt R and Entem D R 2011 *Phys. Rep.* **503** 1–75
[7] Pieper S C and Wiringa R B 2001 *Annu. Rev. Nucl. Part. S.* **51** 53–90
[8] Pieper S C, Varga K and Wiringa R B 2002 *Phys. Rev.* C **66** 044310
[9] Pieper S C 2005 *Nucl. Phys.* A **751** 516–32
[10] Navrátil P, Vary J P and Barrett B R 2000 *Phys. Rev.* C **62** 054311
[11] Caurier E, Navrátil P, Ormand W E and Vary J P 2002 *Phys. Rev.* C **66** 024314
[12] Navrátil P, Quaglioni S, Stetcu I and Barrett B R 2009 *J. Phys. G: Nucl. Part. Phys.* **36** 083101
[13] Feldmeier H, Neff T, Roth R and Schnack J 1998 *Nucl. Phys.* A **632** 61–95
[14] Neff T and Feldmeier H 2003 *Nucl. Phys.* A **713** 311–71
[15] Roth R, Neff T and Feldmeier H 2010 *Prog. Part. Nucl. Phys.* **65** 50–93
[16] Roth R and Navrátil P 2007 *Phys. Rev. Lett.* **99** 092501
[17] Roth R, Hergert H, Paar N and Papakonstantinou P 2007 *Nucl. Phys.* A **788** 12–19
[18] Roth R, Gour J R and Piecuch P 2009 *Phys. Rev.* C **79** 054325
[19] Honma M, Mizusaki T and Otsuka T 1995 *Phys. Rev. Lett.* **75** 1284–7
[20] Mizusaki T, Honma M and Otsuka T 1996 *Phys. Rev.* C **53** 2786
[21] Honma M, Mizusaki T and Otsuka T 1996 *Phys. Rev. Lett.* **77** 3315–8
[22] Otsuka T, Honma M and Mizusaki T 1998 *Phys. Rev. Lett.* **81** 1588–91
[23] Otsuka T, Honma M, Mizusaki T, Shimizu N and Utsuno Y 2001 *Prog. Part. Nucl. Phys.* **47** 319–400
[24] Utsuno Y, Otsuka T, Mizusaki T and Honma M 1999 *Phys. Rev.* C **60** 054315
[25] Shimizu N, Otsuka T, Mizusaki T and Honma M 2001 *Phys. Rev. Lett.* **86** 1171–4
[26] Mizusaki T, Otsuka T, Honma M and Brown B A 2001 *Phys. Rev.* C **63** 044306
[27] Utsuno Y, Otsuka T, Mizusaki T and Honma M 2001 *Phys. Rev.* C **64** 011301
[28] Shimizu N, Utsuno Y, Mizusaki T, Otsuka T, Abe T and Honma M 2010 *Phys. Rev.* C **82** 061305
[29] Abe T, Maris P, Otsuka T, Shimizu N, Utsuno Y and Vary J P 2011 *AIP Conf. Proc.* **1355** 173
[30] Abe T, Maris P, Otsuka T, Shimizu N, Utsuno Y and Vary J P 2012 *Phys. Rev.* C **86** 054301
[31] Liu L, Otsuka T, Shimizu N, Utsuno Y and Roth R 2012 *Phys. Rev.* C **86** 014302
[32] McCutchan E A, Lister C J, Wiringa R B *et al* 2009 *Phys. Rev. Lett.* **103** 192501
[33] McCutchan E A, Lister C J, Pieper S C *et al* 2012 *Phys. Rev.* C **86** 014312
[34] Timmer G A 1976 Ph.D. thesis University of Utrecht
[35] Davydov A S and Filippov G F 1958 *Nucl. Phys.* **3** 237

Nuclear spectroscopy with fast exotic beams: News on $N = 28$ from recent NSCL measurements

Alexandra Gade

National Superconducting Cyclotron Laboratory and Department of Physics and Astronomy, Michigan State University, East Lansing, MI 48824, USA

E-mail: gade@msu.edu

Abstract. The nuclear potential and resulting shell structure are well established for the valley of stability, however, dramatic modifications to the familiar ordering of single-particle orbitals in rare isotopes with a large imbalance of proton and neutron numbers have been found: new shell gaps emerge and conventional magic numbers are no longer valid. Current efforts in nuclear structure physics are aimed at unraveling the driving forces behind this structural evolution, which was found most dramatic in neutron-rich species. This manuscript will outline some of the recent efforts at NSCL aimed at shedding light on the shell evolution around neutron number $N = 28$ in neutron-rich Ar, Cl and Si isotopes.

1. Introduction

The goal of a theoretical description of the atomic nucleus with predictive power is driving experimental and theoretical research programs around the world. The nuclear shell model is one of the cornerstones for a comprehensive understanding of nuclei. Strong efforts in the field are aimed at unraveling the driving forces behind structural departures from the well-established, traditional shell model, which have been observed mostly in nuclei with a large neutron excess [1, 2, 3].

Among the robust driving forces of shell evolution identified in neutron-rich nuclei so far are the spin-isospin parts of the nucleon-nucleon NN interaction [4], in particular the monopole part of the tensor force [5]. The important roles of $3N$ forces and the central part have been pointed out recently as well [6, 7]. Toward the nucleon driplines, the density dependence of the spin-orbit force and couplings to the particle continuum were shown to become important [8]. Examples for the shell evolution in neutron-rich nuclei driven by the tensor force are the breakdown of $N = 20$ as a magic number in neutron-rich nuclei in the vicinity of ^{32}Mg and the emergence of new shell gaps at $N = 14, 16$ in the neutron-rich *sd* shell, respectively [4, 5].

In terms of shell evolution, the neutron magic number $N = 28$ has attracted much attention in recent years. On the neutron-rich side of the nuclear chart, below doubly magic $^{48}\mathrm{Ca}_{28}$, the $N = 28$ shell closure was shown to disappear progressively below $Z = 20$ in $^{44}\mathrm{S}_{28}$ [9] and $^{42}\mathrm{Si}_{28}$ [10], however, with conflicting experimental data on $^{46}\mathrm{Ar}_{28}$ [11, 12, 13]. Most of the initial spectroscopic information mentioned above comes from measurements of the 2^+_1 energy, $E(2^+_1)$, and the absolute $B(E2; 0^+_1 \to 2^+_1) \equiv B(E2\uparrow)$ quadrupole excitation transition strength in even-even nuclei. These observables have been used extensively as indicators signaling the breakdown or persistence of a shell gap, with a high 2^+_1 energy and low $B(E2\uparrow)$ excitation strength at a magic number (reduced collectivity) and vice versa mid-shell (collective character or deformation). The properties of odd-A or odd-odd nuclei have not been exploited as extensively. The following sections will summarize recent measurements performed at NSCL [14] that used Coulomb excitation and in-beam γ-ray spectroscopy following secondary fragmentation reactions to benchmark shell-model calculations for very neutron-rich Ar, Cl and Si isotopes.

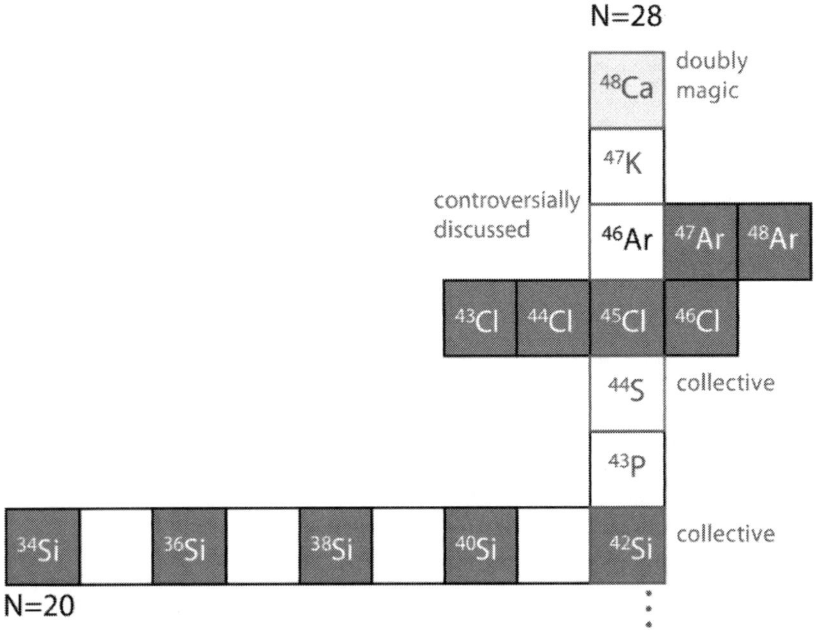

Figure 1. (Color online) Rapid shell evolution has been observed along the $N = 28$ isotone line south of doubly magic ^{48}Ca as ^{42}Si is approached. Ar, Cl and Si isotopes recently studied at NSCL are shaded.

2. In-beam γ-ray spectroscopy at NSCL

Experimental approaches aimed at tracking the changes in nuclear shell structure are manifold. One way of probing nuclear structure in quantitative ways is the application of nuclear reactions that selectively probe specific degrees of freedom, e.g. single-particle or collective degrees of freedom. Coulomb excitation of nuclei has long been used to investigate collectivity. Intriguing possibilities arise along isotonic and isotopic chains of nuclei where the onset of collectivity or the migration of single-particle levels can be tracked consistently as function of the neutron number. Other less selective reactions, as for example secondary fragmentation of fast projectiles, populate the low-lying excited states of nuclei and can provide a first glimpse at the excitation level schemes of exotic nuclei [3].

At in-flight facilities, exotic medium-mass and light nuclei furthest away from stability can be efficiently produced by fragmentation of stable beams impinging upon stable targets at high energy. The resulting secondary beams of rare isotopes are then available for experiments at velocities typically exceeding $v/c = 0.3$. Well-established experimental techniques used for decades to explore the structure of stable nuclei are not applicable at the low beam rates encountered for the shortest-lived species. Powerful new experimental techniques have been developed to enable in-beam spectroscopy studies of fast rare-isotope beams with intensities several orders of magnitude less than needed for the established low-energy techniques.

The intensities of rare-isotope beams are typically lower than stable-beam rates by orders of magnitude. The experimental technique of in-beam γ-ray spectroscopy compensates for the reduced intensities by employing thick reaction targets, taking advantage of the high beam velocity, and enabling experiments at luminosities comparable to stable-beam experiments but with projectile rates of up to a factor of 10000 less. At NSCL, rare-isotope induced reactions, for example intermediate-energy Coulomb excitation, nucleon knockout or secondary fragmentation reactions, are induced by thick reaction targets (several hundred mg/cm^2) and with the detection of γ rays for the identification of the reaction residue's final state or to tag the inelastic process in Coulomb excitation measurements [3]. Contrary to barrier-energy Coulomb excitation, intermediate-energy Coulomb excitation of exotic projectile beams selectively excites the first and higher-lying 2^+ states and possibly 3^- states *in a single step* and therefore possesses a different and unique sensitivity compared to its lower-energy counter part.

Since the γ-ray emission from the reaction residue occurs *in flight*, the γ-ray detection systems have to be granular to enable an angle-dependent, event-by-event reconstruction of the Doppler-shifted γ-ray energies into the rest frame of the emitter. The choice of the target material is dictated by the desired reaction: secondary fragmentation and one- and two-nucleon knockout reactions [15] are typically induced by light targets, for example ^{9}Be or ^{12}C, while intermediate-energy Coulomb excitation [16, 17] requires high-Z targets like gold, lead or bismuth. For the measurements described in this manuscript, the reacted beam emerging from the target was identified with the large acceptance S800 magnetic spectrograph [18] to uniquely select the reaction channel of interest. The γ-ray spectroscopy was performed with NSCL's highly segmented Ge detector array SeGA [19] or the high-efficiency CsI(Na) scintillator array CAESAR [20]. Similar in-beam γ-ray spectroscopy programs with fast fragmentation beams are pursued at GSI [21] in Germany, at RIBF/RIKEN [22] in Japan and at GANIL [23] in France.

3. Coulomb excitation of the very neutron-rich ^{47,48}Ar

The role of the Ar isotopes around $N = 28$ is of great interest. They are, with element number $Z = 18$, located between doubly-magic ^{48}Ca and the already collective S isotopes ($Z = 16$) on the path toward ^{42}Si, which has the lowest-lying 2_1^+ state along the $N = 28$ isotonic chain. Pioneering intermediate-energy Coulomb excitation measurements [11, 9] showed that ^{40,42,44}S are collective with high $B(E2\uparrow)$ excitation strengths while, in two independent measurements, the $B(E2\uparrow)$ value for ^{46}Ar was found low as one may expect for a persistence of an $N = 28$ shell gap [11, 12]. Shell-model calculations were unable to explain the reduced collectivity in ^{46}Ar [24, 25] and already predict the breakdown of $N = 28$ as a magic number and an onset of collectivity in the chain of Ar isotopes. A recent, low-statistics lifetime measurement for the 2_1^+ state extracted a much higher $B(E2\uparrow)$ value in agreement with the shell-model description [13] but contradicts the two previous measurements.

Motivated by the pivotal location of the Ar chain, the first study of quadrupole collectivity in the $N = 30, 29$ isotopes ^{48,47}Ar was performed at the NSCL with intermediate-energy Coulomb excitation using SeGA in conjunction with the S800 spectrograph [26]. $B(E2; 0_1^+ \to 2_1^+)$ and $B(E2; 3/2_1^- \to J)$ values were determined from Coulomb excitation cross sections and compared to state-of-the-art shell model calculations. The following paragraphs briefly summarize the findings reported in [26].

A consistent description of the collectivity at $N = 28$ in the isotopic chains of sulfur ($Z = 16$) and silicon ($Z = 14$) has been a challenge for large-scale shell-model calculations and was realized by Nowacki and Poves by devising two effective interactions (SDPF-U) for the *sdpf* model space, one valid for $Z \leq 14$ and one valid for $Z > 14$ [25]. In a recent shell-model work by Kaneko *et al.*, the extended pairing plus quadrupole-quadrupole force with inclusion of a monopole interaction (EPQQM) was proposed to provide a consistent description of the breakdown of $N = 28$ across the $Z = 20$ to $Z = 14$ isotopic chains [27]. In Fig. 2, the shell model

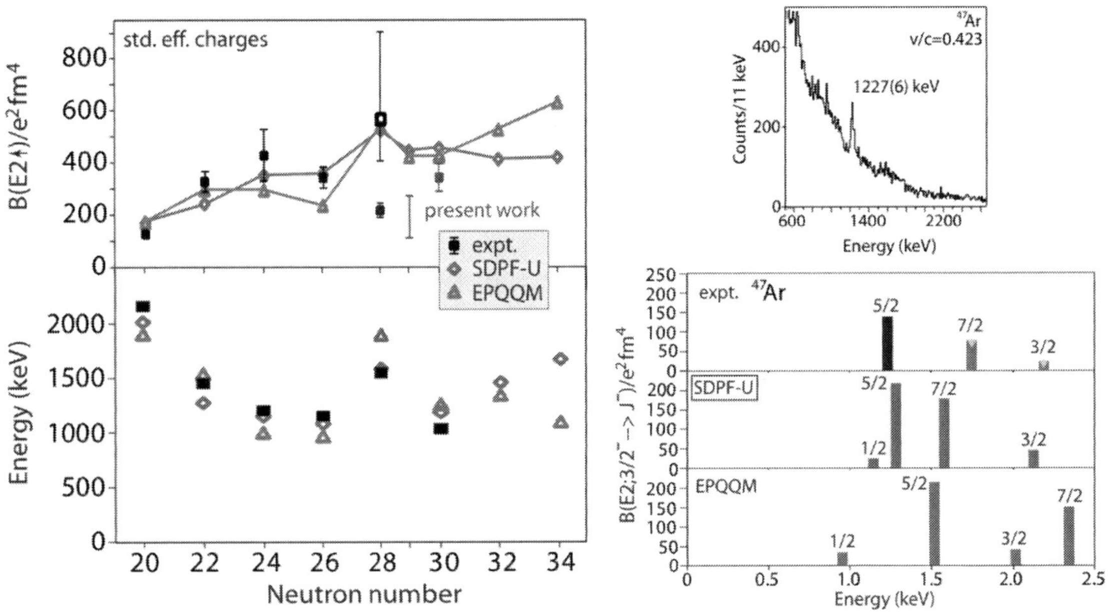

Figure 2. (Color online) Left: $B(E2)$ values and 2^+ excitation energies for the chain of Ar isotopes. Upper right: γ-ray spectrum of ^{47}Ar, showing the decay from the alleged $5/2^-$ states to the $3/2^-$ ground state. Lower right: $B(E2)$ values in ^{47}Ar. Figure adapted from [26].

$B(E2; 0_1^+ \rightarrow 2_1^+)$ values calculated with the two different effective interactions are compared to the available measured values in the chain of Ar isotopes. The two calculations describe the trend of the data on the even-even isotopes well only if one neglects the two consistent, low $B(E2\uparrow)$ values at $N = 28$ [11, 12] and assumes that the low-statistics excited-state lifetime measurement for ^{46}Ar stands. For ^{48}Ar the calculation is within 2-σ of the experimental value for the SDPF-U effective interaction. For ^{48}Ar, the EPQQM calculation is within 1.5-σ of the measured value. At $N = 28$, both calculations predict high $B(E2\uparrow)$ and $E(2_1^+)$ values, deviating from established Grodzins systematics that rather suggests their anti-correlation as observed for ^{38}Ar$_{20}$, for example.

With conflicting experimental results at ^{46}Ar and rather robust shell-model calculations at $N = 28$, it is interesting to turn to the odd-N neighbor ^{47}Ar. For ^{47}Ar, the coupling of the odd $p_{3/2}$ proton to the ^{46}Ar 2_1^+ state would give rise to a quartet of states with spin values of $1/2^-$, $3/2^-$, $5/2^-$, and $7/2^-$ at about $E(2_1^+(^{46}\mathrm{Ar})) = 1555$ keV [12] and with a total $E2$ excitation strength of $\Sigma_J B(E2; 3/2^- \rightarrow J) = B(E2\uparrow)_{^{46}\mathrm{Ar}}$. The extracted $B(E2; 3/2^- \rightarrow J^-)$ strength distribution for the lowest-lying excited $1/2^- - 7/2^-$ states is shown in Fig. 2. The experimental values for $J = 7/2$ and $3/2$ are upper limits as described in detail in [26]. The total $\Sigma_J B(E2; 3/2^- \rightarrow J^-)$ strength concentrated in these states is included in the $B(E2)$ systematics of Fig. 2 at $N = 29$ for experiment as well as theory. The comparison shows that both shell-model effective interactions significantly overpredict the $B(E2)$ strength concentrated in the low-lying states of ^{47}Ar. The experimental $B(E2)$ range given in the figure indicates the possible span of the total $B(E2)$ strength ranging from the sole excitation of the $5/2^-$ states to the maximum strength observed in case the upper limits for the excitation of the $7/2^-$ and $3/2^-$ members of the quartet should be realized. This discrepancy is consistent with the situation at $N = 28$ if the measured low $B(E2\uparrow)$ value for ^{46}Ar should be proven correct. This result for ^{47}Ar indicates that challenges remain for the most modern shell-model effective interactions in this area of rapid structural change and it also demonstrates the rarely exploited, unique benchmark posed by measurements of collectivity in odd-A nuclei.

4. In-beam γ-ray spectroscopy of neutron-rich Cl isotopes

In order to inform theoretical descriptions of the region around $N = 28$ more, in-beam γ-ray spectroscopy following secondary fragmentation of a ^{48}K beam was used to investigate the low-energy structure of $^{43-46}$Cl [28]. One new transition in ^{43}Cl, three in ^{44}Cl, six in ^{45}Cl, and eight new transitions in ^{46}Cl, for which no previous spectroscopic data exist, were observed in a spectroscopy measurement using the S800 spectrograph and SeGA. γ-γ coincidences registered with SeGA were used to produce experimental level schemes up to 2 MeV for ^{43}Cl and ^{45}Cl, although no spin-parity assignments could be made based on the measured data. These level schemes were compared to large-scale shell model calculations, showing generally excellent agreement with the SDPF-U interaction. The results for ^{45}Cl are displayed in Fig. 3 as an example. Additionally, the lifetime of the first excited state in ^{45}Cl was obtained through a γ-ray lineshape analysis, yielding a lifetime one order of magnitude shorter than that predicted by the SDPF-U interaction, and four orders of magnitude shorter than that predicted by the EPQQM interaction. In [28], this observation is discussed in the framework of tensor corrections to the effective M1 operator for ℓ-forbidden transitions. For the odd-odd nuclei ^{44,46}Cl, the two shell-model effective interactions differ widely, demonstrating the power of the spectroscopy of odd-odd nuclei to benchmark shell model, however, firm spin assignments are needed and have to remain a challenge for future experimental studies.

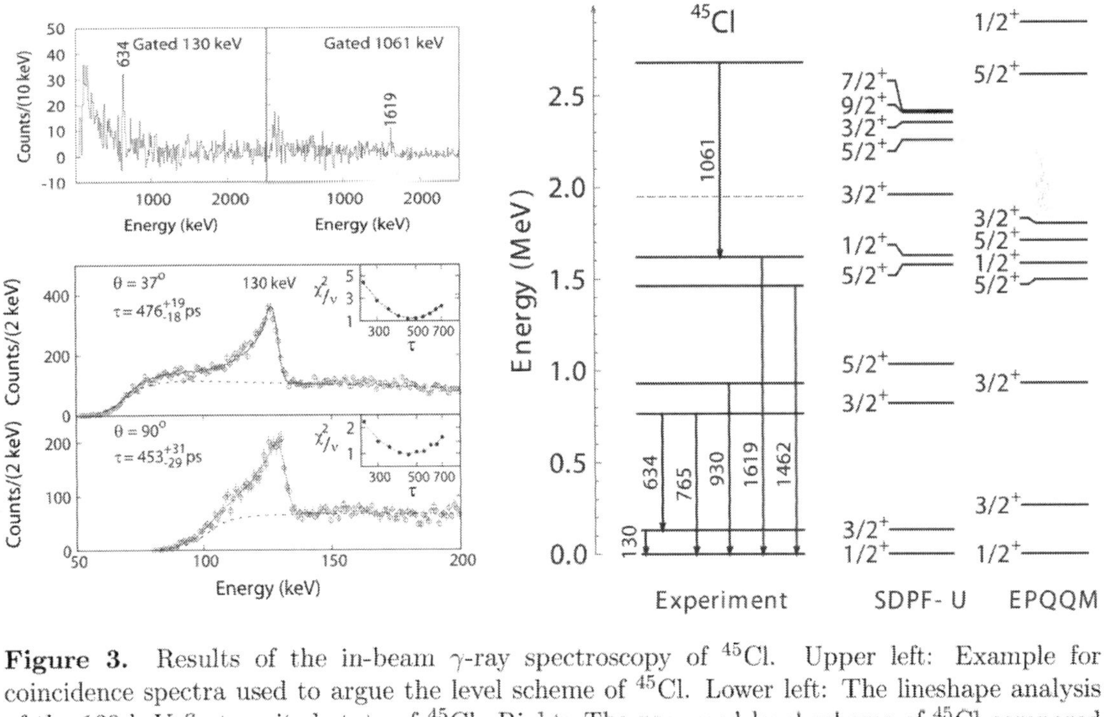

Figure 3. Results of the in-beam γ-ray spectroscopy of ^{45}Cl. Upper left: Example for coincidence spectra used to argue the level scheme of ^{45}Cl. Lower left: The lineshape analysis of the 130-keV first excited state of ^{45}Cl. Right: The proposed level scheme of ^{45}Cl compared to calculations with the two shell-model effective interactions introduced in Section 3. Figures adapted from [28].

5. Collectivity along the chain of Si isotopes

The $B(E2;0_1^+ \rightarrow 2_1^+)$ excitation strength for ^{34,36,40,42}Si were measured at NSCL using the high-efficiency scintillator array CAESAR [20] in conjunction with the S800 spectrograph. Preliminary results indicate that the novel SDPF-MU effective interaction [29] that explicitly includes contributions from the tensor force describes the collectivity in the chain of Si isotopes best. The results are presently being finalized [30].

6. Summary

Neutron-rich nuclei around $N = 28$ have provided an exceptional insight into the phenomenon of shell evolution. Complementary experimental methods are employed around the world to probe various aspects of nuclear structure and serve as valuable benchmarks for nuclear theory. At NSCL, neutron-rich Ar [26], Cl [28] and Si [30] isotopes were studied experimentally and confronted with large-scale shell model calculations using two recent effective interactions. The important and under-utilized role of the spectroscopy of odd-odd, odd-A, and in particular the measurement of quadrupole collectivity in odd-A nuclei was shown to pose sensitive benchmarks for state-of-the-art shell models.

Acknowledgments

This work was supported by the National Science Foundation under Grants No. PHY-0606007, No. PHY-1102511, and No. PHY-1068217. Over the many years of collaboration, countless inspiring discussions with Takaharu Otsuka are acknowledged that shaped the in-beam γ-ray spectroscopy program at NSCL.

References

[1] Brown BA 2001 *Prog. Part. Nucl. Phys.* **47** 517
[2] Sorlin O and Porquet MG 2008 *Prog. Part. Nucl. Phys.* **61** 602
[3] Gade A and Glasmacher T 2008 *Prog. Part. Nucl. Phys.* **60** 161
[4] Otsuka T et al. 2001 *Phys. Rev. Lett.* **87**, 082502
[5] Otsuka T et al. 2005 *Phys. Rev. Lett.* **95** 232502
[6] Otsuka T et al. 2010 *Phys. Rev. Lett.* **104** 012501
[7] Otsuka T et al. 2010 *Phys. Rev. Lett.* **105** 032501
[8] Dobaczewski J et al. 2007 *Prog. in Part. and Nucl. Phys.* **59** 432
[9] Glasmacher T et al. 1997 *Phys. Lett.* B **395** 163
[10] Bastin B et al. 2007 *Phys. Rev. Lett.* **99** 022503
[11] Scheit H et al. 1996 *Phys. Rev. Lett.* **77** 3967
[12] Gade A et al. 2003 *Phys. Rev.* C **68** 014302
[13] Mengoni D et al. 2010 *Phys. Rev.* C **82** 024308
[14] Gade A and Gelbke CK 2010 *Scholarpedia* **5(1)** 9651 and http://www.nscl.msu.edu/
[15] Hansen PG and Tostevin JA 2003 *Annu. Rev. Nucl. Part. Sci.* **53** 221
[16] Glasmacher T 1998 *Annu. Rev. Nucl. Part. Sci.* **48** 1
[17] Motobayashi T et al. 1995 *Phys. Lett.* B **346** 9
[18] Bazin D et al. 2003 *Nucl. Instrum. Methods in Phys. Res., Sect.* B **204** 629
[19] Mueller WF et al. 2001 *Nucl. Instrum. and Methods in Phys. Res., Sect.* A **466** 492
[20] Weisshaar D et al. 2010 *Nucl. Instrum. and Methods in Phys. Res., Sect.* A **624** 615
[21] http://www.gsi.de/portrait/index
_e.html
[22] http://www.rarf.riken.go.jp/Eng/index.html
[23] Gales S 2010 *Scholarpedia* **5(7)** 9670
[24] Nummela S et al. 2009 *Phys. Rev.* C **63** 044316
[25] Nowacki F and Poves A 2009 *Phys. Rev.* C **79** 014310
[26] Winkler R et al. 2012 *Phys. Rev. Lett.* **108** 182501
[27] Kaneko K et al. 2011 *Phys. Rev.* C **83** 014320
[28] Stroberg SR et al. 2012 *Phys. Rev.* C **86** 024321
[29] Utsuno Y et al. 2012 *Phys. Rev.* C **86** 051301
[30] Ratkiewicz A et al. 2013, to be submitted

Ground state moments and spins of neutron-rich K and Al isotopes as probes for changes in nuclear structure

G Neyens

KU Leuven, Instituut voor Kern- en Stralingsfysica, B-3001 Heverlee, Belgium

E-mail: gerda.neyens@fys.kuleuven.be

Abstract. This contribution focusses on recent measurements of nuclear moments and spins using two complementary methods. The ground state magnetic moments and spins of the exotic isotopes ^{49,51}K have been measured at the ISOLDE facility at CERN using bunched-beam high-resolution collinear laser spectroscopy. The re-inversion of the ground state spin from $I = 1/2$ in ^{47,49}K back to the normal $I = 3/2$ in ^{51}K has been established. At GANIL (Caen, France) the quadrupole moment of the ^{33}Al ground state has been measured using the continuous-beam β-nuclear magnetic resonance method applied to a spin-polarized beam produced at the LISE fragment separator. The large value establishes a very mixed wave function with about equal amounts of normal and neutron particle-hole excited configurations contributing to its ground state wave function. This illustrates the transitional nature of isotopes at the border of the island-of-inversion.

1. Introduction

Static moments of nuclear ground states are ideal probes to study their wave function: the magnetic moment provides information on the nature of the unpaired single particle nature of the wave function while the quadrupole moment is also sensitive to deformation and core-polarization effects. Experimental techniques to study these properties for exotic nuclei are constantly optimized to reach out to the most exotic regions.

At ISOLDE-CERN the collinear laser spectroscopy method is used to study in a single experiment the spin, the magnetic dipole moment, the electric quadrupole moment and the mean square charge radius of long-lived isomeric states and ground states in a model-independent way [1]. In the last years, the implementation of an ion cooler-buncher [2] has lead to an improvement of the experimental sensitivity by a factor 500. This allow experiments on beams of a few 10^4 ions/s using the bunched-beam collinear laser spectroscopy technique [3, 4]. To study the exotic potasium isotopes up to ^{51}K, produced at less than 5000/s, a further improvement in sensitivity of almost an order of magnitude was needed. This was achieved by building a dedicated light collection region with lenses for enhancing the geometrical efficiency and dedicated shielding for reducing background photon counting [5].

Not all isotopes are produced efficiently at an ISOL-type facility and other methods are needed to study moments of nuclei produced e.g. in fragmentation reactions and selected as fast beams using an in-flight spectrometer. The β-nuclear magnetic resonance (β-NMR) method is ideally suited to study the magnetic dipole moment or electric quadrupole moment, by implanting a

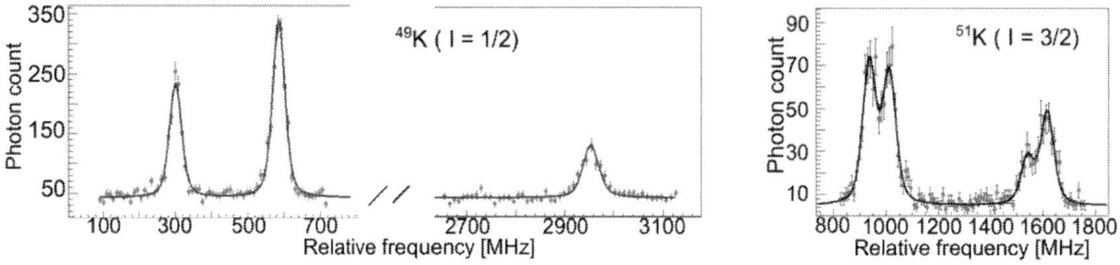

Figure 1. Measured hyperfine structures for ^{49}K and ^{51}K.

polarized beam in a suitable crystal [6]. The polarized ensemble of exotic nuclei is obtained by making a proper selection in the transverse and longitudinal momentum distributions of the fragments [7, 8, 9]. With this method polarized beams of Al isotopes up to ^{34}Al are produced at rates of a few 1000/s at the LISE fragment separator at GANIL.

In this paper we present results from both types of experiments, and we concentrate on the nuclear structure information that is deduced from the measured spin, magnetic moment or quadrupole moment.

2. Ground state spins and magnetic moments of ^{49,51}K.

The potassium isotopes have 19 protons and the properties of the ground state of the odd-A isotopes is therefore dominated by the single hole in the magic calcium isotopes. The protons occupy the positive parity sd shell while the neutrons occupy the sd shell below $N = 20$ and the pf shell from $N = 21$ onwards. Since many years it is known that the energy difference between the $\pi 1d_{3/2}$ and $\pi 2s_{1/2}$ single particle levels is strongly decreasing from ^{40}Ca to ^{48}Ca, which is illustrated by the lowering of the $1/2^+$ first excited state in the odd-K isotopes. This leads to an inversion of the ground state spin from 3/2 in $^{39-45}$K to 1/2 in ^{47}K [10]. The migration of the proton single particle levels has been attributed to the isospin-dependent monopole residual interaction, which is strongest between levels with the same number of nodes and changes with increasing number of neutrons occupying a certain orbit [11, 12, 13]. For the K-isotopes between $N = 20$ and $N = 28$, it is the attractive $\pi 1d_{3/2} - \nu 1f_{7/2}$ interaction that induces the lowering of the $\pi 1d_{3/2}$ level with respect to the $\pi 2s_{1/2}$ level. Beyond $N = 28$ the $\nu 2p_{3/2}$ level is filled and in these isotopes the monopole interaction between the orbits the ($\pi 2s_{1/2}$ and $\nu 2p_{3/2}$) is the dominant one (as orbits with the same radial quantum number are interacting more strongly). Different theoretical models predict a different evolution of the single particle levels beyond $N = 28$ and therefore experimental data are crucial to test and improve the parametrization of the models.

With this aim we have studied the ground state spins and magnetic moments of the neutron-rich ^{49,51}K isotopes, with respectively $N = 30, 32$. From the measured hyperfine structures in the $^2S_{1/2} \rightarrow ^2P_{1/2}$ transition, a spin $I = 1/2$ is clearly established for the ^{49}K ground state, as only three transitions are observed (figure 1) [5]. The ground state of ^{51}K has spin $I = 3/2$ or larger, considering that four transitions are seen in the hyperfine structure. Its spin is determined by comparing the measured magnetic moment to that of the calculated $3/2^+$ ground state in different large scale shell model calculations with effective interactions for this mass region (details in [5]). The SDPF-NR [14, 15] and its recently upgraded SDPF-U interaction [16] have monopole matrix elements that were tuned by fitting experimental spectra of isotopes in this region, from O to Ca. The SDPF-MU interaction [17] is based on the recently developed monopole-based universal interaction V_{MU} [13] and involves therefore fewer fitted parameters.

Calculations are performed with protons restricted to the *sd* shell and neutrons in the full *pf* shell.

A very good agreement is found between the experimental magnetic moment and all theoretical values for the calculated $I = 3/2^+$ ground states of ^{51}K (figure 2). This establishes the re-inversion to a spin 3/2 ground state in ^{51}K [5]. None of the magnetic moments calculated for the ^{49}K, $I^\pi = 1/2^+$, state agrees very well with the observed value. This suggests that the $\pi 2s^{-1}_{1/2}$ configuration is very likely significantly mixed with the $(\pi 1d^{-1}_{3/2}\nu(2^+))_{1/2^+}$ configuration. Indeed, the magnetic moment of such a mixed wave function agrees with the experimental value for a 25% mixing, as illustrated in figure 3.

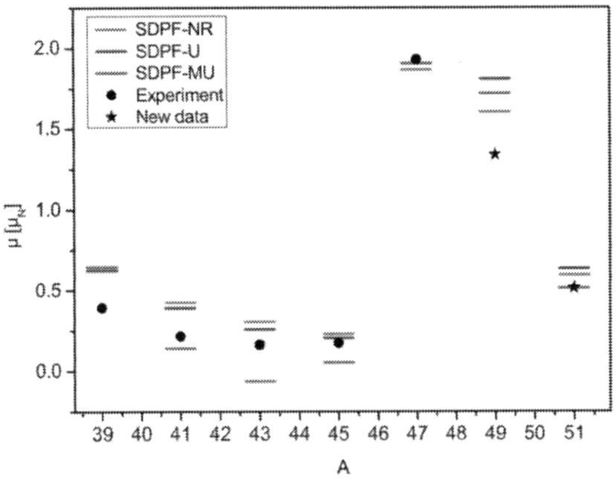

Figure 2. Experimental magnetic moments compared to shell model calculations (see [5] for details).

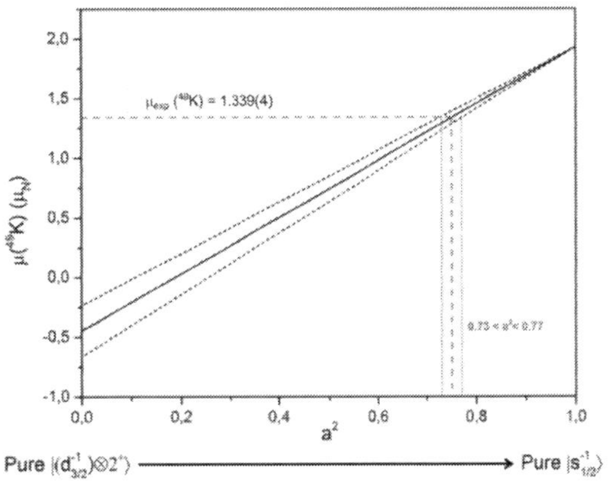

Figure 3. Calculated magnetic moment of the ^{49}K $1/2^+$ state for a mixed wave function.

In figure 4, we compare the experimental $(3/2^+ - 1/2^+)$ energy differences with the calculated energy differences. Those calculated with the fitted interactions are better in agreement with the data than those calculated with the SDPF-MU. However, considering that the number of fitted parameters is much less in the latter, the agreement is also satisfactory. The re-inversion of the energy difference beyond $N = 28$ is somewhat overestimated with the SDPF-MU interaction. Therefore the present data, and in particular the magnetic moment of ^{49}K, will be helpful in further investigating the role of the different components in the monopole interaction. As can

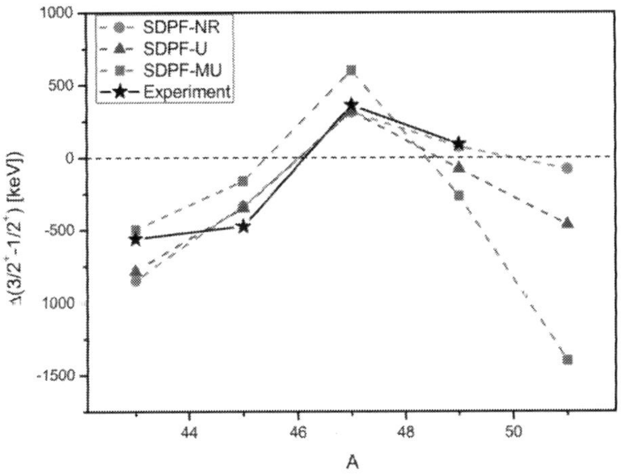

Figure 4. Experimental and calculated energy differences between the $3/2^+$ and $1/2^+$ states in odd K-isotopes. See text and ref. [5] for details.

be seen in figure 4, establishing the energy of the first excited state in ^{51}K is crucial for further improving all of these effective interactions, which predict very different excitation energies for the $1/2^+$ state.

3. The static quadrupole moment of ^{33}Al

The ^{33}Al isotope, with 13 protons and 20 neutrons, is located at the border of the 'island-of-inversion'. From early experiments, the island-of-inversion was suggested to contain neutron-rich Na and Mg isotopes with a deformed ground state [18, 19, 20, 21, 22]. First theoretical studies suggested that the isotopes of Na and Mg with 20-22 neutrons [23] belong to the island of inversion. It was later shown that these isotopes have a ground state wave function that is dominated by the excitation of neutrons from the *sd* shell into the *pf* shell, across a reduced $N = 20$ shell gap (e.g. [24]). Such configurations are called 'intruder' configurations, as they appear lower in the energy spectrum than normally expected. Nuclear moments measurements on Mg isotopes with $N = 19, 20, 21$ and Na isotopes with $N = 19, 20$ have confirmed that these isotopes have an intruder ground state, with two neutrons excited into the *pf* shell [25, 26, 27]. Thus the island-of-inversion extends down to $N = 19$ for the Na and Mg isotopes.

Measured magnetic moments of Al isotopes, occurring at the Z-border of the island-of-inversion, suggested that the ground states of ^{31,32}Al have a normal wave function [28, 29], while ^{33}Al($N = 20$) and ^{34}Al($N = 21$) have a mixed ground state wave function, with a significant contribution of intruder configurations [29, 30]. This was the first signature for the island-of-inversion to extend beyond $Z = 12$.

Quadrupole moments are more sensitive to the occurrence of intruder configurations in the wave function because such configurations are related to an increased deformation. Therefore, experimental campaigns at GANIL and RIKEN were started to study this observable in order to shed more light on the wave function of these isotopes. The production of exotic Al isotopes is higher in fragmentation reactions than at ISOL-type facilities. Furthermore, using a fragmentation reaction for the production of these isotopes has the advantage that a spin-polarized ensemble can be selected [31]. By implanting spin-polarized Al-fragments into a suitable crystal with an axially symmetric electric field gradient, their quadrupole interaction frequency can be measured using the multiple-rf nuclear quadrupole resonance method. Two types of measurements have been performed: the adiabatic fast passage technique using a pulsed fragment beam [32, 33] and the continuous-rf magnetic resonance (MR) technique using the continuous implantation of a fragment beam [34]. The former technique has the advantage that if the proper magnetic resonance conditions are fulfilled, the resonance signal registers the inversion of the ensemble spin polarization. This leads to a signal with the double amplitude as compared to the continuous-rf MR method where the spin-polarization is not inverted, but simple reduced to zero. However, the full spin-inversion is only obtained if all rf-conditions are fully controlled, which is rarely the case. The continuous-rf method has the advantage that the full duty-cycle can be used, while the adiabatic technique requires a pulsed beam and therefore a duty cycle which is typically less than 50%.

The quadrupole moments of the Al isotopes have been investigated by implanting them into an Al_2O_3 (corundum) single crystal. From the ratio of the measured quadrupole frequencies, the ratio of quadruole moments is determined. Thus all quadrupole moments are determined with respect to a reference value, which is chosen to be the stable ^{27}Al isotope. The measured quadrupole frequency of ^{27}Al implanted in the same crystal structure under similar conditions at 77 K is $\nu_Q = 2389(2)$ kHz [35] (used for adiabatic fast passage measurements), while the frequency at room temperature is 2402.5(1.7) kHz [36] (for measurements using the continuous-rf method). As a reference quadrupole moment for ^{27}Al, Q=146.6(10) emb is used [37].

Comparison of the measured quadrupole moments of the less exotic Al isotopes to shell model calculations with and without excitations of neutrons from the *sd* to the *pf* shell, revealed a

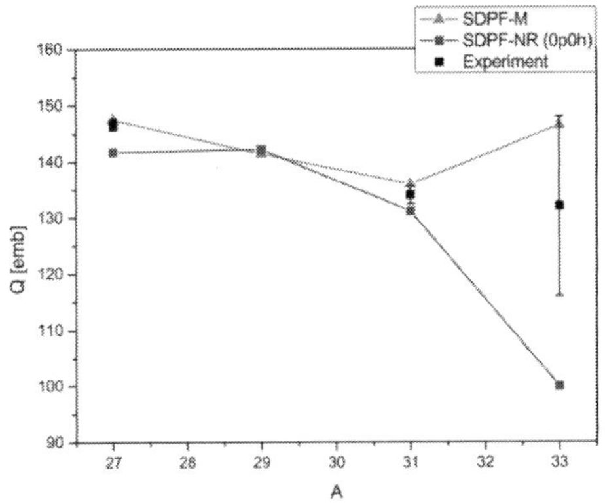

Figure 5. Experimental quadrupole moments of odd-A Al isotopes compared to shell model calculations with protons and neutrons restricted to the sd-shell (SDPF-NR 0p0h) and with neutrons in the full $sd-pf$ shell (SDPF-M). The amount of intruder configurations in the latter calculation for ^{33}Al is 63%.

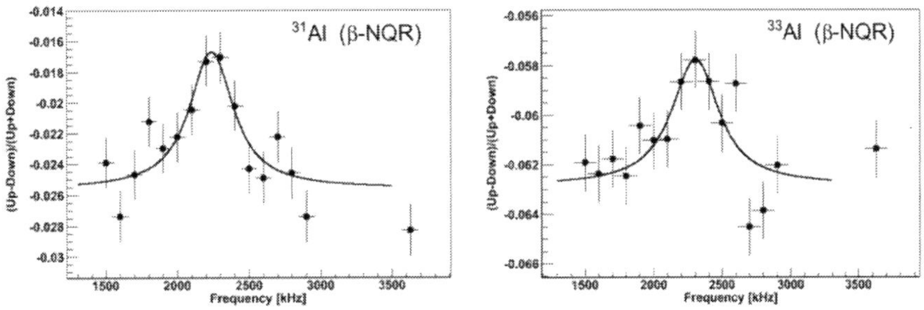

Figure 6. Measured quadrupole resonances for ^{31}Al and ^{33}Al using the continuous-rf nuclear magnetic resonance method, from which the ratio of quadrupole moments is determined to a precision of better than 2% [41].

normal sd wave function for the isotopes of ^{31}Al [38] and ^{32}Al [39], in agreement with the conclusion from the magnetic moments measurements. That is illustrated in figure 5, where the experimental quadrupole moments of the odd-Al isotopes are compared to shell model calculations. The SDPF-M calculations have been performed with the Monte Carlo Shell Model code, with protons restricted to the sd shell and neutrons occupying the full pf shell. The SDPF-NR calculations are performed with the ANTOINE shell model code, with protons and neutrons both restricted to the sd shell. Note that these calculations give a similar result as with the USD-interactions, shown in [40]. The effective charges are respectively 1.1e for protons and 0.5e for neutrons, as used by De Rydt et al. [38]. A similar result is found when the isospin-dependent effective charges are used, as can be seen in the paper by Shimada et al. [40].

The quadrupole moment of the ^{33}Al ground state could be measured recently using the adiabatic fast passage β-NMR method at GANIL. Although the obtained value has a rather large error bar (figure 5), the deviation of the experimental value from the SDPF-NR value (restricting neutrons to the sd shell) is clear and from the good agreement with the SDPF-M value a mixing with intruder configurations in its ground state has been confirmed [40]. A new measurement using the continuous-rf NMR method, also performred at GANIL, allowed to measure the quadrupole resonance frequency with a much higher precision, resulting in a ratio of the ^{33}Al to ^{31}Al quadrupole moment with a precision of less than 2%. The measured resonances from this experiment are shown in figure 6. The results from this work will be published soon [41].

4. Conclusion

In this presentation we have shown some recent results from measurements of nuclear moments and spins, and how these observables are good tests for the nuclear ground state wave functions and shell model calculations.

Acknowledgments

The author wants to thank the COLLAPS collaboration at ISOLDE-CERN and the β-NMR/LISE collaboration at GANIL for their support. This work was financed through several grants from the FWO-Vlaanderen, the IAP projects P6/23 and P7/12 of the OSTC Belgium and a GOA-project of the KU Leuven.

References

[1] Neugart R and Neyens G 2006 Nuclear Moments *Lect. Notes Phys.* vol 700 (Berlin Heidelberg: Springer-Verlag) pp 135-189
[2] Mane E *et al.* 2009 *Eur. Phys. J.* A **42** 503
[3] Vingerhoets P *et. al.* 2010 *Phys. Rev.* C **82** 064311
[4] Flanagan KT *et al.* 2009 *Phys. Rev. Lett.* **103** 142501
[5] Papuga J *et al.* 2013 *submitted to Phys. Rev. Lett.*
[6] Neyens G 2003 *Rep. Prog. Phys.* **66** 633 and 1251
[7] Okubo H *et al.* 1994 *Phys. Lett.* B **335** 29
[8] Borremans D *et al.* 2002 *Phys. Rev.* C **66** 054601
[9] Turzo K *et al.* 2006 *Phys. Rev.* C **73** 044313
[10] Touchard F *et al.* 1982 *Phys. Lett.* B **108** 169
[11] Otsuka T *et al.* 2001 *Phys. Rev. Lett.* **87** 082502
[12] Otsuka T *et al.* 2005 *Phys. Rev. Lett.* **95** 232502
[13] Otsuka T *et al.* 2010 *Phys. Rev. Lett.* **104** 012501
[14] Retamosa J *et al.* 1997 *Phy. Rev.* C **55** 1266
[15] Nummela S *et al.* 2001 *Phys. Rev.* C **64** 054313
[16] Nowacki F and Poves A 2009 *Phys. Rev.* C **79** 014310
[17] Utsuno Y *et al.* 2012 *Phys. Rev.* C **86** 051301(R)
[18] Thibault C *et al.* 1975 *Phys. Rev.* C **12** 644
[19] Lunney D *et al.* 2006 *Eur. Phys. J.* A **28** 129
[20] Motobayashi T *et al.* 1995 *Phys. Lett.* B **346** 9
[21] Niedermaier O *et al.* 2005 *Phys. Rev. Lett.* **94** 172501
[22] Pritychenko BV *et al.* 2000 *Phys. Rev.* C **63** 011305(R)
[23] Warburton EK, Becker JA and Brown BA 1990 *Phys. Rev.* C **41** 1147
[24] Brown BA 2001 *Prog. Part. Nucl. Phys.* **47** 517
[25] Keim M *et al.* 1998 *Proc. Int. Conf. on Exotic Nuclei and Atomic Masses (ENAM98)* (AIP Conf. Proc. vol 455) ed BM Sherrill *et al.* (New York: AIP) p 50
[26] Utsuno Y *et al.* 2004 *Phys. Rev.* C **70** 044307
[27] Neyens G 2011 *Phys. Rev.* C **84** 064310
[28] Borremans D *et al.* 2002 *Phys. Lett.* B **537** 45
[29] Himpe P *et al.* 2006 *Phys. Lett.* B **643** 257
[30] Himpe P *et al.* 2008 *Phys. Lett.* B **658** 203
[31] Borremans D *et al.* 2002 *Phys. Rev.* C **66** 054601
[32] Hardy CJ, Edelstein WA and Vatis D 1986 *J. Mag. Reson.* **66** 470
[33] Minamisona T *et al.* 1993 *Hyp. Int.* **80** 1315
[34] Borremans D *et al.* 2005 *Phys. Rev.* C **72** 044309
[35] Gravina SJ *et al.* 1990 *J. Mag. Reson.* **89** 515
[36] Filsinger B *et al.* 1997 *J. Mag. Reson.* **125** 280
[37] Kello V *et al.* 1999 *Chem. Phys. Lett.* **304** 414
[38] De Rydt M *et al.* 2009 *Phys. Lett.* B **678** 344
[39] Kameda D *et al.* 2007 *Phys. Lett.* B **647** 93
[40] Shimada K *et al.* 2012 *Phys. Lett.* B **714** 246
[41] De Rydt M *et al.* 2013 *in preparation*

International Symposium on Exotic Nuclear Structure From Nucleons (ENSFN 2012) IOP Publishing
Journal of Physics: Conference Series **445** (2013) 012008 doi:10.1088/1742-6596/445/1/012008

Towards unified description of shell evolution — Takaharu Otsuka's achievements

Y Utsuno

Advanced Science Research Center, Japan Atomic Energy Agency, Tokai, Ibaraki 319-1195, Japan

E-mail: utsuno.yutaka@jaea.go.jp

Abstract. Recent studies on exotic nuclei have rapidly advanced the understanding of the evolution of shell structure, to which Takaharu Otsuka has made remarkable contributions particularly in terms of the nuclear force. This paper presents a brief survey of his achievements, focused on the concept of shell evolution due to the tensor force and the three-nucleon force and on related shell-model works.

1. Introduction

Evolution of the shell structure, often called shell evolution, has been attracting much theoretical and experimental interest in the physics of exotic nuclei, as found in many talks devoted to this subject in this conference. Among various recent findings, Professor Takaharu Otsuka has greatly contributed to unveiling the mechanism of the shell evolution (see [1, 2] for overviews by himself). It is an honor for me, as a former student and a close collaborator of Prof. Otsuka, to have an opportunity to survey his remarkable achievements at the conference celebrating his 60th birthday. This paper concentrates on how he has advanced the idea of shell evolution from personal point of view.

Since Mayer and Jensen accounted for the origin of the magic number by introducing strong spin-orbit coupling more than 60 years ago [3], the shell structure has constituted a robust basis of the nuclear structure. It has been widely accepted that the Woods-Saxon potential consisting of a central and a spin-orbit *one-body* term gives a global property of the shell structure as far as the regions of stable nuclei are concerned [4]. A characteristic behavior of the shell structure created by one-body potentials is a gradual and monotonic evolution of the shell structure: difference of neighboring single-particle energies changes rather flat and increases or decreases monotonically with the mass number. As a result, the conventional magic numbers are quite stable.

In exotic nuclei, on the other hand, several phenomena that cannot be accounted for by the conventional picture are found. An early example is the so-called parity inversion in ^{11}Be. Talmi and Unna proposed [5] that the inversion can be explained by a linear extrapolation of observed single-particle spectra of the $N = 7$ isotones $^{12}_{5}$B and $^{13}_{6}$C into $Z = 4$ and that this linearity is due to the *two-body* force. This particular two-body term causing the shell evolution is often called the monopole interaction acting between nucleons in the orbits i and j whose matrix element is defined with the j-j coupled two-body matrix elements as

$$v_{i,j}^{m} = \frac{\sum_J (2J+1)\langle ijJ|V|ijJ\rangle}{\sum_J (2J+1)}, \tag{1}$$

where J runs over possible couplings of angular momentum with a nucleon in i and one in j. The mean (i.e., center-of-mass) energy of a configuration specified by a set of the occupation numbers $\{n_j\}$ is evaluated with the monopole interaction as

$$E(\{n_j\}) = \sum_j \varepsilon_j n_j + \sum_j \frac{1}{2} n_j(n_j-1)v_{j,j}^{m} + \sum_{j\neq j'} n_j n_{j'} v_{j,j'}^{m}, \tag{2}$$

where ε_j is the single-particle energy of the orbit j. Hence, the monopole interaction $v_{i,j}^{m}$ acts as the bond energy between nucleons in i and j. When one defines the effective single-particle energy (ESPE) of an orbit i as a nucleon separation energy from this orbit, it leads to

$$\bar{\varepsilon}_i \equiv \frac{\partial E(\{n_j\})}{\partial n_i} \approx \varepsilon_i + \sum_j v_{i,j}^{m} n_j. \tag{3}$$

The linearity of the shell evolution during the occupation of j is thus obtained. Equation (3) can also cause a non-monotonic evolution of a shell gap $\bar{\varepsilon}_i - \bar{\varepsilon}_{i'}$ depending on the sign of $v_{i,j}^{m} - v_{i',j}^{m}$ for different j. It should be noted that the contribution of the monopole interaction to the effective shell gap is magnified with the occupation number.

Consequently, the monopole interaction plays an essential role in the evolution of the shell structure and also in the resulting many-body properties. For instance, while the $N = 28$ magic structure is rather fragile with the G-matrix interactions, it can be stabilized by tuning the monopole terms [6]. The Strasbourg-Madrid group has demonstrated with large-scale shell-model calculations that an appropriate choice of the monopole interaction leads to a successful description of a wide range of nuclei [7, 8].

Thus, the monopole interaction is one of the most important but undetermined parts of the effective interaction. The empirical monopole shift works well only for the regions with plenty of experimental data to fix monopole matrix elements. Universal understanding of the monopole interaction, or shell evolution, is strongly needed for constructing a general theory of shell formation and for enhancing predictive power for unexplored regions. Otsuka has proposed several novel mechanisms of the shell evolution in terms of the nuclear forces which is severely tested by experiment and microscopic theory. In the following sections, I attempt to display those ideas as comprehensively as possible. In Sect. 2, the importance of the tensor force is demonstrated. It was overlooked for a long time but is now regarded as one of the dominant proton-neutron interactions causing the shell evolution. Section 3 is focused on the three-nucleon force. It gives a solution to the long-standing issue why effective interactions based on microscopic theories failed to describe neutron-rich nuclei due to an incorrect shell structure. Section 4 is directed to shell-model studies to show how they play an indispensable role in clarifying the shell evolution.

2. Shell evolution due to the tensor force

2.1. Importance of spin-isospin dependence

At the beginning of the 2000s, Otsuka has found a phenomenology of shell evolution common to various major shells [9]. Figure 1 shows how this evolution occurs schematically. Here, the case of light nuclei is considered where the β-stability line lies around $N = Z$. For stable nuclei shown in Fig. 1 (a), the splitting of the orbits belonging to the same harmonic oscillator quantum number

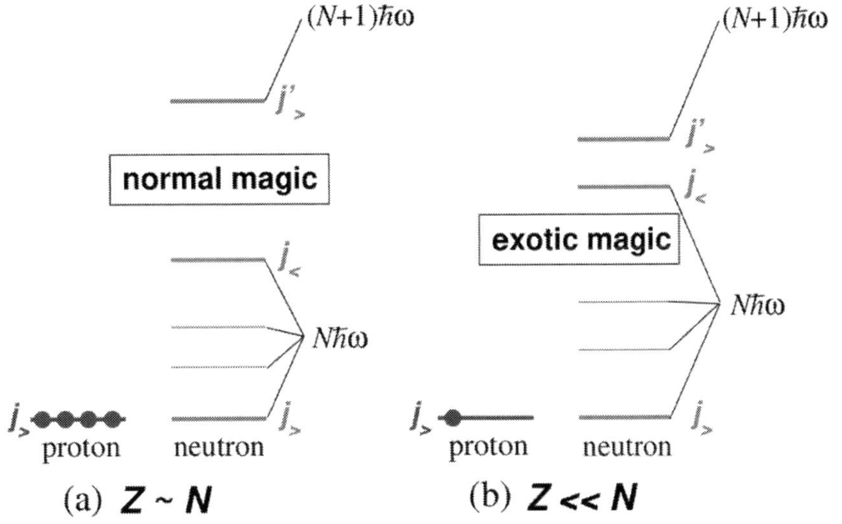

Figure 1. Phenomenological shell evolution compared between (a) stable nuclei with $Z \sim N$ and (b) exotic nuclei with $Z \ll N$.

is not so large that the normal magic number is made up between the harmonic-oscillator shell gap. For exotic nuclei illustrated in Fig. 1 (b), on the other hand, the shell structure seems rather different. The highest orbit of the $N\hbar\omega$ shell denoted as $j_<$ is located too high to stabilize the normal magic number. A new magic number can emerge below $j_<$ instead. This accounts for the appearance of a new magic number $N = 16$ in oxygen isotopes pointed out earlier [12, 13]. When applied to the $0p$ shell, this leads to the parity inversion in ^{11}Be. When one moves to the $0f$-$1p$ shell, a new magic number $N = 34$ could appear around calcium isotopes [14], attracting much experimental interest.

The difference between Fig. 1 (a) and (b) is the proton number. In any case mentioned above, the $j_<$ orbit is sharply lowered especially when protons occupy the $j_>$ orbit, i.e., $Z = 2$-6 for the $0p$ shell, $Z = 8$-14 for the $0d$-$1s$ shell, and $Z = 20$-28 for the $0f$-$1p$ shell. This indicates that the proton-neutron monopole interaction $v^m_{k_1,k_2}$ with $k_1 = j_>$ and $k_2 = j_<$ is quite a strongly attractive one. This effect includes the so-called Federman-Pittel mechanism causing deformation in medium-heavy nuclei [15]. Otsuka has noticed that the interaction depending on spin and isospin is responsible for this property. While the spin-isospin dependent central interaction $(\tau \cdot \tau)(\sigma \cdot \sigma)f(r)$ was introduced originally [9], he proposed the tensor force later [16] as the source of the strong proton-neutron monopole interaction between $j_>$ and $j_<$.

It is worthwhile mentioning here the possibility of the $N = 34$ magic number in further detail. Since a drastic lowering of $0f_{5/2}$ in moving from calcium ($Z = 20$) [10] to nickel ($Z = 28$) [11] is observed on top of the $N = 28$ core, the mechanism discussed above must be valid also for the $0f$-$1p$ shell. It is thus a matter of the neutron-neutron interaction, which is discussed in Sect. 3, whether the new $N = 34$ magic number really survives or not in going from $N = 28$ to 34.

2.2. *Tensor force*
The tensor force is well known to play a crucial role in the binding of nuclei. However, before Otsuka's work, it was almost ignored as the *effective* interaction as exemplified by the absence of the tensor force from standard Skyrme and Gogny interactions. This is possibly because the short-range tensor force, which is rather important in the gain of binding energy, serves predominantly as the effective central force for the independent particle picture. Namely, the

expectation value of the tensor force vanishes when a spin-saturated Slater determinant such as an L-S closure ^{16}O is taken, and then a strong effective central force is required for keeping the binding energy. Indeed, according to an analysis of realistic interactions [17] by using the spin-tensor decomposition [18], the magnitude of the effective tensor force is much smaller than that of the effective central force.

Hence, Otsuka's finding [16] that the tensor force is vital for the shell evolution, a sort of the mean-field effect, was a surprise. The significance of the tensor force in spite of being relatively small comes from its monopole property that is opposite between $j_>$ and $j_<$. The monopole matrix elements of any tensor operator satisfy the equation

$$(2j_> + 1)v^m_{j_>,j'} + (2j_< + 1)v^m_{j_<,j'} = 0 \tag{4}$$

for j' not equal to $j_>$ or $j_<$. This represents that the net effect from a spin saturated wave function ($2j_> + 1$ nucleons in $j_>$ plus $2j_< + 1$ nucleons in $j_<$) on the ESPE of j' vanishes. As a result, $v^m_{j_>,j'}$ and $v^m_{j_<,j'}$ are opposite in sign. As illustrated in his paper (see Fig. 2 of [16]), the realistic tensor force gives, on average, attraction between $j_<$ and $j'_>$ and thus repulsion between $j_>$ and $j'_>$.

Therefore, the tensor force changes the spin-orbit splitting as depicted in Fig. 1 of [16]. The shift of the magic number illustrated in Fig. 1 is one of the manifestations of the shell evolution due to the tensor force. As presented in [16], many experimental data are described in this manner with a single choice of the tensor force of the $\pi + \rho$ exchange potential with cutoff at 0.7 fm, including a famous example of the energy levels in antimony isotopes [19]. Otsuka has successfully introduced the tensor force to the mean-field interaction soon after [20], as the pioneering work of the same kind.

2.3. Renormalization persistency of the tensor force and monopole-based universal interaction

The success of the "bare" $\pi + \rho$ tensor force in the description of the shell evolution should not be taken for granted because the effective interaction is in general modified from the bare interaction due to short-range correlation and rather limited model space. Otsuka has demonstrated [21, 22] that the monopole part of the tensor force is almost unchanged due to the renormalization procedures involving the short-range correlation and also involving the reduction of the model space. This property is referred to as renormalization persistency [22].

Since the $\pi + \rho$ tensor force is thus justified microscopically, it makes sense to examine the remaining part of the effective interaction. From analyses of empirical interactions for the *sd* and *pf* shells, Otsuka has found [21] that the monopole interaction after subtracting the tensor force has a very simple structure: rather weak dependence on spin and strong dependence on node. The $T = 0$ monopole matrix elements are much strongly attractive between the orbits of the same node than between those of different nodes. This property is well simulated by the Gaussian central force. It is amazing that despite the complexity of the effective central force comprising various effects it is expressed so simply at least as far as the monopole part is concerned. As a result, a simple and comprehensive effective interaction, called monopole-based universal interaction (V_{MU}) [21], is obtained for describing the shell evolution in a unified way. The V_{MU} is very useful particularly for regions where a reliable interaction is not available such as for those far from stability. Some applications of the V_{MU} to exotic nuclei are shown in Sect. 4.

3. Shell evolution due to the three-nucleon force

As recognized earlier [6, 7], effective interactions derived from microscopic NN forces are not useful as a *universal* interaction in many cases. This is attributed largely to the $T = 1$ monopole interaction. From comparison to empirically fitted interactions, the $T = 1$ monopole matrix elements of microscopic effective interactions for the pf shell are known to be systematically too attractive [14]. This applies also to the sd shell [21]. From comparison to experiment, for instance, the $N = 28$ magic number for the $\nu(0f_{7/2})^8$ configuration collapses with a microscopic effective interaction because its $T = 1$ monopole interaction between $0f_{7/2}$ and $1p_{3/2}$ is too attractive to stabilize the shell gap between them. Another example is found in the drip line of oxygen isotopes. Microscopic effective interactions predicts the drip line of oxygen isotopes to be located at $N = 20$ or beyond contrary to the experimental drip line located at $N = 16$. The discrepancy for heavy oxygen isotopes is due to a bound neutron $0d_{3/2}$ orbit affected by $T = 1$ attraction from other occupied neutron sd orbits. Thus, correct understanding of the $T = 1$ monopole interaction is essential for describing the neutron shell structure evolving as the neutron number increases.

As an origin of the overbinding of the $T = 1$ monopole interaction, Otsuka has proposed a new mechanism due to the three-nucleon force (3NF) [23]. According to his explanation, the dominant source of the $T = 1$ repulsion is the Fujita-Miyazawa (FM) 3NF in which a virtual excitation of a Δ-particle nucleon-hole occurs as an intermediate state. The 3NF of this type among two valence nucleons and one core nucleon is reduced to a two-body interaction between valence nucleons by summing up all the contributions of the core states. Since this reduced two-body interaction is repulsive as illustrated in [23], the inclusion of this 3NF process gives rise to a repulsive shift of the $T = 1$ monopole interaction as desired. Otsuka has demonstrated also that the FM 3NF accounts for the main part of the repulsive shift by including other types of the chiral 3NFs.

As a result, he has succeeded in reproducing the drip line of oxygen isotopes in a microscopic way without any empirical tuning. This feature has recently been confirmed by a more elaborate coupled-cluster calculation [24] with the inclusion of the chiral 3NFs. A similar effect is demonstrated to occur also in calcium isotopes [25].

4. Shell-model calculations for probing the shell evolution

Although the shell structure is often crucial for the nuclear structure, it does not appear in general through pure single-particle states but through strongly correlated many-body states. Thus, solving a many-body problem to a good accuracy is needed to trace the nuclear property back to the shell evolution. The shell model is one of the most suitable approaches for this purpose. Otsuka has led the shell-model group of the University of Tokyo, producing substantial important results. Among them is advancing large-scale shell-model calculations mainly by developing the Monte Carlo shell model (MCSM) [26, 27]. An overview of this aspect is presented in this conference by Shimizu [28].

Shell-model studies conducted by Otsuka have been useful for clarifying the shell evolution. A unified pf-shell interaction, called GXPF1(A) [14, 29], was completed at the beginning of the 2000s with the help of the development of large-scale shell-model calculations. This interaction is used as one of the standard pf-shell interactions, and its monopole properties provide a basis for the shell-evolution picture as a reference interaction. Studies of the "island of inversion" with the SDPF-M interaction [30] are also strongly connected to constructing the shell evolution. Its feature that the magic number is shifted from $N = 16$ to 20 was phenomenologically introduced [30], and now is regarded as a manifestation of the shell evolution predominantly due to the tensor force for its Z dependence and due to the three-nucleon force for its N dependence.

The concept of the shell evolution has been exported to the construction of effective interactions, which provide good tests for the concept. Spin-isospin properties such as the Gamow-Teller transition in the p-sd shell region are well described by a new interaction [31] incorporating stronger spin-isospin dependent monopole interaction of [9]. The $V_{\rm MU}$ interaction [21] has recently been applied to shell-model interactions with success [32, 33]. For instance, a large deformation in ^{42}Si with the $N = 28$ magic number arises naturally as a consequence of the tensor-force driven Jahn-Teller effect [33].

Towards shell-model descriptions of medium-heavy nuclei, the barium region [34] and the $28 \leq N(Z) \leq 50$ region [35] were explored already, and the neutron-rich nickel region is under investigation [27]. This direction should be accompanied by the development of large-scale calculations as well as reliable theory of shell evolution.

Acknowledgments
I would express my deepest gratitude to Professor Takaharu Otsuka for a long-standing collaboration and for many fruitful discussions. I also acknowledge many colleagues, in particular Drs. T. Mizusaki, M. Honma and N. Shimizu, for collaborations under the CNS-RIKEN joint research project on large-scale nuclear-structure calculations. This contribution is in part supported by MEXT KAKENHI Grant Number 21740204.

References
[1] Otsuka T and Schwenk A 2012 *Nuclear Physics News* vol 22 (London: Taylor & Francis) pp 12-17
[2] Otsuka T 2013 *Phys. Scr.* **T152** 014007
[3] Mayer M G 1949 *Phys. Rev.* **75** 1969; Haxel O, Jensen J H D and Suess H E 1949 *Phys. Rev.* **75** 1766
[4] Bohr A and Mottelson B R 1969 *Nuclear Structure* vol 1 (New York: Benjamin)
[5] Talmi I and Unna I 1960 *Phys. Rev. Lett.* **4** 469
[6] McGrory J B, Wildenthal B H and Halbert E C 1970 *Phys. Rev.* C **2** 186
[7] Poves A and Zuker A 1981 *Phys. Rep.* **70** 235
[8] Caurier E, Martínez-Pinedo G, Nowacki F, Poves A and Zuker A. P. 2005 *Rev. Mod. Phys.* **77** 427
[9] Otsuka T, Fujimoto R, Utsuno Y, Brown B A, Honma M and Mizusaki T 2001 *Phys. Rev. Lett.* **87** 082502
[10] Uozumi Y *et al.* 1994 *Nucl. Phys.* A **576** 123
[11] Rehm K E *et al.* 1998 *Phys. Rev. Lett.* **80** 676
[12] Utsuno Y, Otsuka T, Mizusaki T and Honma M 1999 *Phys. Rev.* C **60** 051315
[13] Ozawa A, Kobayashi T, Suzuki T, Yoshida K and Tanihata I 2000 *Phys. Rev. Lett.* **84** 5493
[14] Honma M, Otsuka T, Brown B A and Mizusaki T 2004 *Phys. Rev.* C **69** 034335; Honma M, Otsuka T, Brown B A and Mizusaki T 2002 *Phys. Rev.* C **65** 061301(R)
[15] Federman P and Pittel S 1977 *Phys. Lett.* **69B** 385
[16] Otsuka T, Suzuki T, Fujimoto R, Grawe H and Akaishi Y 2005 *Phys. Rev. Lett.* **95** 232502
[17] Brown B A, Richter W A, Julies R E and Wildenthal B H 1988 *Ann. Phys.* **182** 191
[18] Kirson M W 1973 *Phys. Lett.* **47B** 110
[19] Schiffer J P *et al.* 2004 *Phys. Rev. Lett.* **92** 162501
[20] Otsuka T, Matsuo T and Abe D 2006 *Phys. Rev. Lett.* **97** 162501
[21] Otsuka T, Suzuki T, Honma M, Utsuno Y, Tsunoda N, Tsukiyama K and Hjorth-Jensen M 2010 *Phys. Rev. Lett.* **104** 012501
[22] Tsunoda N, Otsuka T, Tsukiyama K and Hjorth-Jensen M 2011 *Phys. Rev.* C **84** 044322
[23] Otsuka T, Suzuki T, Holt J D, Schwenk A and Akaishi Y 2010 *Phys. Rev. Lett.* **105** 032501
[24] Hagen G, Hjorth-Jensen M, Jansen G R, Machleidt R and Papenbrock T 2012 *Phys. Rev. Lett.* **108** 242501
[25] Holt J D, Otsuka T, Schwenk A and Suzuki T 2012 *J. Phys.* G **39** 085111
[26] Otsuka T, Honma M, Mizusaki T, Shimizu N and Utsuno Y 2001 *Prog. Part. Nucl. Phys.* **47** 319
[27] Shimizu N, Abe T, Tsunoda Y, Utsuno Y, Yoshida T, Mizusaki T, Honma M and Otsuka T 2012 *Prog. Theor. Exp. Phys.* **2012** 01A205
[28] Shimizu N *et al.* contribution to this symposium
[29] Honma M, Otsuka T, Brown B A and Mizusaki T 2005 *Eur. Phys. J.* A **25** Suppl. 1 499
[30] Utsuno Y, Otsuka T, Mizusaki T and Honma M 1999 *Phys. Rev.* C **60** 05431; Utsuno Y, Otsuka T, Glasmacher T, Mizusaki T and Honma M 2004 *Phys. Rev.* C **70** 044307
[31] Suzuki T, Fujimoto R and Otsuka T 2003 *Phys. Rev.* C **67** 044302
[32] Yuan C, Suzuki T and Otsuka T 2012 *Phys. Rev.* C **84** 064324
[33] Utsuno Y, Otsuka T, Brown B A, Honma M, Mizusaki T and Shimizu N 2012 *Phys. Rev.* C **86** 051301(R)
[34] Shimizu N, Otsuka T, Mizusaki T and Honma M 2001 *Phys. Rev. Lett.* **86** 1171
[35] Honma M, Otsuka T, Mizusaki T and Hjorth-Jensen M 2009 *Phys. Rev.* C **80** 064323

International Symposium on Exotic Nuclear Structure From Nucleons (ENSFN 2012) IOP Publishing
Journal of Physics: Conference Series **445** (2013) 012009 doi:10.1088/1742-6596/445/1/012009

Three-nucleon forces and nuclei at the extremes

A Schwenk

Institut für Kernphysik, Technische Universität Darmstadt, 64289 Darmstadt, Germany and ExtreMe Matter Institute EMMI, GSI Helmholtzzentrum für Schwerionenforschung GmbH, 64291 Darmstadt, Germany

E-mail: schwenk@physik.tu-darmstadt.de

Abstract. Neutron-rich nuclei become increasingly sensitive to three-nucleon forces. These components of nuclear forces are at the forefront of theoretical developments based on effective field theories of quantum chromodynamics. We discuss our understanding of three-nucleon forces and their impact on exotic nuclei, and show how new measurements test and constrain them. Three-nucleon forces therefore provide an exciting link between theoretical and experimental nuclear physics frontiers.

1. Chiral effective field theory (EFT) and three-nucleon (3N) forces

Chiral EFT is based on the symmetries of quantum chromodynamics and is applicable at momentum scales of order of the pion mass $Q \sim m_\pi$, where pions are included as explicit degrees of freedom and build up the long-range parts of strong interactions. In chiral EFT, nucleons interact via pion exchanges and shorter-range contact interactions [1]. The resulting nuclear forces are organized in a systematic expansion in powers of Q/Λ_b, where $\Lambda_b \sim 500\,\text{MeV}$ denotes the breakdown scale. As shown in Fig. 1, at a given order this includes contributions from one- or multi-pion exchanges and from contact interactions, with short-range couplings that are fit to data and thus capture all short-range effects relevant at low energies. In addition, nuclear forces depend on a resolution scale, where the evolution is governed by the renormalization group (RG). The RG decouples low and high momenta and leads to universal low-momentum interactions with greatly enhanced convergence in few- and many-body systems [2].

Chiral EFT opens up a systematic path to investigate many-body forces and their impact on neutron-rich nuclei and neutron-rich matter [3]. This results from the consistency of NN and 3N interactions, which predicts the two-pion-exchange c_1, c_3, c_4 parts of 3N forces at N^{2}LO, leaving only two low-energy couplings c_D, c_E that encode pion interactions with short-range NN pairs and short-range three-body physics. Moreover, all 3N and 4N forces at the next order, N^{3}LO, are predicted [1]. For systems of only neutrons, the c_D, c_E parts do not contribute because of the Pauli principle and the coupling of pions to spin [4]. Therefore, chiral EFT predicts all three-neutron and four-neutron forces to N^{3}LO. At the same time, 3N forces are a frontier in the physics of nuclei and nucleonic matter in stars. This leads to a forefront connection of 3N forces with the exploration of exotic nuclei at rare isotope beam facilities worldwide.

2. Three-nucleon forces and exotic nuclei

Three-nucleon forces play a key role for understanding and predicting exotic nuclei and for the formation and evolution of shell structure. As shown in Fig. 2, chiral 3N forces (fit only to ^{3}H

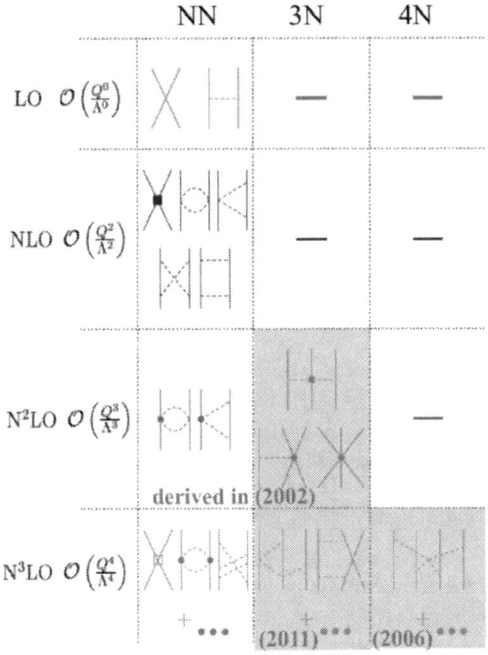

Figure 1. Chiral EFT for nuclear forces, where the different contributions at successive orders are shown diagrammatically [1]. Many-body forces are highlighted in orange including the year they were derived. For neutrons, 3N and 4N forces are predicted parameter-free to N^{3}LO.

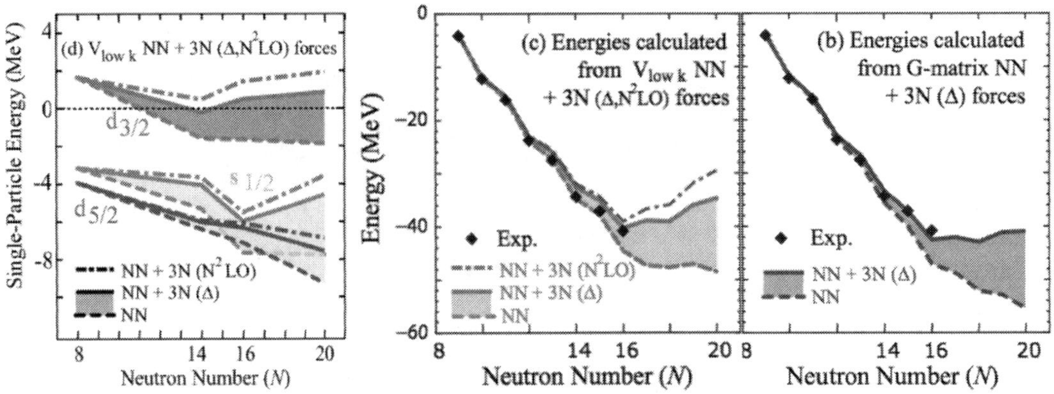

Figure 2. Left panel: Single-particle energies in the oxygen isotopes as a function of neutron number. Results are shown based on NN forces only (RG-evolved to low-momentum interactions $V_{\text{low}\,k}$) and with N^{2}LO 3N forces (NN+3N). The changes due to the single-Δ contribution to 3N forces are highlighted by the shaded areas in all panels. Right panels: Ground-state energies of the neutron-rich oxygen isotopes relative to ^{16}O, compared to the experimental energies of the bound isotopes $^{17-24}$O. The middle panel shows the results corresponding to the left panel. The right most panel is for a G matrix. For details see Ref. [5].

and ^{4}He) lead to repulsive interactions between valence neutrons that change the location of the neutron dripline from ^{28}O (with NN forces only) to the experimentally observed ^{24}O [5, 6].

Journal of Physics: Conference Series **445** (2013) 012009 doi:10.1088/1742-6596/445/1/012009

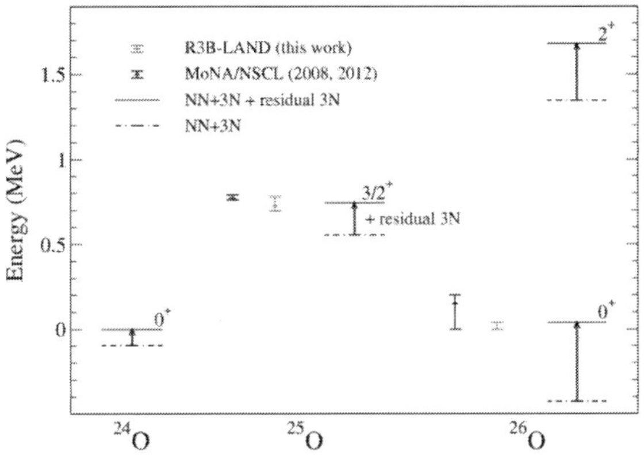

Figure 3. Comparison of the experimental ^{25,26}O energies from MoNA [11] and R3B-LAND [12] with theoretical shell-model calculations based on NN+3N forces and including residual 3N forces. The impact of residual 3N forces is highlighted by the arrows. For details see Ref. [12].

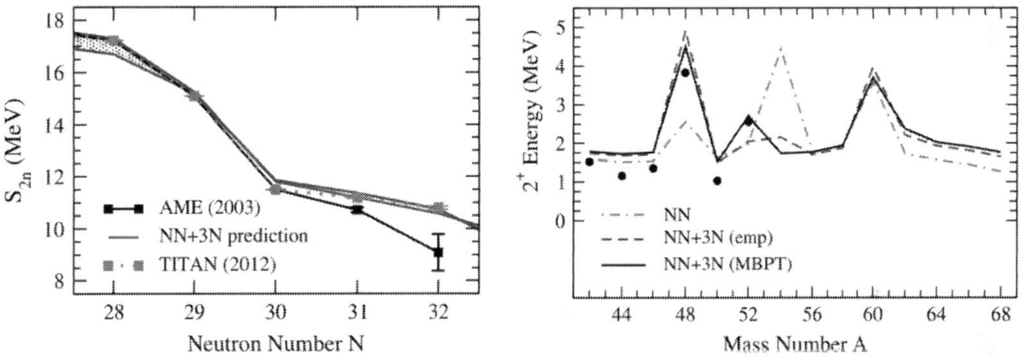

Figure 4. Left panel: Two-neutron separation energy S_{2n} of the neutron-rich calcium isotopes, with experimental energies from the AME 2003 atomic mass evaluation. We also show the new precision mass measurements for ^{51,52}Ca from TITAN, which disagree significantly with the indirectly measured masses of AME 2003. Our predictions based on NN+3N forces are in excellent agreement with these masses and with the flat S_{2n} behavior from ^{50}Ca to ^{52}Ca. For details see Ref. [15]. Right panel: 2^+ energy in the even calcium isotopes with and without 3N forces compared with experiment (dots from ENSDF). The excitation energies are calculated to ^{68}Ca in an extended $pfg_{9/2}$ valence space, using both empirical (emp) and calculated (MBPT) single-particle energies. For details see Ref. [16].

The position of the neutron dripline is driven by the location of the $d_{3/2}$ orbital, which remains unbound with 3N forces. This presents the first explanation of the oxygen anomaly based on nuclear forces. The 3N-force mechanism is dominated by the single-Δ contribution (see the shaded areas in Fig. 2) and was recently confirmed in large-space calculations [7, 8, 9].

The interactions between valence neutrons are dominated by 3N forces between two valence neutrons and one nucleon in the core of ^{16}O. This is expected for normal Fermi systems [10], where the contributions from residual three-valence-nucleon interactions are small compared to the normal-ordered two-body part. In the shell model, the impact of residual 3N forces increases with the number of valence nucleons, so they are amplified in the most neutron-rich

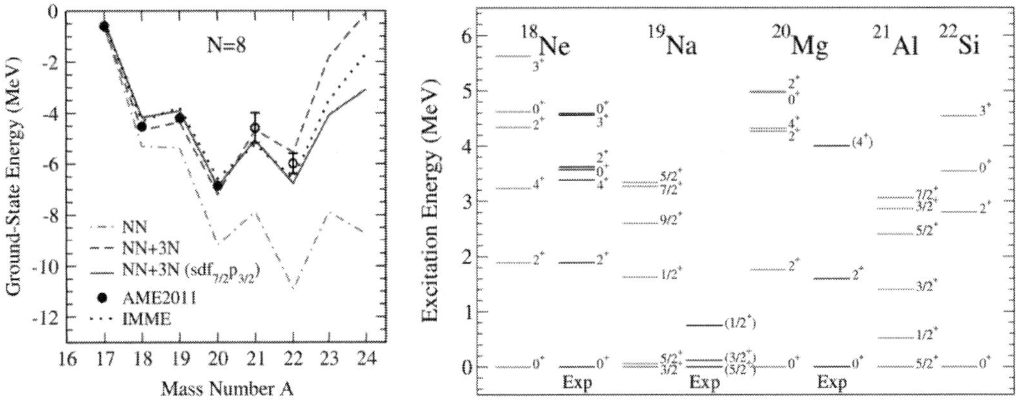

Figure 5. Left panel: Ground-state energies of $N = 8$ isotones relative to ^{16}O. Experimental energies (AME2011 with extrapolations as open circles) and energies from the isobaric multiplet mass equation (IMME) are shown. NN-only results in the *sd* shell are compared to calculations based on NN+3N forces in both *sd* and $sdf_{7/2}p_{3/2}$ valence spaces with calculated (MBPT) single-particle energies. Right panel: Excitation energies of $N = 8$ isotones calculated with NN+3N forces in the $sdf_{7/2}p_{3/2}$ valence space, compared with experimental data where available. For details see Ref. [17].

^{25,26}O isotopes studied at MoNA [11] and R3B-LAND [12], as demonstrated in Fig. 3.

While the magic numbers $N = 2, 8, 20$ are generally well understood, $N = 28$ is the first standard magic number that is not reproduced in microscopic theories with NN forces only. In studies for calcium isotopes [13, 14], it was shown that 3N forces are key to explain the $N = 28$ magic number, leading to a high 2^+ excitation energy. Moreover, chiral 3N forces improve the agreement with experimental masses, and as shown in Fig. 4 predicted a flat behavior of the two-neutron separation energy from ^{50}Ca to ^{52}Ca, in excellent agreement with new precision TITAN Penning-trap mass measurements [15]. The 2^+ excitation energy in the even calcium isotopes with and without 3N forces, based on the same calculations as for the masses, is shown in Fig. 4 [16]. This predicts a 2^+ energy in ^{54}Ca of $1.7 - 2.2$ MeV (with 3N forces to first order only, the 2^+ energy is higher [13]). The first measurement of the 2^+ energy in ^{54}Ca was achieved at RIKEN and presented by D. Steppenbeck at this symposium. Finally, we have presented first results with 3N forces for the ground and excited states of proton-rich nuclei along the $N = 8$ (see Fig. 5) and $N = 20$ isotones to the proton dripline [17].

3. Neutron matter and neutron stars

The physics of neutron-rich matter ranges from universal properties at low densities and in ultracold atoms to the densest matter we know to exist in neutron stars. The same chiral 3N forces of the previous section are repulsive in neutron matter and dominate the theoretical uncertainties of the neutron-matter energy [4]. The predicted energy range provides tight constraints for the symmetry energy (see Fig. 6) and predicts the neutron skin thickness of ^{208}Pb to 0.17 ± 0.03 fm [18], in excellent agreement with a recent determination from the complete electric dipole response [19]. In addition, our calculations based on chiral EFT interactions constrain the properties of neutron-rich matter below nuclear densities to a much higher degree than is reflected in current neutron star modeling [18]. These constraints have been recently explored for the gravitational wave signal in neutron star mergers [20].

To improve our understanding of neutron matter further, we have performed the first complete

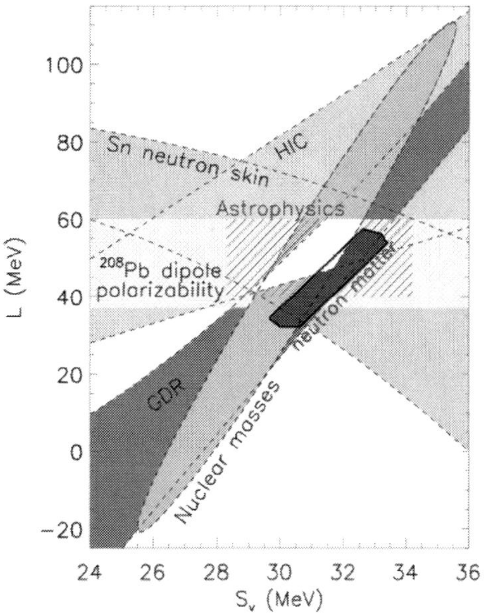

Figure 6. Constraints for the symmetry energy S_v and its density derivative L. The blue region shows our neutron-matter constraints, in comparison to bands from different empirical extractions (the white area gives the overlap region) [21]. For details see Ref. [22].

N^3LO calculation of neutron matter including NN, 3N and 4N forces [23, 24]. The resulting energy is shown in the left panel of Fig. 7. This leads to bands consistent with the RG-evolved results [4]. Many of the equations of state for core-collapse supernova simulations are inconsistent with the N^3LO neutron-matter band. Combined with the heaviest $2M_\odot$ neutron star [25], our results shown in Fig. 7 constrain the radius of a typical $1.4M_\odot$ star to $R = 9.7 - 13.9\,\text{km}$ [22, 24] (the same relative uncertainty as the neutron skin). The predicted radius range is due, in about equal amounts, to the uncertainty in 3N forces and to the extrapolation to high densities. The radius range is also consistent with astrophysical results obtained from modeling X-ray burst sources [26]. The physics of 3N forces therefore connects the heaviest neutron-rich nuclei with the heaviest neutron stars.

Acknowledgments
All the very best for your 60th birthday, Taka, lots of good health, happiness and continued success! It has been a great pleasure working together on 3N forces and exotic nuclei and I look forward to exciting future work. I would also like to thank J. Dilling, A. T. Gallant, K. Hebeler, J. D. Holt, T. Krüger, J. M. Lattimer, J. Menéndez, C. J. Pethick, J. Simonis, T. Suzuki, and I. Tews, who contributed to the results presented in this talk. This work was supported by the BMBF under Contract No. 06DA70471, the DFG through Grant SFB 634, the ERC Grant No. 307986 STRONGINT, and the Helmholtz Alliance HA216/EMMI.

References
[1] Epelbaum E, Hammer H-W and Meißner U-G 2009 *Rev. Mod. Phys.* **81** 1773
[2] Bogner S K, Furnstahl R J and Schwenk A 2010 *Prog. Part. Nucl. Phys.* **65** 94
[3] Hammer H-W, Nogga A and Schwenk A 2013 *Rev. Mod. Phys.* **85** 197
[4] Hebeler K and Schwenk A 2010 *Phys. Rev. C* **82** 014314
[5] Otsuka T, Suzuki T, Holt J D, Schwenk A and Akaishi Y 2010 *Phys. Rev. Lett.* **105** 032501

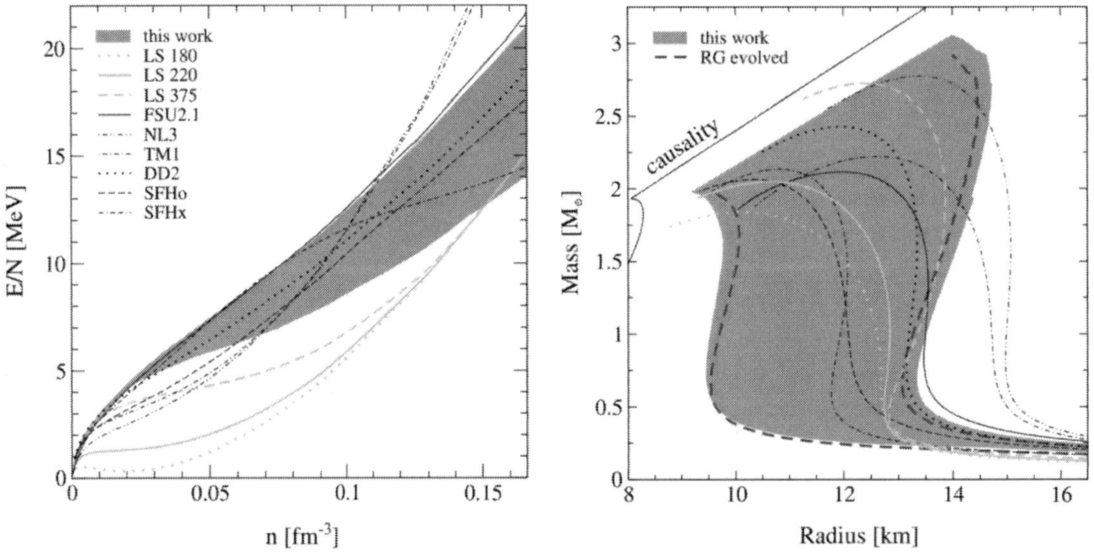

Figure 7. Comparison of the neutron-matter energy at N^{3}LO (red band) with equations of state for core-collapse supernova simulations provided by Lattimer-Swesty (LS with different incompressibilities 180, 220, and 375 MeV), G. Shen (FSU2.1, NL3), Hempel (TM1, SFHo, SFHx), and Typel (DD2). Right panel: Constraints on the mass-radius diagram of neutron stars based on the neutron-matter results of the left panel and following Ref. [22] for the extension to neutron-star matter and to high densities (red band), in comparison to the constraints based on RG-evolved interactions (thick dashed blue lines) [22]. We also show the mass-radius relations obtained from the equations of state for core-collapse supernova simulations shown in the left panel (the same legend applies). For details and references see Ref. [24].

[6] Holt J D, Menéndez J and Schwenk A 2013 *Eur. Phys. J.* A **49** 39
[7] Hagen G, Hjorth-Jensen M, Jansen G R, Machleidt R and Papenbrock T 2012 *Phys. Rev. Lett.* **108** 242501
[8] Hergert H, Binder S, Calci A, Langhammer J and Roth R 2013 *Preprint* arXiv:1302.7294
[9] Cipollone A, Barbieri C and Navrátil P 2013 *Preprint* arXiv:1303.4900
[10] Friman B and Schwenk A 2011, *From Nuclei to Stars: Festschrift in Honor of Gerald E. Brown*, Ed. S Lee (World Scientific) p 141 (*Preprint* arXiv:1101.4858)
[11] Hoffman C R et al. 2008 *Phys. Rev. Lett.* **100** 152502; Lunderberg E et al. 2012 *Phys. Rev. Lett.* **108** 142503
[12] Caesar C, Simonis J et al. 2012 *Preprint* arXiv:1209.0156
[13] Holt J D, Otsuka T, Schwenk A and Suzuki T 2012 *J. Phys.* G **39** 085111
[14] Hagen G, Hjorth-Jensen M, Jansen G R, Machleidt R and Papenbrock T 2012 *Phys. Rev. Lett.* **109** 032502
[15] Gallant A T et al. 2012 *Phys. Rev. Lett.* **109** 032506
[16] Holt J D, Menéndez J and Schwenk A 2013 *Preprint* arXiv:1304.0434
[17] Holt J D, Menéndez J and Schwenk A 2013 *Phys. Rev. Lett.* **110** 022502
[18] Hebeler K, Lattimer J M, Pethick C J and Schwenk A 2010 *Phys. Rev. Lett.* **105** 161102
[19] Tamii A et al. 2011 *Phys. Rev. Lett.* **107** 062502
[20] Bauswein A and Janka H-T 2012 *Phys. Rev. Lett.* **108** 011101; Bauswein A, Janka H-T, Hebeler K and Schwenk A 2012 *Phys. Rev.* D **86** 063001
[21] Lattimer J M and Lim Y 2012 *Preprint* arXiv:1203.4286
[22] Hebeler K, Lattimer J M, Pethick C J and Schwenk A 2013 *Preprint* arXiv:1303:4662
[23] Tews I, Krüger T, Hebeler K and Schwenk A 2013 *Phys. Rev. Lett.* **110** 032504
[24] Tews I, Krüger T, Hebeler K and Schwenk A 2013 *Preprint* arXiv:1304.2212
[25] Demorest P B, Pennucci T, Ransom S M, Roberts M S E and Hessels J W T 2010 *Nature* **467** 1081
[26] Steiner A W, Lattimer J M and Brown E F 2010 *Astrophys. J.* **722** 33

Pairing and shell gaps in nuclei

B Alex Brown

National Superconducting Cyclotron Laboratory and Department of Physics and Astronomy, Michigan state University, East Lansing, Michigan 48824-1321, USA

E-mail: brown@nscl.msu.edu

Abstract. The pairing contribution to the odd-even oscillations in the nuclear binding energies is considered in the framework on the nuclear shell model. Schematic and realistic Hamiltonians are used to understand the trends related to pairing and shell gaps. Detailed results are shown and discussed for the calcium isotopes. Results are also shown for carbon, oxygen and neon as well as the overall trends for all nuclei.

1. Introduction

One of the most robust signatures of pairing in nuclei is the odd-even oscillation in binding energies (BE) as a function of neutron or proton number. This is illustrated in Fig. 1 which shows the energies, E, and one-neutron separation energies, S_n, for the calcium isotopes. The figure shows the results of a shell-model calculation in the $(0f_{7/2}, 0f_{5/2}, 1p_{3/2}, 1p_{1/2})(fp)$ model space with the GX1A Hamiltonian (also referred to as GXPF1A in the literature [1]) compared with experiment. We will discuss the oscillation in the one-neutron separation energies as a function of neutron number in terms of the energy differences

$$D_n(N) = (-1)^{N+1}[S_n(Z,N+1) - S_n(Z,N)] = (-1)^N[2\mathrm{BE}(Z,N) - \mathrm{BE}(Z,N-1) - \mathrm{BE}(Z,N+1)] \quad (1)$$

where $S_n(N) = \mathrm{BE}(Z,N) - \mathrm{BE}(Z,N-1)$ is the one-neutron separation energy (BE = -E). N is the number of neutrons and Z is the number of protons. This quantity turns out to be always positive and reflects the fact that the even nuclei are always more bound on the average than the neighboring odd nuclei. We will distinguish the results for even and odd N value by, D_{ne} and D_{no}, respectively. In the literature one commonly finds the related quantity $\Delta_n(N) = \frac{D_n(N)}{2}$. We use D rather than Δ because its values are more directly connected to simple underlying quantities associated with pairing and shell gaps. Equivalent equations as a function of proton number are obtained by fixing N and varying Z.

Fig. 2 shows values of D_n for the calcium isotopes ($N > 20$) obtained from experiment and from two commonly used Hamiltonians in the fp model space, KB3G [2] and GX1A [1]. We also show the excitation energies for the lowest 2^+ states of the even nuclei. The trends observed in Fig. 2 will be understood on the basis of both simple and realistic models. This will lead to an understanding of the results for all nuclei.

One observes in Fig. 2 a rather complicated experimental trend that is rather well reproduced by the calculations. The results for the two Hamiltonians are similar up to $N = 33$ where the experimental data are known. Beyond $N = 33$ there are differences, in particular for $N = 34$

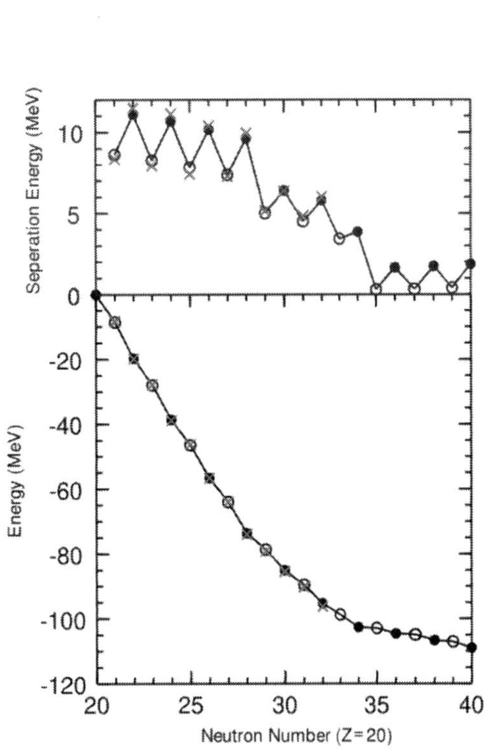

Figure 1. The bottom panel shows the ground-state energies for the calcium isotopes obtained from the pf shell-model calculation with the GX1A Hamiltonian relative to ^{40}Ca with filled circles even N and open circles for odd N, all connected by a line. The crosses are the experimental data. The top panel shows the one-neutron separation energies for GX1A and experiment.

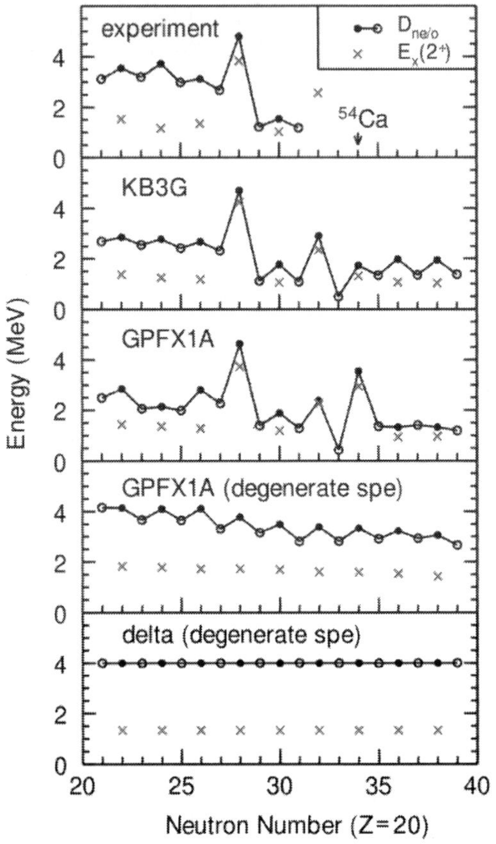

Figure 2. D and $E_x(2^+)$ for the calcium isotopes as a function of neutron number. D values are shown by the filled circles for even N and open circles for odd N, all connected by a line. The $E_x(2^+)$ values are shown by the crosses.

(^{54}Ca), where the GX1A result for the D and $E_x(2^+)$ values are nearly two times larger than for those for KB3G.

Fig. 2 also shows the results obtained for GX1A when all of the single-particle energies are set equal to each other. The result does not depend on the value of the single-particle energy since in this degenerate case the one-body part of the Hamiltonian does not contribute differences of Eq. (1). The result is a very regular oscillating pattern. What we learn from this is that the larger irregularities observed in the upper panels of Fig. 2 are due to the finite spacing of the single-particle energies related to the shell gaps.

In this contribution we show how the results for D in Fig. 2 and other nuclei arise from schematic and realistic interactions between nucleons. An essential ingredient will be the pairing part of the interaction, but the consequences of pairing will be influenced by shell gaps and other components of the interactions. We start with the schematic surface-delta interaction in the next few sections, and then come to more realistic interactions. In the last sections we examine the results for D in light nuclei out to the neutron-drip line, and also show the experimental results for all nuclei. The experimental masses were taken from the intermediate evaluation of Audi and Wang [3] supplemented by more recent data on ^{51,52}Ca [4] and ^{26}O [5].

2. The surface-delta interaction model

To get some insight into the reasons for the patterns observed in Fig. 2, we start with the simple "surface-delta-function" (SDI) model for V [6]. The two-body matrix elements for a delta function $V(\mid \vec{r}_1 - \vec{r}_2 \mid) = A\delta(\mid \vec{r}_1 - \vec{r}_2 \mid)$ can be obtained analytically as a product of angular and radial matrix elements, $< V > = < V >_{\rm ang} < V >_{\rm rad}$ where

$$< V >_{\rm rad} = \int R_a(r) R_b(r) R_c(r) R_d(r) r^2 dr. \qquad (2)$$

where a, b, c and d label the single-particle states with quantum numbers (n, ℓ, j), and $R(r)$ are the radial wavefunctions. For the SDI one evaluates this integral at the surface, $r = r_o$, and assumes that
$\mid R_a(r_o) \mid = \mid R_b(r_o) \mid = \mid R_c(r_o) \mid = \mid R_d(r_o) \mid = R$. Then $< V >_{\rm rad} = (-1)^{n_a+n_b+n_c+n_d} R^4 r_o^2 = (-1)^{n_a+n_b+n_c+n_d} C$, where the phase factor is used for the radial wavefunction convention of being positive near the origin. The SDI two-body matrix elements for $J = 0$ are

$$< aaJ = 0 \mid {\rm SDI} \mid bbJ = 0 > = (A'/2)\sqrt{(2j_a + 1)(2j_b + 1)}. \qquad (3)$$

The results obtained with SDI for the fp model space with degenerate single-particle energies is shown in Fig. 3. We have chosen the constant $A' = 4\pi A(-1)^{n_a+n_b+n_c+n_d} C = 0.4$ MeV in order to make the value $D = 4$ MeV similar to that obtained for the GX1A Hamiltonian when the single-particle energies are degenerate.

For the SDI the D and $E_x(2^+)$ are constant. The SDI with degenerate single-particle energies has some very simple and interesting properties [7]. The $J = 0$ pairing properties are determined only by strength of the interaction and by the total number of possible pairs, $m/2 = 10$ in our example, where $m = 20$ is the maximum number of valence neutrons. Exactly the same results for D are obtained if there were a single orbital with $j = 19/2$, but the value of $E_x(2^+)$ depends upon the j values. Use of a delta function with the radial integral evaluated with oscillator or Woods-Saxon radial wavefunctions gives results that are close to SDI (within about 5%), but the properties are more complicated but analytic (oscillator) or non analytic (Woods-Saxon).

In addition to D and $E_x(2^+)$ we show in Fig. 3 the energies obtained when the degenerate single-particle energies are zero (if the energies were not zero there would be an additional linear dependence to the total energies). The total energy for SDI [line (a) in Fig. 3] as a function of the number of valence neutrons, n, is very simple

$$E(n) = \frac{nV_o}{2} \text{ (even } n\text{)}, \quad \text{and} \quad E(n) = \frac{(n-1)V_o}{2} \text{ (odd } n\text{)}, \qquad (4)$$

where V_o is the interaction energy for two-nucleons in the $J = 0$ paired state. We also have $D = -V_o$. All of the odd N nuclei have four degenerate states with $J^\pi = 1/2^-$, $3/2^-$, $5/2^-$ and $7/2^-$.

One can also make a "mean-field" approximation for the total energy [line (b) in Fig. 3] with $\bar{E}(n) = n(n-1)\bar{V}/2$, where the average (monopole) energy, $\bar{V}$ is the total closed-shell energy divided by the total number of m state two-particle combinations ($m = 20$ in this case). We can define the pairing correlation energy as the difference $E^{\rm corr}(n) = E(n) - \bar{E}(n)$. This is shown by curve (c) in Fig. 3.

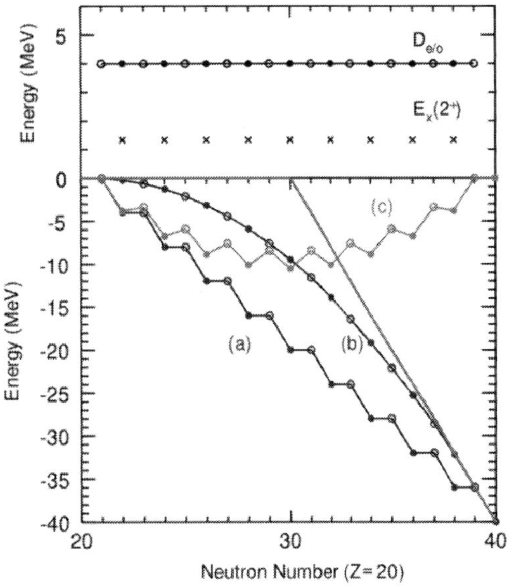

Figure 3. Results obtained with the SDI interaction. The bottom panel shows the total energy (a), the monopole energy (b) and the correlation energy (c).

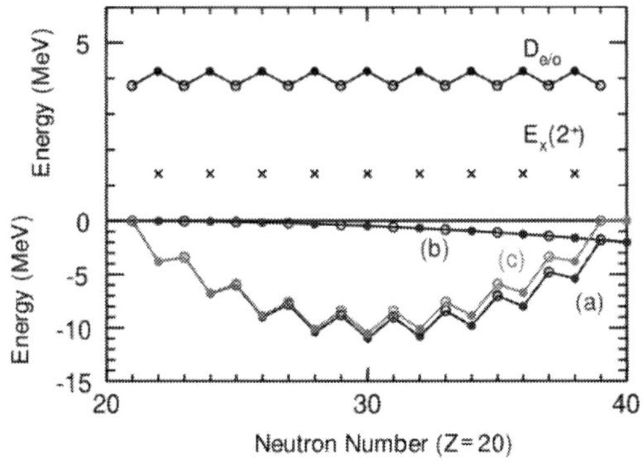

Figure 4. Results obtained with the MSDI interaction. The bottom panel shows the total energy (a), the monopole energy (b) and the correlation energy (c).

3. The modified surface-delta interaction model

One observes two differences between the results in the two lower panels of Fig. 2. One is that the GX1A result tends to fall off as a function of the number of neutrons whereas the SDI result is flat. A part of this is due to the fact that GX1A contains an assumed smooth mass dependence of the form $(A/42)^{-0.3}$ that is meant to take into account a smooth increase the harmonic oscillator parameters $\hbar\omega$ as a function of A. This was not included in the SDI. The main reason for the fall off is that the average interactions, $\bar{V}_{ab}$, differ from SDI and lead to a splitting in the single-hole energies for ^{60}Ca that reduces the effective pairing for larger N values.

The other feature is that the GX1A result for D oscillates whereas the SDI result does not. In the 1960's this was recognized as a basic failure of the delta and SDI Hamiltonians. It was empirically fixed by adding a constant to the interaction to make the so-called modified-delta interaction [8] of the form $V(\mid \vec{r}_1 - \vec{r}_2 \mid) = A\delta(\mid \vec{r}_1 - \vec{r}_2 \mid) + B$ where B is the added constant. In terms of two-body interaction it is infinitely long range. With the SDI form of the radial integral this becomes the so-called modified-surface-delta interaction (MSDI). The modern interpretation of this constant is that it comes from core-polarization corrections and three-body interactions. The energy and D values obtained with $B = 0.2$ MeV are shown in Fig. 4. This constant simply adds a term $n(n-1)B/2$ to the all energies, and does not change the correlation energy. It gives an oscillation in D; $D_o = -V_o - B$ for odd n and $D_e = -V_o + B$ for even n. Half the sum of neighboring even and odd D gives the pairing contribution: $D_a = \frac{1}{2}[D_e(N) + D_o(N-1)] = -V_o$, and half of the difference gives quadratic dependence: $D_b = \frac{1}{2}[D_e(N) - D_o(N-1)] = B$. Fig. 2-5 in Bohr and Mottelson [9] is based upon the D_a combination ($\Delta = D_a/2$).

The Coulomb interaction between protons behaves like the addition of a long-range monopole term ($B \approx 0.31$ MeV for Z=20). There is also a small anti-pairing addition to D_a of (about +0.17 MeV for $Z = 20$). This is important for understanding the trends in $D_p(Z)$ compared to those for $D_n(N)$.

A microscopic calculation of these quadratic monopole corrections requires an explicit consideration of core-polarization the real and induced three-body interactions (those between two-valence nucleons and one in the core). It has been proposed that the density-dependent interactions empirically included in the Skyrme interaction can be used to obtain B values for the valence shell-model interactions [10]. This method improves the relative binding energies obtained for shell-model calculations in heavy nuclei.

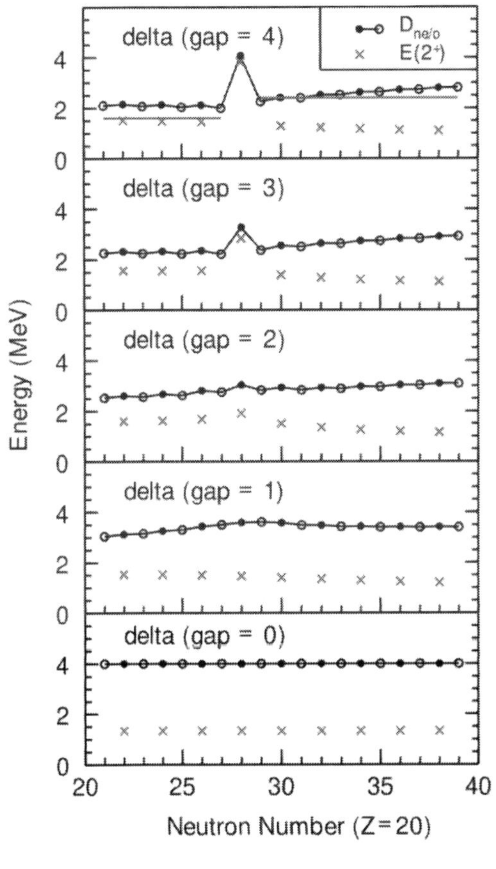

Figure 5. Results obtained with the SDI interaction as a function of the shell gap between the $0f_{7/2}$ and $(0f_{5/2}, 1p_{3/2}, 1p_{1/2})$ set of orbitals. The top panel also show the results for D (thin lines) obtained when the shell gap is infinite.

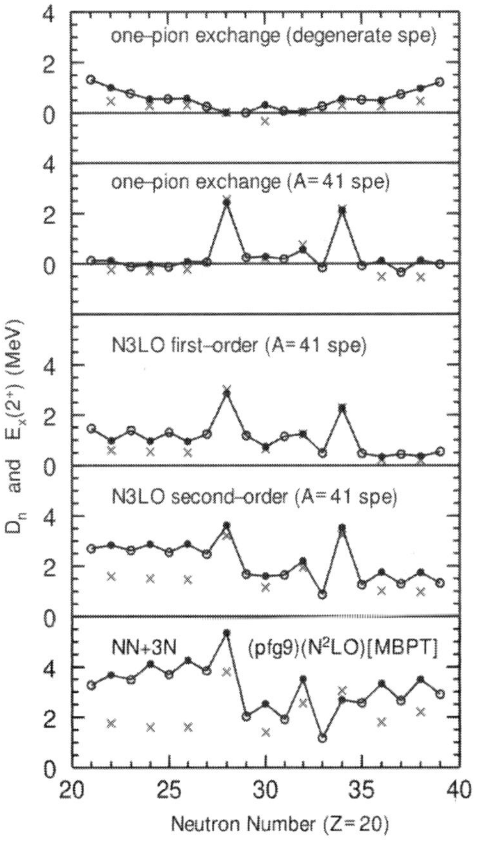

Figure 6. Values for D and $E_x(2^+)$ obtained from the one-pion exchange (top two panels), the N^{3}LO interaction (middle panels), and by including three-nucleon interactions [11] (bottom panel).

4. The surface-delta interaction model with a shell gap

Next we show how D changes when the single-particle energies are not degenerate. This requires a computational experiment with the SDI applied to the fp orbits divided into two groups, $f = (0f_{7/2})$, and $r = (0f_{5/2}, 1p_{3/2}, 1p_{1/2})$ with a variable gap in between them, so that the second group is raised in energy relative to the the $0f_{7/2}$. The result is shown in Fig. 5 where the gap starts at zero and is increased up to four MeV. Four MeV is approximately equal to the experimental shell gap found for ^{48}Ca ($N = 28$). When the gap is small (one MeV), the energies of the 2^+ states remain flat, and the pairing correlation starts to fall off away from the break at $N = 28$. When the gap is two MeV there is a small rise in the 2^+ energy and a small increase in D at $N = 28$. When the gap is above two MeV there is an obvious increase in the 2^+ energy and D and $D \approx E_x(2^+)$ at the gap. Thus, the relative values of D and $E_x(2^+)$ provide a confirming measure of shell gaps. For the calcium isotopes the calculations are carried out with only neutrons in the fp model space and the 2^+ energies are those for neutrons. The experimental states used for comparison are also dominated by the neutron configuration. In other cases when the gap is between orbitals with difference parity, the excitation energy of the relevant negative parity state (like 3^-) should be used. When experimental values of $E_x(2^+)$ are used one should confirm that the 2^+ state in question has the appropriate neutron or proton configuration. Away from semi-magic nuclei the 2^+ states usually have strongly mixed proton and neutron configurations.

One can truncate the full fp calculation by considering only the $J = 0$ part of the Hamiltonian [12]. This method provides a good approximation to ground-state energies for identical-particle systems with realistic Hamiltonians that includes the affects of both pairing and shell gaps.

In the limit of an infinite gap, the results are those shown by the thin line in the top panel of Fig. 5. In this case the total pairing energy that starts out at $V_o = -4$ MeV in the degenerate cases gets divided into two parts, $V_o^f = -1.6$ for the $0f_{7/2}$ below $N = 28$ and $V_o^r = -2.4$ for the other three orbitals above $N = 28$. Unlike the D_e which shows a large increase at the shell gap at some even N, the D_o for odd N does not show any direct evidence of the shell gap. As in this example, the D_o values often change across a gap because the number of m states that are involved in the pairing change.

5. Features for the calcium isotopes

We can now understand the features of the calcium isotopes shown in Fig. 2. With the GX1A Hamiltonian there are two large shell gaps at $N = 28$ and $N = 34$, and a smaller shell gap at $N = 32$, related to preferential filling of the orbitals in the order $0f_{7/2}$, $1p_{3/2}$, $1p_{1/2}$ and $0f_{5/2}$. The results for KB3G are similar except that above $N = 32$ the $1p_{1/2}$ and $f_{5/2}$ orbitals are closer together and there is no shell gap at $N = 34$. The smallest D_{no} value occurs for ^{53}Ca ($N = 33$) (0.43 MeV for GX1A). If confirmed experimentally this would be smallest D_{no} value in all nuclei. It is related to the fact that there is some shell gap on both sides of the $1p_{1/2}$ orbital which has the smallest possible pairing matrix element [see Eq. (3)]. Experimental confirmation of this is important.

The results for D and $E_x(2^+)$ obtained with the bare one-pion exchange potential are shown at the top of Fig. 6. One-pion exchange provides almost no pairing. The middle of Fig. 6 shows the results obtained with the N^{3}LO potential [13] in first order and in second order. Almost half of the pairing comes from the second-order corrections, and most of this comes from the bubble diagrams mediated by the tensor interaction. This is also seen in calculations for the heavy nuclei [10]. Also the second-order correction changes the sign of the quadratic term (the sign of the odd-even oscillations) to bring the result into close agreement with experiment. The bottom panel of Fig. 6 shows the results obtained with the inclusion of three-nucleon interactions from [11]. The three-nucleon interactions improve the agreement with experiment for the shell gap at $N = 28$, but at the expense of giving an upward trend in the D_n values as one approaches $N = 28$ that does not agree very well with experiment. The empirical interactions KB3G and GX1A reproduce the shell gap at $N = 28$ due to monopole adjustments of the two-body matrix elements. We can conclude that these type of monopole adjustments take into account the (missing) three-nucleon interactions.

Journal of Physics: Conference Series **445** (2013) 012010 doi:10.1088/1742-6596/445/1/012010

6. Light nuclei toward the neutron drip line

How do the D values change as we approach the neutron drip line? The heaviest isotope for which the neutron drip line has been reached is oxygen where it is located at $N = 16$ (^{24}O). Recent experiments on neutron decay have established the energies of the lowest neutron-unbound state in ^{25}O [14] and ^{26}O [5]. These results together with other mass data have been used for Fig. 7 that shows the D_n values for carbon, oxygen and neon. For oxygen large shell gaps are observed for $N = 6$ (a) and 8 (b) with two smaller ones at 14 (c) and 16 (d). ^{20}Ne with $N = 10$ shows the Wigner-cusp at (e) due to proton-neutron pairing ($T = 0$ and $T = 1$) [15]. ^{12}C at $N = 6$ (a) should also have a Wigner-cusp, but it is on top of the shell gap at $Z = 6$. The 2^+ states for neutron-rich carbon and oxygen are dominated by the the neutron configuration and can be compared to the D values. The 2^+ states for carbon and neon have mixed proton-neutron configurations.

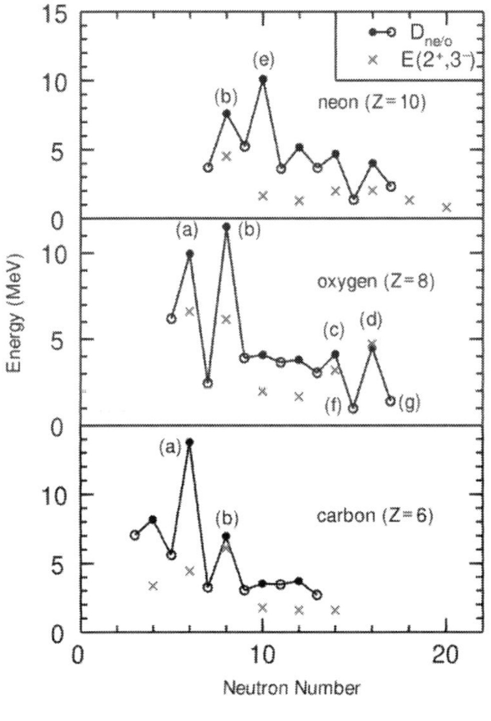

Figure 7. Experimental values for D and $E_x(J^\pi)$ for the carbon, oxygen and neon isotopes. J^π is 2^+ except for ^{14}C, ^{16}O and ^{18}Ne where it is the lowest observed negative parity state.

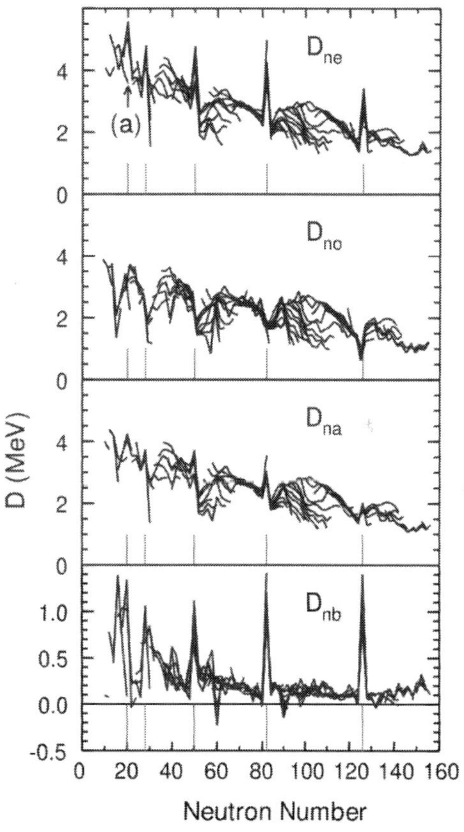

Figure 8. D_n for all nuclei. Values for for even Z are given as a function of the number of neutrons ($N > Z$) connected by lines. D_{ne} are for even N, D_{no} are for odd N values, D_{na} are the averages and D_{nb} are the differences given in Sec. 3. The vertical lines show the location of the magic numbers 20, 28, 50, 82 126.

The small D_{no} values at $N = 15$ (f) is related to the filling of the $1s_{1/2}$ orbital at this point; ^{25}O and ^{27}Ne have $1/2^+$ ground states. Shell-model calculations such as those in the $p - sd$ model space with the WBP Hamiltonian [16], and in the *sd* model space with the USDA/B Hamiltonians [17] are in general good agreement with these data. The main exception is that WBP strength needs to be reduced by about 25% in order to reproduce the experimental energies for the neutron-rich carbon [18]. For WBP one assumed that the interaction (including pairing) is the same for oxygen and carbon. A similar situation exists for a reduction of the interaction strength in the silicon isotopes ($Z = 14$) compared to calcium [19]. Two possible reasons for this reduction are: (1) The radial matrix for the bare interaction elements become smaller for more loosely bound neutrons. This can be qualitatively modeled by evaluating Eq. (2) with Woods-Saxon or Hartree-Fock radial wavefunctions. (2) The core-polarization corrections to the matrix elements are reduced.

Both of these effects have been considered in Ref [20], where it was found that (1) is most important for silicon. Thus we may expect pairing to be reduced as one approaches the drip line simply due to the fact that the neutrons are on the average further apart. This is confirmed by the small value of D_{no} for ^{25}O ($N = 17$) (g) where the $0d_{3/2}$ and the $1p_{3/2}$ orbitals start to fill. Understanding the situation near and beyond the neutron drip line will require an explicit consideration of the continuum part of the wavefunctions.

The examples we have discussed are part of the complete set of data shown in Fig. 8. It is interesting to compare the experimental systematics with those obtained from various mean-field (e.g. [21]) and configuration-mixing models. In the next decade we expect a tremendous growth in the data for masses of neutron-rich nuclei from radioactive beam facilities. These are essential for understanding how the pairing and shell gap properties evolve as one approaches the neutron drip line.

Acknowledgments
We acknowledge support from NSF grant PHY-1068217.

References
[1] Honma M, Otsuka T, Brown B A and Mizusaki T 2002 *Phys. Rev.* C **65** 061301(R); 2004 *Phys. Rev.* C **69** 034335; 2005 *Euro. Phys. Jour.* A **25 Suppl. 1** 499
[2] Poves A, Sanchez-Solano J, Caurier E and Nowacki F 2001 *Nucl. Phys.* A **694** 157
[3] Audi G and Wang M 2011 private communication
[4] Gallant A T, et al. 2012 *Phys. Rev. Lett.* **109** 032506
[5] Lunderberg E, et al. 2012 *Phys. Rev. Lett.* **108** 142503
[6] Plastino A, Arvieu R and Moszkowski S A 1966 *Phys. Rev.* **145** 837
[7] Talmi I 1993 *Simple Models of Complex Nuclei, Contemporary Concepts in Physics* vol 7 (Harwood Academic Publishers)
[8] Glaudemans P W M, Brussaard P J and Wildenthal B H 1967 *Nucl. Phys.* A **102** 593
[9] Bohr A and Mottelson B R 1998 *Nuclear Structure* vol I (World Scientific Publishing Company)
[10] Brown B A, Signoracci A and Hjorth-Jensen M 2011 *Phys. Lett.* B **695** 507
[11] Holt J D, Otsuka T, Schwenk A and Suzuki T 2012 *J. Phys.* G **39** 085111
[12] Volya A, Brown B A and Zelevinsky V 2001 *Phys. Lett.* B **509** 37
[13] Entem D R and Machleidt R 2003 *Phys. Rev.* C **68** 041001
[14] Hoffman C R, et al. 2008 *Phys. Rev. Lett.* **100** 152502
[15] Senkov R, Bertsch G F, Brown B A, Luo Y L and Zelevinsky V 2008 *Phys. Rev.* C **78** 044304
[16] Warburton E K and Brown B A 1992 *Phys. Rev.* C **46** 923
[17] Brown B A and Richter W A 2006 *Phys. Rev.* C **74** 034315
[18] Stanoiu M et al. 2008 *Phys. Rev.* C **78** 034315
[19] Nowacki F and Poves A 2009 *Phys. Rev.* C **79** 014310
[20] Signoracci A, Brown B A and Hjorth-Jensen M 2011 *Phys. Rev.* C **83** 024315
[21] Bertsch G F, Bertulani C A, Nazarewicz W, Schunck N and Stoitsov M V 2009 *Phys. Rev.* C **79** 034306

Mean-field and RPA approaches to stable and unstable nuclei with semi-realistic NN interactions

H Nakada

Department of Physics, Graduate School of Science, Chiba University
Yayoi-cho 1-33, Inage, Chiba 263-8522, Japan

E-mail: `nakada@faculty.chiba-u.jp`

Abstract. Important roles of the NN interaction in the shell evolution, which have been pointed out by Otsuka and his collaborators, can be investigated by the self-consistent mean-field approaches with the semi-realistic interactions. We show that the shell evolution in the $Z = 50$ and $Z = 20$ nuclei is appropriately described by the semi-realistic interactions that include the realistic tensor force. The $M1$ strengths in ^{208}Pb are also reproduced by the semi-realistic interactions in the HF+RPA, which are another manifestation of the tensor force in nuclear structure.

1. Introduction

Experimental data on unstable nuclei have disclosed that the nuclear shell structure has significant Z- and N-dependence, even altering magic numbers [1]. This Z- and N-dependence of the shell structure is often called *shell evolution* [2]. Although the ℓ-dependence of the single-particle (s.p.) energies in the loosely bound orbitals could influence the shell evolution [3], there has been no clear evidence that the ℓ-dependence of the s.p. energies gives rise to appearance or disappearance of magic numbers. On the other hand, significant roles of several parts of the effective nucleonic interaction have been pointed out. The central spin-isospin channel of the NN interaction was first considered [4], although it seems only subsidiary in many cases. In contrast, it has been established that the tensor channels are important in the shell evolution [5]. As discarded in the conventional mean-field approaches, tensor-force effects are one of the current hot topics in nuclear structure physics. In addition to the NN interaction, roles of the NNN interaction have also been investigated [6]. Because these mechanisms of the shell evolution according to the NN and NNN interactions were first proposed by Otsuka, together with his collaborators, I would call them *Otsuka scheme* in this paper.

The mean-field (MF) theories, the Hartree-Fock (HF) and the Hartree-Fock-Bogolyubov (HFB) theories in particular, give us a framework to describe nuclear shell structure from the nucleonic degrees of freedom in a self-consistent manner. On top of the MF theories, the random-phase approximation (RPA) is useful in describing properties of excited states at relatively low energy. The only input to these theories is effective interactions (or energy density functionals), in principle. In the non-relativistic MF approaches, the Skyrme interaction [7] has most widely been employed, in which all terms of the interactions are restricted to zero range, though certain part of finite-range effects is pretended by the derivative terms. The Gogny interaction [8] has also been applied systematically, in which the central channels have finite range. The author has developed effective interactions [9, 10, 11] based on the Michigan 3-range Yukawa (M3Y) interaction [12]. The M3Y interaction was derived by fitting the Yukawa functions to the G-matrix. Whereas the original M3Y interaction cannot reproduce the saturation and the spin-orbit splitting within the MF regime, this is remedied by adding a density-dependent contact term and multiplying the LS channels by an enhancement factor. This type of interactions has been called *semi-realistic* interactions. They are contrasted to the Skyrme and the Gogny interactions whose parameters have usually been determined in a phenomenological manner.

In harmony with the Otsuka scheme, the MF approaches with semi-realistic interactions are expected to be useful in investigating the shell evolution. The M3Y-type semi-realistic interactions contain the central channels of the one-pion exchange potential (OPEP), and thereby have reasonable spin-isospin channels. They also have the realistic tensor force originating from the G-matrix. Though with restricted form and fully phenomenological treatment at present, the density-dependent terms may simulate effects of the NNN interaction. Using the recently developed algorithm [13, 14, 15, 16], semi-realistic interactions have been applied to the MF and RPA calculations of medium- to heavy-mass spherical nuclei. In this paper we shall show some of the results, primarily focusing on the tensor-force effects on nuclear structure. It is commented that, with attempts of application to nuclear reactions [17] and neutron-star problems [18] as well as nuclear structure studies, the semi-realistic interactions may give a step toward unified description of all these problems.

2. M3Y-type semi-realistic interactions

The M3Y-type semi-realistic interactions have been developed [9, 10, 11], with aiming at high predictive power and wide applicability. The M3Y-Pn parameter-sets ($n = 1, 2, \cdots$, *etc.*) have been obtained from the M3Y-Paris interaction [19], by replacing a part of short-range repulsion by density-dependent contact term so as to reproduce the saturation, and by tuning some of the parameters. In the M3Y-Pn interactions with $n \geq 5$, the tensor channels ($v^{(\mathrm{TN})}$) of the M3Y-Paris interaction have been kept unchanged, as well as the longest-range central channels that are fixed to be those of the OPEP ($v^{(\mathrm{C})}_{\mathrm{OPEP}}$). While these channels are closely connected to the chiral symmetry breaking, they have not explicitly been taken into account in the conventional MF approaches. It is remarked that these channels in the M3Y-type interactions provide realistic basis corresponding to roles of the spin-isospin channels and the tensor channels in the Otsuka scheme, and that, owing to explicit and realistic $v^{(\mathrm{C})}_{\mathrm{OPEP}}$ and $v^{(\mathrm{TN})}$, we obtain reasonable spin properties with the M3Y-type interactions. For instance, M3Y-Pn gives the Landau-Migdal parameter g_0' consistent with the data [20], to which contribution of $v^{(\mathrm{C})}_{\mathrm{OPEP}}$ is significant. Thus the M3Y-type semi-realistic interactions are suitable to investigating their effects on the shell evolution within a self-consistent MF regime. For the NNN interaction, the density-dependent terms may carry certain part of its effects, although these terms may also represent medium effects. Since the density-dependent terms are currently determined in a phenomenological manner, its connection to the microscopic NNN interaction is obscure and we shall not argue their effects in this paper.

Density profile of the symmetry energy $a_t(\rho)$ may be important in nuclear reaction problems and in neutron stars. Although the M3Y-P5 interaction [10] has been applied to these problems [17, 18], $a_t(\rho)$ is not well constrained in M3Y-P5, typified by instability of the symmetric nuclear matter at $\rho \approx 4\rho_0$ (ρ_0 is the saturation density). In the newly explored M3Y-P6 and P7 interactions [21], several parameters have been fitted to the neutron matter energy obtained from microscopic calculations of Refs. [22] (M3Y-P6) and [23] (M3Y-P7). This improves $a_t(\rho)$ to substantial extent, and the spin-saturated symmetric matter is found stable in wide range of ρ, in contrast to many phenomenological interactions including the tensor force [24]. However, in the subsequent sections we mainly focus on the shell evolution and on the spin excitation, by employing the M3Y-P5 or P5$'$ interaction.

Table 1. Binding energies $-E$ (MeV) and rms matter radii $\sqrt{\langle r^2\rangle}$ (fm) of several doubly magic nuclei.

		SLy5	D1S	M3Y-P5′	M3Y-P7	Exp.
^{16}O	$-E$	128.6	129.5	124.1	125.9	127.6
	$\sqrt{\langle r^2\rangle}$	2.59	2.61	2.60	2.57	2.61
^{40}Ca	$-E$	344.3	344.6	331.7	334.3	342.1
	$\sqrt{\langle r^2\rangle}$	3.29	3.37	3.37	3.35	3.47
^{48}Ca	$-E$	416.0	416.8	411.5	414.9	416.0
	$\sqrt{\langle r^2\rangle}$	3.44	3.51	3.51	3.49	3.57
^{90}Zr	$-E$	782.4	785.9	775.7	780.8	783.9
	$\sqrt{\langle r^2\rangle}$	4.22	4.24	4.23	4.22	4.32
^{208}Pb	$-E$	1635.2	1639.0	1635.7	1635.5	1636.4
	$\sqrt{\langle r^2\rangle}$	5.52	5.51	5.51	5.51	5.49

All the numerical calculations in this paper are implemented by applying the Gaussian expansion method [13, 14, 15, 16]. The Hamiltonian is $H = H_N + V_C - H_{\rm c.m.}$, where $H_N = \sum_i \boldsymbol{p}_i^2/2M + \sum_{i<j} v_{ij}$ with the effective NN interaction v_{ij}, V_C represents the Coulomb interaction among protons, and $H_{\rm c.m.} = \boldsymbol{P}^2/2AM$. Note that finite-range nature of V_C and $H_{\rm c.m.}$ is handled exactly, up to the pairing channel in the HFB case.

The spherical HF results for the binding energies and the rms matter radii of the doubly-magic nuclei with the M3Y-P5′ and P7 interactions are compared to those with Skyrme SLy5 [25] and Gogny D1S [26] as well as to the measured values [27, 28, 29], in Table 1. We view that the M3Y-type semi-realistic interactions are adjusted to the properties of the doubly magic nuclei comparably well with the conventional MF interactions.

3. Shell evolution in medium-mass nuclei

Since the shell structure is a fundamental concept in nuclear structure physics, the shell evolution is a highly important subject. A significant role of the tensor force, which is a central issue in the Otsuka scheme, was first pointed out for the N-dependence of $\varepsilon_p(0h_{11/2}) - \varepsilon_p(0g_{7/2})$ just above $Z = 50$ [5]. We here view the s.p. energies of these orbits relative to $p1d_{5/2}$, $\Delta\varepsilon_p(j) = \varepsilon_p(j) - \varepsilon_p(1d_{5/2})$. In Fig. 1, $\delta\Delta\varepsilon_p(0h_{11/2})$ and $\delta\Delta\varepsilon_p(0g_{7/2})$, where $\delta\Delta\varepsilon_p(j)$ indicates the difference between $\Delta\varepsilon_p(j)$ at a given N and at $N = 64$, are plotted. The HF results with M3Y-P5′ are compared to those with D1S and to the experimental data [30], for which we use the energies of the lowest states in Sb having appropriate J^π. It is noted that the N-dependence of $\varepsilon_p(0h_{11/2}) - \varepsilon_p(0g_{7/2})$ evaluated from the lowest states is similar to that obtained after averaging the known states with the weight of the spectroscopic factors [31]. Unlike D1S that does not explicitly include tensor channels, the M3Y-type interactions including the realistic tensor force well describe the N-dependence of $\delta\Delta\varepsilon_p(j)$, particularly $\delta\Delta\varepsilon_p(0h_{11/2})$, as pointed out in Ref. [10]. This result confirms the Otsuka scheme both qualitatively and quantitatively. It is remarked that the realistic tensor force gives reasonable slope of $\delta\Delta\varepsilon_p(j)$.

Fragmented over certain energy range, the s.p. energies obtained in the MF approaches should better correspond to the averaged energy of the states weighted by the spectroscopic factors, rather than energies of individual observed levels. For the proton $1s_{1/2}$ and $0d_{3/2}$ hole states in ^{40}Ca and ^{48}Ca, sum of the measured spectroscopic factor exceeds 90% [32, 33], by which inversion of the $p1s_{1/2}$ and $p0d_{3/2}$ levels is indicated as N increases from ^{40}Ca to ^{48}Ca. This

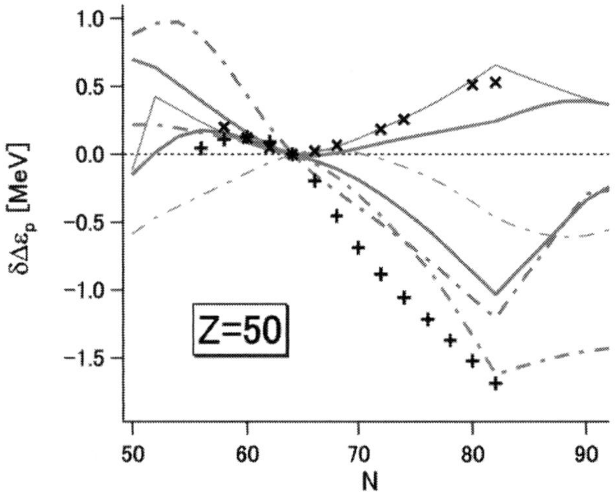

Figure 1. $\delta\Delta\varepsilon_p(0g_{7/2})$ (dot-dashed lines) and $\delta\Delta\varepsilon_p(0h_{11/2})$ (solid lines) in the $Z = 50$ nuclei obtained from the spherical HF calculations with D1S (blue lines) and M3Y-P5′ (thick red lines), in comparison with the observed levels of the Sb nuclei (+ for $0g_{7/2}$ and × for $0h_{11/2}$) [30]. Thin red lines are the $v^{(\mathrm{TN})}$ contribution in the M3Y-P5′ results.

$p1s_{1/2}$-$p0d_{3/2}$ difference and their inversion can be a good test of the MF effective interactions. We have found that M3Y-P5′ reproduces the N-dependence of $\Delta\varepsilon_p = \varepsilon_p(1s_{1/2}) - \varepsilon_p(0d_{3/2})$ from ^{40}Ca to ^{48}Ca remarkably well (Fig. 2). A crucial role of the realistic tensor force in the slope of $\Delta\varepsilon_p$ is obvious, as $\Delta\varepsilon_p$ becomes almost parallel to that of D1S when the tensor force is excluded. This consequence is qualitatively consistent with those from phenomenological tensor force [34, 35], though $\Delta\varepsilon_p$ has not been reproduced quantitatively in those studies. Further details including possibility of the bubble structure are discussed in Ref. [36].

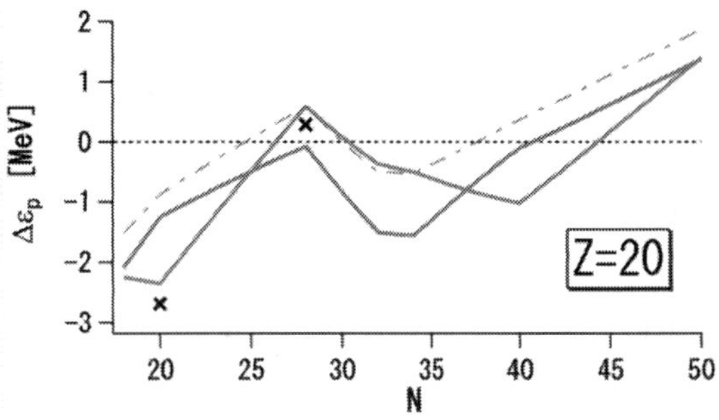

Figure 2. $\Delta\varepsilon_p$ in the $Z = 20$ nuclei obtained from the spherical HF calculations with D1S (blue lines) and M3Y-P5′ (thick red lines), in comparison with the experimental values in ^{40,48}Ca (×) [32, 33]. Thin red dot-dashed line shows the values by excluding the $v^{(\mathrm{TN})}$ contribution from the M3Y-P5′ results, with shifting so as to coincide at ^{48}Ca.

Influence of $v^{(\mathrm{C})}_{\mathrm{OPEP}}$ on the shell evolution, which forms another issue in the Otsuka scheme, has not been so clear. However, in certain cases it can give additional effects to those of $v^{(\mathrm{TN})}$, as illustrated by the $N = 32$ nuclei in Refs. [10, 37].

Although our MF approaches with the semi-realistic interactions have realized the Otsuka scheme on the shell evolution, the tensor-force effects in particular, they have given different consequence on the magicity from those predicted by Otsuka and his collaborators for several cases. Even though contribution of the tensor force is similar in qualitative respect, interplay of the tensor force and the other channels of the effective NN interaction can make sizable difference. Such an example is found in $N = 34$ magicity around ^{54}Ca, which has been claimed in Ref. [38] but not supported by the self-consistent MF calculations with the semi-realistic interactions [10, 11]. Another is erosion of the $Z = 28$ magicity at ^{78}Ni, which is not predicted in the MF approaches with the semi-realistic interactions [37] as with the conventional interactions. Experiments on these points are awaited.

Table 2. Comparison of $M1$ strength in ^{208}Pb, quoted from Ref. [39].

			M3Y-P5		Exp.
			$\hat{v} - \hat{v}^{(\mathrm{TN})}$	$\hat{v}$	
"IS"	E_x	(MeV)	6.87	5.85	5.85
	$B(M1)\uparrow$	(μ_N^2)	4.7	2.4	2.0
"IV"	E_x	(MeV)	9.2 – 10.9	9.2 – 10.9	7.1 – 8.7
	$\sum B(M1)\uparrow$	(μ_N^2)	16.3	19.4	16.3 or 18.2

4. $M1$ excitations of ^{208}Pb

The tensor-force effects are expected to appear in excitations involving spin degrees of freedom. However, self-consistent calculations within the HF+RPA including the tensor force have not been performed until recently. An earliest calculation in this line was implemented for the $M1$ excitation of ^{208}Pb with the semi-realistic interaction [39], and the results were compared to those of the high-precision measurements. By adopting the $M1$ operator including the two-particle-two-hole ($2p$-$2h$) as well as the meson-exchange current corrections [40], we have successfully reproduced the separation of the isoscalar (IS) dominant $M1$ strength at low energy and the isovector (IV) dominant ones at higher energies.

In ^{208}Pb, the $M1$ excitation strengths are basically carried by the $(p0h_{11/2})^{-1}(p0h_{9/2})$ and the $(n0i_{13/2})^{-1}(n0i_{11/2})$ configurations. The residual interaction mixes these two configurations. The excitation energy and the $M1$ strength of the IS-dominant state are sensitive to the residual interaction (see arguments in Ref. [41]). We have found that the repulsive IV spin-isospin interaction in the central channel is not necessarily sufficient to give rise to the separation as large as viewed in the measurement, and that the additive effect of the tensor force well accounts for the excitation energy and the $M1$ strength of the IS-dominant state, as presented in Table 2.

5. Summary

The M3Y-type semi-realistic NN interactions have been developed and applied to describing nuclear structure in the mean-field and RPA framework. They are useful in investigating roles of the NN interaction in the shell evolution within the self-consistent mean-field approaches, in harmony with the mechanisms proposed by Otsuka and his collaborators (Otsuka scheme). We have shown that the shell evolution in the $Z = 50$ and $Z = 20$ nuclei predicted by the semi-realistic interactions is in good agreement with the experimental data, owing primarily to the realistic tensor force. The realistic tensor force gives important and appropriate contribution also to the $M1$ strengths in ^{208}Pb, as shown within the HF+RPA. These results exemplify that the semi-realistic effective NN interactions provide us with a promising tool to understand nuclear properties on a microscopic basis.

Acknowledgments

I would express my gratitude to Prof. Otsuka because I learned a lot from him during fruitful collaboration and instructive discussions, particularly at early stages of my career. This work is financially supported as Grant-in-Aid for Scientific Research (C) No. 22540266 by JSPS, and as Grant-in-Aid for Scientific Research on Innovative Areas No. 24105008 by MEXT, Japan.

References

[1] Sorlin O and Porquet M-G 2008 *Prog. Part. Nucl. Phys.* **61** 602

[2] Otsuka T 2008 *Eur. Phys. J* ST **156** 169
[3] Ozawa A, *et al.* 2000 *Phys. Rev. Lett.* **84** 5493
[4] Otsuka T, *et al.* 2001 *Phys. Rev. Lett.* **87** 082502
[5] Otsuka T, Suzuki T, Fujimoto R, Grawe H and Akaishi Y 2005 *Phys. Rev. Lett.* **95** 232502
[6] Otsuka T, Suzuki T, Holt J D, Schwenk A and Akaishi Y 2010 *Phys. Rev. Lett.* **105** 032501
[7] Vautherin D and Brink D M 1972 *Phys. Rev.* C **5** 626
[8] Dechargé J and Gogny D 1980 *Phys. Rev.* C **21** 1568
[9] Nakada H 2003 *Phys. Rev.* C **68** 014316
[10] Nakada H 2008 *Phys. Rev.* C **78** 054301; 2010 *ibid.* **82** 029902(E)
[11] Nakada H 2010 *Phys. Rev.* C **81** 027301; 2010 *ibid.* **82** 029903(E)
[12] Bertsch G, Borysowicz J, McManus H and Love W G 1977 *Nucl. Phys.* A **284** 399
[13] Nakada H and Sato M 2002 *Nucl. Phys.* A **699** 511; 2003 *ibid.* **714** 696
[14] Nakada H 2006 *Nucl. Phys.* A **764** 117; 2008 *ibid.* **801** 169
[15] Nakada H 2008 *Nucl. Phys.* A **808** 47
[16] Nakada H, Mizuyama K, Yamagami M and Matsuo M 2009 *Nucl. Phys.* A **828** 283
[17] Khoa D T private communication
[18] Loan D T, Tan N H, Khoa D T and Margueron J 2011 *Phys. Rev.* C **83** 065809
[19] Anantaraman N, Toki H and Bertsch G F 1983 *Nucl. Phys.* A **398** 269
[20] Suzuki T and Sakai H 1999 *Phys. Lett.* B **455** 25
[21] Nakada H 2013 *Phys. Rev.* C **87** 014336
[22] Friedman B and Pandharipande V R 1981 *Nucl. Phys.* A **361** 502
[23] Akmal A, Pandharipande V R and Ravenhall D G 1998 *Phys. Rev.* C **58** 1804
[24] Cao L-G, Colò G and Sagawa H 2010 *Phys. Rev.* C **81** 044302
[25] Chabanat E, Bonche P, Haensel P, Meyer J and Schaeffer R 1998 *Nucl. Phys.* A **635** 231
[26] Berger J F, Girod M and Gogny D 1991 *Comp. Phys. Comm.* **63** 365
[27] Audi G, Wapstra A H and Thibault C 2003 *Nucl. Phys.* A **729** 337
[28] Khoa D T, Than H S and Grasso M 2003 *Nucl. Phys.* A **722** 92c
[29] Alkhazov G D, Belostotsky S L and Vorobyov A A 1978 *Phys. Rep.* **42** 89
[30] http://www.nndc.bnl.gov/ensdf/.
[31] Schiffer J P, *et al.* 2004 *Phys. Rev. Lett.* **92** 162501
[32] Doll P, Wagner G J, Knöpfle K T and Mairle G 1976 *Nucl. Phys.* A **263** 210
[33] Ogilvie C A, *et al.* 1987 *Nucl. Phys.* A **465** 445
[34] Grasso M, Ma Z Y, Khan E, Margueron J and Van Giai N 2007 *Phys. Rev.* C **76** 044319
[35] Wang Y Z, Gu J Z, Zhang X Z and Dong J M 2011 *Phys. Rev.* C **84** 044333
[36] Nakada H, Sugiura K and Margueron J 2013 *Phys. Rev.* C **87** 067305
[37] Nakada H 2010 *Phys. Rev.* C **81** 051302
[38] Honma M, Otsuka T, Brown B A and Mizusaki T 2005 *Eur. Phys. J.* A **25** s01 499
[39] Shizuma T, *et al.* 2008 *Phys. Rev.* C **78** 061303(R)
[40] Towner I S 1987 *Phys. Rep.* **155** 263
[41] Nakada H 2012 *Prog. Theor. Phys. Suppl.* **196** 371

Journal of Physics: Conference Series **445** (2013) 012012 doi:10.1088/1742-6596/445/1/012012

Investigating the strength of the $N = 34$ subshell closure in ^{54}Ca

D Steppenbeck[1], S Takeuchi[2], N Aoi[3], P Doornenbal[2], J Lee[2], M Matsushita[1], H Wang[2,4], H Baba[2], N Fukuda[2], S Go[1], M Honma[5], K Matsui[6], S Michimasa[1], T Motobayashi[2], D Nishimura[7], T Otsuka[1,6], H Sakurai[2,6], Y Shiga[8], P-A Söderström[2], T Sumikama[9], H Suzuki[2], R Taniuchi[6], Y Utsuno[10], J J Valiente-Dobón[11] and K Yoneda[2]

[1] Center for Nuclear Study, University of Tokyo, 2-1, Hirosawa, Wako, Saitama 351-0198, Japan
[2] RIKEN Nishina Center, 2-1, Hirosawa, Wako, Saitama 351-0198, Japan
[3] Research Center for Nuclear Physics, Osaka University, Ibaraki, Osaka 567-0047, Japan
[4] School of Physics and State Key Laboratory of Nuclear Physics and Technology, Peking University, Beijing 100871, People's Republic of China
[5] Center for Mathematical Sciences, University of Aizu, Aizu-Wakamatsu, Fukushima 965-8580, Japan
[6] Department of Physics, University of Tokyo, Bunkyo, Tokyo 113-0033, Japan
[7] Department of Physics, Tokyo University of Science, Noda, Chiba 278-8510, Japan
[8] Department of Physics, Rikkyo University, Toshima, Tokyo 172-8501, Japan
[9] Department of Physics, Tohoku University, Sendai, Miyagi 980-8578, Japan
[10] Japan Atomic Energy Agency, Tokai, Ibaraki 319-1195, Japan
[11] Istituto Nazionale di Fisica Nucleare, Laboratori Nazionali di Legnaro, Legnaro I-35020, Italy

E-mail: `steppenbeck@cns.s.u-tokyo.ac.jp`

Abstract. Excited states in the exotic $N = 34$ isotope ^{54}Ca have been measured for the first time via in-beam γ-ray spectroscopy with proton-knockout reactions from ^{55}Sc and ^{56}Ti radioactive beams on a Be reaction target at the Radioactive Isotope Beam Factory, Japan. A strong candidate for the transition from the first 2^+ state to the 0^+ ground state has been identified, in addition to several other weaker transitions. The structure of the $N = 33$ isotope ^{53}Ca has also been investigated from the same data. Preliminary γ-ray energy spectra for ^{54}Ca and ^{53}Ca will be presented and the significance of the $N = 34$ subshell closure will be examined. Predictions of several shell-model interactions performed in the fp model space will be discussed in light of the new results.

1. Introduction

Over recent years, the evolution of nuclear shell structure in exotic, neutron-rich nuclei has attracted much attention on both the experimental and theoretical fronts. The theoretical studies of Otsuka *et al* [1, 2], for example, has highlighted the role of the tensor and central forces in changing the traditional ordering of nuclear single-particle orbitals that can, in some instances, lead to the disappearance of magic numbers while other new ones appear. Such

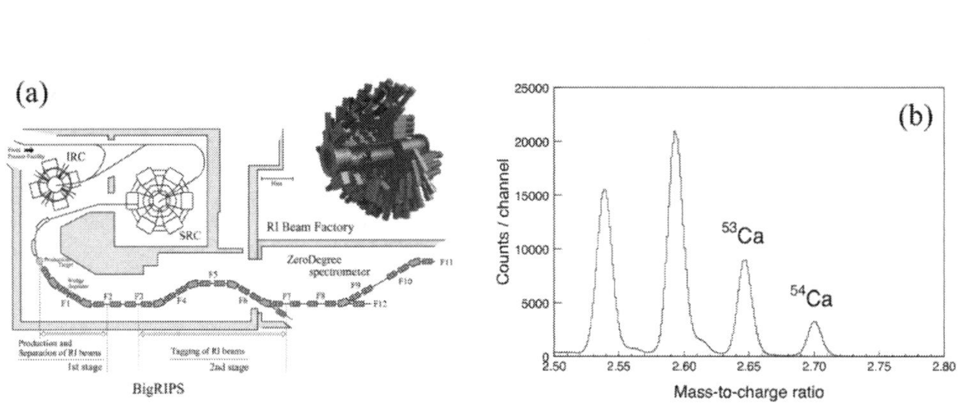

Figure 1. (Colour online) (a) Layout of the RIBF facility at RIKEN Nishina Center. The configuration of the DALI2 array used in the present work is indicated in the top-right corner. (b) Particle identification plot for Ca isotopes measured by ZDS. See text for details.

examples include the onset of an $N = 16$ subshell closure in neutron-rich oxygen [3] and the quenching of the $N = 20$ shell closure in the well-known "island of inversion" around $Z = 10-12$ [4]. In neutron-rich isotopes of Ca [5, 6], Ti [7, 8] and Cr [9, 10], much recent experimental work has focused on the onset of a new subshell gap at $N = 32$, which is reproduced well by the GXPF1 [11] and KB3G [12] shell-model Hamiltonians. In the framework of Refs. [1, 2], the onset of the $N = 32$ closure results as a direct consequence of a sizable $\nu p_{3/2} - \nu p_{1/2}$ gap, which presents itself as the $\nu f_{5/2}$ orbital moves up in energy due to a weakening of the attractive proton-neutron $\pi f_{7/2} - \nu f_{5/2}$ interaction as protons are removed from the $\pi f_{7/2}$ orbital. The GXPF1 Hamiltonian also predicts the onset of a significant $N = 34$ gap, which appears because the $\nu f_{5/2}$ orbital lies relatively high above the $\nu p_{1/2}$ orbital in that interaction. In fact, the size of the $N = 34$ gap predicted by GXPF1 is comparable to the one at $N = 28$ along the Ca and Ti isotopic chains. However, in the case of the Ti isotopes, recent experimental studies [8, 13] indicate that no subshell closure exists at $N = 34$. This observation led to a modified version of the GXPF1 Hamiltonian, labelled GXPF1A [14], in order to better reproduce the experimental systematics of the $N \approx 34$ isotopes. The updated interaction was found to account for the energy of the first 2^+ state [$E(2_1^+)$] in ^{56}Ti and states with $J > 6\hbar$ in ^{54}Ti [7], which correspond to neutron excitations above $N = 32$, in a more satisfactory manner. The new interaction was also found to show a systematic improvement in predictions of $E(2_1^+)$ in the Cr isotopes, and the recent assignment of $J^\pi = 1/2^-$ to the ^{55}Ti ground state [15] supports the prediction of the GXPF1A interaction, indicating that the $\nu f_{5/2}$ orbital lies above $\nu p_{1/2}$ in Ti, albeit not high enough to result in a new subshell closure. Despite the imposed modifications, a significant $N = 34$ subshell closure resides in the predictions of the GXPF1A interaction for ^{54}Ca, where the energy of the 2_1^+ state is predicted to lie at about 3 MeV, which is larger than, though comparable to, the size of $E(2_1^+)$ in ^{52}Ca [5, 6]. The latest addition to the GXPF1 family of interactions, the GXPF1B Hamiltonian [16], predicts the 2_1^+ state in ^{54}Ca to lie at 2.6 MeV. It must be stressed, however, that no significant $N = 34$ subshell closure is predicted by other shell-model effective interactions that are commonly used in the fp shell, such as the KB3G Hamiltonian. Thus, the measurement of $E(2_1^+)$ is important not only because it may highlight the doubly magic nature of ^{54}Ca, but also because it will bear important consequences for nuclear shell-model Hamiltonians, irrespective of the strength of the $N = 34$ gap.

2. Experimental details

The structures of ^{54}Ca and ^{53}Ca were investigated using in-beam γ-ray spectroscopy at the Radioactive Isotope Beam Factory [17], operated by RIKEN Nishina Center and Center for

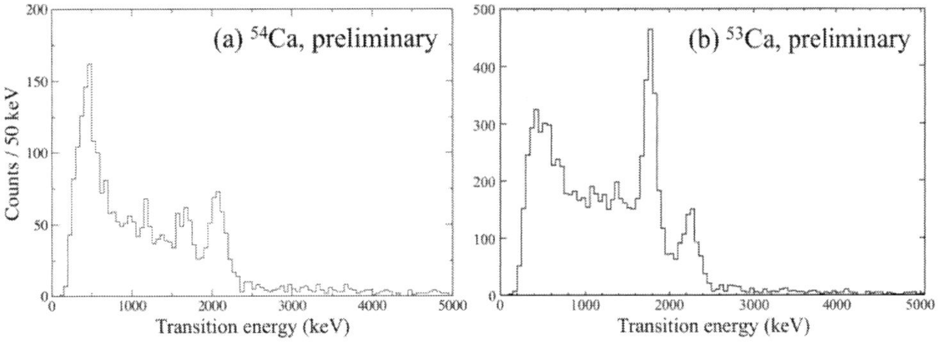

Figure 2. Preliminary Doppler-corrected γ-ray energy spectra for (a) ^{54}Ca and (b) ^{53}Ca deduced in the present work.

Nuclear Study, University of Tokyo. A ^{70}Zn primary beam at 345 MeV/u was delivered to the BigRIPS separator (BRS) [18], where a radioactive beam containing ^{55}Sc and ^{56}Ti was produced and focused on a 10-mm-thick Be reaction target located at the 8th focal plane of BRS [see Fig. 1(a)]. Typical secondary beam rates were $\sim$ 12 and 125 particles per second, respectively, per 1 pnA of primary beam intensity. The DALI2 array [19] was used to measure γ-ray transitions from excited nuclear states populated via nucleon knockout reactions. The reaction products were identified with the ZeroDegree spectrometer (ZDS) [18] using well established techniques involving measurements of particle magnetic rigidities, time of flights and energy losses in an ionisation chamber on an event-by-event basis (details are provided, for example, in Ref. [20]). A particle identification plot for Ca isotopes in ZDS is presented in Fig. 1(b), indicating the events measured in coincidence with at least one γ ray in DALI2.

3. Results and discussion

Several new γ-ray transitions have been identified in the present work for ^{53,54}Ca. A preliminary spectrum for ^{54}Ca is presented in Fig. 2(a), where ^{54}Ca events in ZDS measured in coincidence with either ^{55}Sc or ^{56}Ti in BRS and a single γ ray have been selected. The peak at 2.07 MeV[1] is assigned as the $2_1^+ \to 0_1^+$ transition based on the relative intensities of the transitions and systematics of even-even nuclei. Two weaker lines at 1.69 and 1.19 MeV are also present in the spectrum of Fig. 2(a). Based on the measured relative intensities, it is likely that the 1.69-MeV transition depopulates a level at 3.76 MeV and feeds the 2_1^+ state discussed above. Placement of the 1.19-MeV γ ray in the ^{54}Ca level scheme is, however, uncertain at present and cannot be deduced by intensity sum rules alone.

Systematics of $E(2_1^+)$ for the Ca isotopic chain are displayed in Fig. 3(a). The result of the present study indicates a reduction of $\sim$ 0.5 MeV in $E(2_1^+)$ relative to ^{52}Ca, suggesting a slightly weaker $N = 34$ subshell closure than the one at $N = 32$ for Ca. Systematics of $E(3_1^-)$ are also plotted in Fig. 3(a). The result for the 3.76-MeV level appears to continue the general trend of the data well and, therefore, a spin-parity assignment of 3^- appears plausible for this state. It must be stressed, however, that spin-parity assignments and $\gamma\gamma$ coincidences will be investigated further in the future. In Fig. 3(b), predictions of $E(2_1^+)$ by several shell-model Hamiltonians are presented as examples. It is clear that while the older GXPF1 and KB3G interactions deviate significantly from the new experimental result, the modified interaction GXPF1B provides a more satisfactory description. Numerous other predictions for the Ca isotopes are available, for example, those performed in the *sdfp* model space [21] and those involving three-body forces [22, 23]. It is noted that $E(2_1^+)$ is reproduced rather well in some cases.

[1] A preliminary uncertainty of ± 0.05 MeV is given for all transitions deduced in the present work.

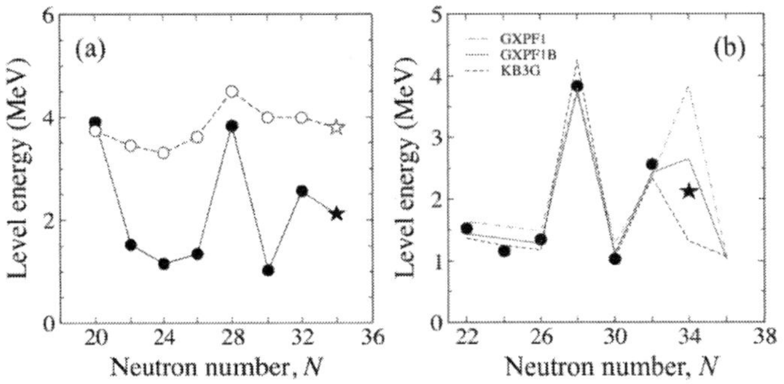

Figure 3. (Colour online) (a) Systematics of $E(2_1^+)$ [filled markers] and $E(3_1^-)$ [open markers] for Ca isotopes. (b) Examples of shell-model predictions of $E(2_1^+)$ along the Ca isotopic chain.

The structure of ^{53}Ca should also provide important information that may help shed light on the nature of the $N = 34$ subshell closure. A preliminary spectrum for ^{53}Ca deduced in the present work is is provided in Fig. 2(b). The 1.77-MeV transition is reported here for the first time, while the line at 2.24 MeV is consistent in energy with a transition previously reported in a β-decay study [24]. The relative intensities of the two γ rays in the spectrum of Fig. 2(b) and the result of Ref. [24] suggest that the transitions do not form a cascade sequence, indicating that an excited state at 1.77 MeV exists in ^{53}Ca in addition to the one at 2.24 MeV. Moreover, the observation of two states at ~ 2 MeV that are separated by ~ 0.5 MeV appears consistent with the predictions of shell-model Hamiltonians, for example, the interactions of Refs. [16, 22], among others. It is stressed, however, that the spins and parities of the two states could not be deduced from the experimental data. In the future, $\gamma\gamma$ coincidences will be investigated to confirm the suggested structure of ^{53}Ca based on the arguments presented above.

References

[1] Otsuka T *et al* 2005 *Phys. Rev. Lett.* **95** 232502
[2] Otsuka T *et al* 2010 *Phys. Rev. Lett.* **104** 012501
[3] Hoffman C R *et al* 2008 *Phys. Rev. Lett.* **100** 152502
[4] Warburton E K, Becker J A and Brown B A 1990 *Phys. Rev.* C **41** 1147
[5] Huck A *et al* 1985 *Phys. Rev.* C **31** 2226
[6] Gade A *et al* 2006 *Phys. Rev.* C **74** 021302(R)
[7] Janssens R V F *et al* 2002 *Phys. Lett.* B **546** 55
[8] Dinca D-C *et al* 2005 *Phys. Rev.* C **71** 041302(R)
[9] Prisciandaro J I *et al* 2001 *Phys. Lett.* B **510** 17
[10] Bürger A *et al* 2005 *Phys. Lett.* B **622** 29
[11] Honma M *et al* 2002 *Phys. Rev.* C **65** 061301(R)
[12] Poves A *et al* 2001 *Nucl. Phys.* A **694** 157
[13] Liddick S N *et al* 2004 *Phys. Rev. Lett.* **92** 072502
[14] Honma M *et al* 2005 *Eur. Phys. J.* A **25** s01 499
[15] Maierbeck P *et al* 2009 *Phys. Lett.* B **675** 22
[16] Honma M, Otsuka T and Mizusaki T 2008 *RIKEN Accel. Prog. Rep.* **41** 32
[17] Yano Y 2007 *Nucl. Instrum. Meth.* B **261** 1009
[18] Kubo T 2003 *Nucl. Instrum. Meth.* B **204** 97
[19] Takeuchi S *et al* 2003 *RIKEN Accel. Prog. Rep.* **36** 148
[20] Ohnishi T *et al* 2008 *J. Phys. Soc. Jpn.* **77** 083201
[21] Kaneko K *et al* 2011 *Phys. Rev.* C **83** 014320
[22] Hagen G *et al* 2012 *Phys. Rev. Lett.* **109** 032502
[23] Holt J D *et al* 2012 *J. Phys. G: Nucl. Part. Phys.* **39** 085111
[24] Perrot F *et al* 2006 *Phys. Rev.* C **74** 014313

Journal of Physics: Conference Series **445** (2013) 012013 doi:10.1088/1742-6596/445/1/012013

Shell-model derivation of the shears mechanism

P Van Isacker

Grand Accélérateur National d'Ions Lourds, CEA/DSM–CNRS/IN2P3
B.P. 55027, F-14076 Caen Cedex 5, France

E-mail: isacker@ganil.fr

Abstract. The geometry of the shears mechanism in nuclei is obtained by taking the limit of large angular momentum of shell-model matrix elements.

1. Introduction

In several lead isotopes, as well as in other regions of the nuclear chart near doubly-magic closures, regular sequences of γ rays are observed resembling those typical of a rotational band [1, 2, 3, 4]. These transitions are of magnetic character, in contrast to the usual quadrupole nature of nuclear rotors [5]. An explanation of this observation invokes the so-called 'shears' mechanism [6]. In a weakly deformed nucleus there exist low-energy configurations in which neutrons and protons combine into stretched structures ('blades') with angular momentum generated by the re-coupling of these blades, resembling the closing of a pair of shears. The breaking of the rotational symmetry in this case originates in an anisotropic distribution of nucleonic current loops (rather than electric charge), and thus the phenomenon is also referred to as magnetic rotation [7].

While an interpretation in terms of the shears mechanism provides an appealing, intuitive picture of these nuclear states, the question remains whether this geometry is borne out by microscopic calculations. A numerical shell-model calculation by Frauendorf *et al.* [8] confirmed the shears picture in the lead region. An analytic derivation of the geometry of the shears mechanism from the shell model is the purpose of the present contribution which presents results complementary to those reported elsewhere [9].

2. General shell-model expressions for the shears matrix element

Consider m nucleons of one type (say neutrons) in particle-like orbits $j_{1\nu}, j_{2\nu}, \ldots,$ and m' nucleons of the other type (protons) in hole-like orbits $j_{1\pi}^{-1}, j_{2\pi}^{-1}, \ldots,$ where $j_{k\rho}$ ($\rho = \nu, \pi$) is used as an abbreviation for the set of quantum numbers $n_{k\rho}$, $\ell_{k\rho}$ and $j_{k\rho}$ of singe-particle levels in a central potential. The m-particle (mp) and m'-hole (m'h) states are represented as $|N\rangle \equiv |j_{1\nu}j_{2\nu}(J_{12\nu})\ldots J_\nu\rangle$ and $|P^{-1}\rangle \equiv |j_{1\pi}^{-1}j_{2\pi}^{-1}(J_{12\pi})\ldots J_\pi\rangle$, where $J_{12\ldots\rho}$ are intermediate angular momenta occurring in some coupling scheme, and J_ν and J_π are the neutron and proton total angular momenta, respectively. The single-particle orbits $j_{k\rho}$ ($k = 1, 2, \ldots$) may or may not be identical and, in the former case, the states $|N\rangle$ and $|P^{-1}\rangle$ are assumed to be Pauli allowed. A shears band consists of the states $|NP^{-1}; J\rangle$ where J results from the coupling of J_ν and J_π.

A general shell-model hamiltonian has a neutron and a proton piece, and a neutron–proton interaction, and therefore can be written as

$$\hat{H} = \hat{H}_\nu + \hat{H}_\pi + \hat{V}_{\nu\pi}.$$

The quantity $\langle NP^{-1};J|\hat{H}|NP^{-1};J\rangle$ will be referred to as the (diagonal mp–m'h) shears matrix element. Due to the scalar character of the hamiltonian, the matrix elements of $\hat{H}_\nu$ and $\hat{H}_\pi$ depend only on the quantum numbers appearing in $|N\rangle$ and $|P\rangle$, respectively; $\hat{H}_\nu$ and $\hat{H}_\pi$ give constant energy contributions to all members of the shears band, irrespective of the total angular momentum J, and are therefore of no interest in the following. Any J dependence originates from the neutron–proton interaction $\hat{V}_{\nu\pi}$, specified by its matrix elements $\langle j_{k\nu}j_{l\pi};R|\hat{V}_{\nu\pi}|j_{k'\nu}j_{l'\pi};R\rangle$. A multipole expansion of $\hat{V}_{\nu\pi}$ leads to the following expression for the shears matrix element:

$$\begin{aligned}&\langle NP^{-1};J|\hat{V}_{\nu\pi}|NP^{-1};J\rangle\\ &= \sum_{klR\lambda}(-)^{j_{k\nu}+j_{l\pi}+J_\nu+J_\pi+J+R}(2R+1)V^R_{j_{k\nu}j_{l\pi}}\left\{\begin{array}{ccc} j_{k\nu} & j_{k\nu} & \lambda \\ j_{l\pi} & j_{l\pi} & R\end{array}\right\}\left\{\begin{array}{ccc} J_\nu & J_\pi & J \\ J_\pi & J_\nu & \lambda\end{array}\right\}\\ &\quad\times\langle N\|(a^\dagger_{j_{k\nu}}\tilde{a}_{j_{k\nu}})^{(\lambda)}\|N\rangle\langle P^{-1}\|(a^\dagger_{j_{l\pi}}\tilde{a}_{j_{l\pi}})^{(\lambda)}\|P^{-1}\rangle,\end{aligned} \tag{1}$$

where $V^R_{j_{k\nu}j_{l\pi}} \equiv \langle j_{k\nu}j_{l\pi};R|\hat{V}_{\nu\pi}|j_{k\nu}j_{l\pi};R\rangle$. The operator $a^\dagger_{j_{k\rho}m_{k\rho}}$ creates a neutron ($\rho=\nu$) or a proton ($\rho=\pi$) in orbit $j_{k\rho}$ with projection $m_{k\rho}$ on the z axis; the corresponding annihilation operator $a_{j_{k\rho}m_{k\rho}}$ is modified to $\tilde{a}_{j_{k\rho}m_{k\rho}} \equiv (-)^{j_{k\rho}+m_{k\rho}}a_{j_{k\rho}-m_{k\rho}}$ to make it behave as a proper tensor under rotations in three dimensions.

The expression (1) for the shears matrix element between mp–m'h states can be further reduced with use of standard techniques of angular momentum. This reduction is carried out here for $m=m'$ with $m=1$ and $m=2$.

2.1. The 1p–1h shears matrix element

For $m=1$, the members of the shears band are written as $|NP^{-1};J\rangle = |j_\nu j_\pi^{-1};J\rangle$ and the expression (1) reduces to one for a particle–hole matrix element,

$$\langle NP^{-1};J|\hat{V}_{\nu\pi}|NP^{-1};J\rangle = -\sum_R(2R+1)V^R_{j_\nu j_\pi}\left\{\begin{array}{ccc} j_\nu & j_\pi & J \\ j_\nu & j_\pi & R\end{array}\right\}.$$

This is nothing but the Pandya relation which expresses a particle–hole matrix element as a sum of particle–particle matrix elements [10]. The object in curly brackets is a $6j$ symbol [11], a quantity which is scalar under rotations, depending on six angular momenta. It is in fact the simplest non-trivial *scalar* $3nj$ symbol, the case $n=1$ being the trivial triangular delta $\{j_1j_2j_3\}$ which is equal to one if j_1, j_2 and j_3 form a triad and zero otherwise [12]. For the purpose of comparing with the case $m=2$ given below, it is of some interest to convert the $6j$ symbol to the standard notation for $3nj$ symbols (see eq. (17.11a) of Yutsis *et al.* [12]), leading to

$$\langle NP^{-1};J|\hat{V}_{\nu\pi}|NP^{-1};J\rangle = -\sum_R(2R+1)V^R_{j_\nu j_\pi}\left\{\begin{array}{ccc} j_\nu & & j_\pi \\ & J & R \\ j_\nu & & j_\pi\end{array}\right\}. \tag{2}$$

2.2. The 2p–2h shears matrix element

For $m=2$, the members of the shears band arise from the two-neutron-particle state $|N\rangle \equiv |j_{1\nu}j_{2\nu};J_\nu\rangle$ and the two-proton-hole state $|P^{-1}\rangle \equiv |j^{-1}_{1\pi}j^{-1}_{2\pi};J_\pi\rangle$ which are coupled to angular momentum J. The two-particle reduced matrix element in eq. (1) equals

$$\langle j_1j_2;J\|(a^\dagger_{j_k}\tilde{a}_{j_l})^{(\lambda)}\|j_1j_2;J\rangle = (-)^{j_1+j_2+J+\lambda}\sqrt{2\lambda+1}(2J+1)\hat{P}_{12}\left\{\begin{array}{ccc} j_1 & j_1 & \lambda \\ J & J & j_2\end{array}\right\}\delta_{k1}\delta_{l1},$$

where $\hat{P}_{12}$ is a symmetrizer defined as

$$\hat{P}_{12} f(j_1, j_2) = f(j_1, j_2) + f(j_2, j_1),$$

for any function f. The corresponding result for holes is obtained from eq. (14.39) of Talmi [11],

$$\langle j_1^{-1} j_2^{-1}; J \| (a_{j_k}^\dagger \tilde{a}_{j_l})^{(\lambda)} \| j_1^{-1} j_2^{-1}; J \rangle = (-)^{j_1+j_2+J+1} \sqrt{2\lambda+1}(2J+1)\hat{P}_{12} \left\{ \begin{array}{ccc} j_1 & j_1 & \lambda \\ J & J & j_2 \end{array} \right\} \delta_{k1}\delta_{l1}.$$

Insertion of these expressions for the two-particle and two-hole reduced matrix elements in eq. (1) leads to the shears matrix element

$$\frac{\langle NP^{-1}; J|\hat{V}_{\nu\pi}|NP^{-1}; J\rangle}{(2J_\nu+1)(2J_\pi+1)} = -\hat{P}_{12}^{\nu\pi} \sum_R (2R+1) V_{j_{1\nu}j_{1\pi}}^R \left\{ \begin{array}{cccc} j_{1\nu} & J_\nu & J_\pi & j_{1\pi} \\ j_{2\nu} & J & j_{2\pi} & R \\ j_{1\nu} & J_\nu & J_\pi & j_{1\pi} \end{array} \right\}, \quad (3)$$

where $\hat{P}_{12}^{\nu\pi} \equiv \hat{P}_{12}^{\nu} \hat{P}_{12}^{\pi}$ now generates four terms according to

$$\begin{aligned} &\hat{P}_{12}^{\nu\pi} f(j_{1\nu}, j_{2\nu}, j_{1\pi}, j_{2\pi}) \\ &= f(j_{1\nu}, j_{2\nu}, j_{1\pi}, j_{2\pi}) + f(j_{1\nu}, j_{2\nu}, j_{2\pi}, j_{1\pi}) + f(j_{2\nu}, j_{1\nu}, j_{1\pi}, j_{2\pi}) + f(j_{2\nu}, j_{1\nu}, j_{2\pi}, j_{1\pi}). \end{aligned}$$

The object in curly brackets in eq. (3) is a $12j$ symbol of the *first* kind, a quantity which is scalar under rotations, depending on twelve angular momenta. The result (3) is obtained by expressing the shears matrix element as a sum over four $6j$ symbols, which can be related to a $12j$ symbol (see eq. (19.1) of Yutsis *et al.* [12]).

If the neutrons and protons are all particle-like (or all hole-like), a result similar to eq. (3) is obtained,

$$\frac{\langle NP; J|\hat{V}_{\nu\pi}|NP; J\rangle}{(2J_\nu+1)(2J_\pi+1)} = \hat{P}_{12}^{\nu\pi} \sum_R (2R+1) V_{j_{1\nu}j_{1\pi}}^R \left[\begin{array}{cccc} j_{1\nu} & J_\nu & J_\pi & j_{1\pi} \\ j_{2\nu} & J & j_{2\pi} & R \\ j_{1\nu} & J_\nu & J_\pi & j_{1\pi} \end{array} \right], \quad (4)$$

where the object in square brackets is a $12j$ symbol of the *second* kind (see eq. (19.2) of Yutsis *et al.* [12]).

Note the striking similarity between the expressions (2) and (3) for the 1p–1h and 2p–2h shears matrix elements, raising the hope for a treatment of the general case with $m > 2$, a problem which is currently under study.

It is now a simple matter to introduce in the sums (2) or (3) values for the neutron–proton interaction matrix elements and to derive the J dependence of the shears matrix element. The geometric significance of these expressions can be understood by the taking the limit of large angular momenta. If these are large in comparison with $\hbar$, this can be considered as the classical limit of the quantum-mechanical expression for the shears matrix element.

3. Classical expressions for the shears matrix element

Classical limits of $3j$ and $6j$ symbols are known since the seminal study of Wigner [13], subsequently refined by Ponzano and Regge [14] whose work was put on a mathematically solid footing by Schulten and Gordon [15]. The classical limit of a $3j$ symbol is associated with the area of a triangle, while that of a $6j$ symbol involves the volume of a tetrahedron, with the lengths of the sides determined by the angular momenta. However, for $9j$ (let alone $12j$) symbols only partial results are known and therefore the classical limit of the expression (3) is difficult to obtain for an arbitrary neutron–proton interaction.

Instead of dealing with a general interaction, consider the modified surface delta interaction (MSDI),

$$\hat{V}^{\mathrm{MSDI}} = \sum_{k<l} \hat{V}^{\mathrm{MSDI}}(k,l), \quad \hat{V}^{\mathrm{MSDI}}(k,l) = -4\pi a'_T \delta(\bar{r}_k - \bar{r}_l)\delta(r_k - R_0) + b'\bar{\tau}_k \cdot \bar{\tau}_l + c',$$

which is known to be a reasonable approximation to the realistic nucleon–nucleon force in terms of the isoscalar and isovector strengths a'_0 and a'_1, and the strengths b' and c' of a charge-exchange and a constant interaction. The neutron–proton matrix element of the MSDI is [16]

$$V^R_{j_\nu j_\pi} = -\frac{(2j_\nu+1)(2j_\pi+1)}{2}\left[a_{01}\begin{pmatrix} j_\nu & j_\pi & R \\ \frac{1}{2} & -\frac{1}{2} & 0 \end{pmatrix}^2 + a_0\begin{pmatrix} j_\nu & j_\pi & R \\ \frac{1}{2} & \frac{1}{2} & -1 \end{pmatrix}^2\right] - b + c, \quad (5)$$

with

$$a_{01} = \frac{a_0 + a_1}{2} - (-)^{\ell_\nu + \ell_\pi + R}\frac{a_0 - a_1}{2},$$

and

$$a_T = a'_T C(R_0), \quad b = b'C(R_0), \quad c = c'C(R_0), \qquad C(R_0) = R^4_{n_\nu \ell_\nu}(R_0)R_0^2 = R^4_{n_\pi \ell_\pi}(R_0)R_0^2.$$

A geometric insight is obtained by introducing the neutron–proton matrix elements of the MSDI into the expressions for the shears matrix elements. The procedure is illustrated with the 1p–1h shears matrix element (2) and results are summarized for the 2p–2h shears matrix element (3).

3.1. The 1p–1h shears matrix element

When the neutron–proton matrix elements of the MSDI (5) are introduced in the shears matrix element (2), one encounters sums of the type

$$\sigma_n^{(\lambda)} \equiv \sum_R (-)^{\lambda R}(2R+1)\begin{pmatrix} j_\nu & j_\pi & R \\ \frac{1}{2} & n-\frac{1}{2} & -n \end{pmatrix}^2 \begin{Bmatrix} j_\nu & j_\pi & J \\ j_\nu & j_\pi & R \end{Bmatrix},$$

for $(\lambda, n) = (0,0)$, (0,1) and (1,0). These reduce to simple expressions in the classical limit. For example, for $\lambda = 0$, the sum can be exactly rewritten as

$$\sigma_n^{(0)} = \begin{pmatrix} j_\nu & j_\pi & J \\ \frac{1}{2} & -n+\frac{1}{2} & n-1 \end{pmatrix}^2.$$

The classical limit of the $3j$ symbol [13] leads to the approximations

$$\sigma_0^{(0)} \approx \frac{2[1+(-)^{j_\nu+j_\pi+J}]}{\pi(2j_\nu+1)(2j_\pi+1)}\frac{1}{\sin\theta_{\nu\pi}},$$

$$\sigma_1^{(0)} \approx \frac{2}{\pi(2j_\nu+1)(2j_\pi+1)}\left(\frac{1}{\sin\theta_{\nu\pi}} + (-)^{j_\nu+j_\pi+J}\frac{1}{\tan\theta_{\nu\pi}}\right),$$

where $\theta_{\nu\pi}$ is the shears angle, that is, the angle between the neutron and proton angular momenta,

$$\theta_{\nu\pi} = \arccos\frac{J(J+1) - j_\nu(j_\nu+1) - j_\pi(j_\pi+1)}{2\sqrt{j_\nu(j_\nu+1)j_\pi(j_\pi+1)}}.$$

For $(\lambda, n) = (1,0)$, a similar approximation is obtained from the relation $\sigma_0^{(1)} = -(-)^J\sigma_1^{(0)}$.

With use of the preceding approximations, the expression for the 1p–1h shears matrix element (2) of a MSDI becomes

$$\langle NP^{-1}; J|\hat{V}^{\rm MSDI}_{\nu\pi}|NP^{-1}; J\rangle \approx (b-c) + \frac{s_1}{2\pi\sin\theta_{\nu\pi}} + \frac{t_1}{2\pi\tan\theta_{\nu\pi}}, \qquad (6)$$

with

$$\begin{aligned} s_1 &= [1+(-)^{j_\nu+j_\pi+J}](a_0+a_1) + 2a_0 + (-)^{\ell_\nu+\ell_\pi+J}(a_0-a_1), \\ t_1 &= 2(-)^{j_\nu+j_\pi+J}a_0 + (-)^{\ell_\nu+\ell_\pi+j_\nu+j_\pi}(a_0-a_1). \end{aligned}$$

3.2. The 2p–2h shears matrix element

A classical limit can also be obtained for the 2p–2h shears matrix element (3). The derivation is more complicated [9] and leads to the result

$$\langle NP^{-1}; J|\hat{V}^{\rm MSDI}_{\nu\pi}|NP^{-1}; J\rangle \approx 4(b-c) + \frac{s_2}{2\pi\sin\theta_{\nu\pi}} + \frac{t_2}{2\pi\tan\theta_{\nu\pi}}, \qquad (7)$$

where $\theta_{\nu\pi}$ is now the angle between $\bar{J}_\nu$ and $\bar{J}_\pi$, and with

$$s_2 = 4(3a_0+a_1), \quad t_2 = 4\varphi(a_0-a_1), \qquad \varphi = \tfrac{1}{4}\hat{P}^{\nu\pi}_{12}(-)^{\ell_{1\nu}+j_{1\nu}+\ell_{1\pi}+j_{1\pi}}.$$

The classical limit of the 2p–2h shears matrix element (7) is remarkably similar to that of the 1p–1h shears matrix element (6). In both expressions the quantity $(b-c)$ appears and the constant as well as the charge-exchange interaction contribute equally to all members of the shears band. The coefficients s_m and t_m are expressed in terms of the isoscalar and isovector strengths a_0 and a_1 of the MSDI but the dependence for $m=1$ is different from that for $m=2$. It would be of interest to find the generalization to $m>2$.

4. Comparison of quantum-mechanical and classical expressions

In Fig. 1 the exact 1p–1h and 2p–2h shears matrix elements of the SDI (*i.e.*, MSDI with $b=c=0$) are compared with their classical approximations. To illustrate the quality of the classical approximation, rather large values for the neutron and proton angular momenta are chosen, $j_\rho = \frac{41}{2}$ for the 1p–1h and $J_\rho = 20$ for the 2p–2h case. Even for such unrealistically high single-particle angular momenta, the 1p–1h matrix element does not display a shears behaviour since the minimum energy is *not* attained for a configuration with orthogonal neutron and proton angular momenta. The 2p–2h matrix element, on the other hand, does behave (qualitatively at least) as expected for the energies of the members of a shears band. In fact, for $t_2=0$ (or $a_0=a_1$) a minimum energy is reached for a shears angle $\theta_{\nu\pi}=90^\circ$ between the neutron and proton angular momentum vectors. Furthermore, it can be shown [9] that the shears picture remains valid for smaller values of the angular momenta, as long as these are made from nearly stretched configurations with $J_\rho \approx j_{1\rho}+j_{2\rho}$.

5. Conclusion

In this contribution 1p–1h and 2p–2h shears matrix elements of a modified surface delta interaction (MSDI) are studied in the limit of large angular momentum. The geometric picture underlying the shears mechanism is confirmed in the latter case and the shears angle at which the 2p–2h matrix element reaches a minimum energy can be related to the isoscalar and isovector strengths of the MSDI.

Future studies are called for along the same lines. They include the extension to more complicated configurations (*i.e.*, mp–m'h and mixed configurations) and to other components of the interaction (*e.g.*, the tensor force), and the study of electromagnetic properties.

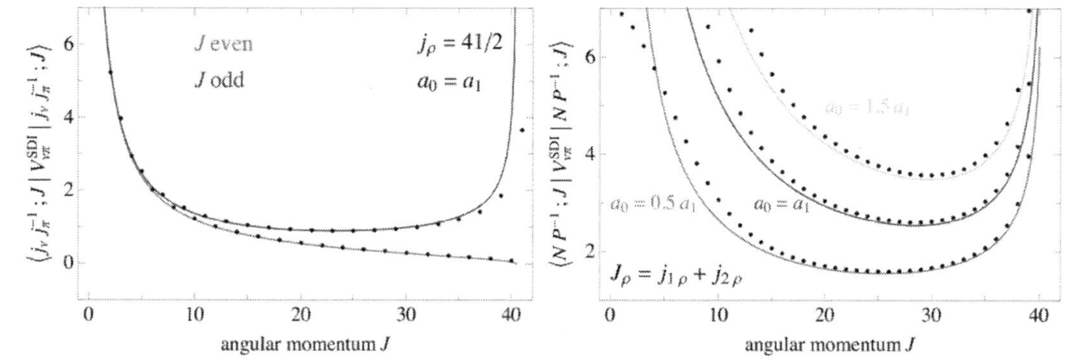

Figure 1. Left: The exact expression (2) for the 1p–1h shears matrix element of the SDI (dots) compared with its classical approximation (6) (lines). The single-particle angular momenta are $j_\rho = \frac{41}{2}$. The matrix element is in units $a_0 = a_1$. Right: The exact expression (3) for the 2p–2h shears matrix element of the SDI (dots) compared with its classical approximation (7) (lines). The single-particle angular momenta are $j_{1\rho} = \frac{19}{2}$ and $j_{2\rho} = \frac{21}{2}$, and the total neutron and proton angular momenta $J_\rho = 20$. Results are shown for three choices of the ratio a_0/a_1 and the matrix element is in units a_1.

Acknowledgements
I wish, on the occasion of his 60th birthday, to express my sincere thanks to Takaharu Otsuka for many years of fruitful and friendly collaboration. Thanks are due to Augusto Macchiavelli in collaboration with whom this work was carried out. This work was supported by FUSTIPEN (French-U.S. Theory Institute for Physics with Exotic Nuclei) under DOE grant No. DE-FG02-10ER41700 and by the exchange programme LIA France–Japan FJ-NSP.

References

[1] Baldsiefen G *et al.* 1992 *Phys. Lett.* B **275** 252
[2] Clark R M *et al.* 1992 *Phys. Lett.* B **275** 247
[3] Clark R M and Macchiavelli A O 2000 *Annual Rev. Nucl. Part. Sci.* **50** 1
[4] Hubel H 2005 *Progr. Part. Nucl. Phys.* **54** 1
[5] Bohr A and Mottelson B R 1975 *Nuclear Structure. II Nuclear Deformations* (New York: Benjamin)
[6] Frauendorf S 1993 *Nucl. Phys.* A **557** 259c
[7] Frauendorf S 2001 *Rev. Mod. Phys.* **73** 463
[8] Frauendorf S, Reif J and Winter G 1996 *Nucl. Phys.* A **601** 41
[9] Van Isacker P and Macchiavelli A O to be published.
[10] Pandya S P 1956 *Phys. Rev.* C **103** 956
[11] Talmi I 1993 *Simple Models of Complex Nuclei. The Shell Model and the Interacting Boson Model* (Chur: Harwood)
[12] Yutsis A P, Levinson I B and Vanagas V V 1962 *The Theory of Angular Momentum* (Jerusalem: Israel Program for Scientific Translations)
[13] Wigner E P 1959 *Group Theory and Its Application to the Quantum Mechanics of Atomic Spectra* (New York: Academic)
[14] Ponzano G and Regge T 1968 *Spectroscopy and Group Theoretical Methods in Physics* (Amsterdam: North-Holland)
[15] Schulten K and Gordon R G 1975 *J. Math. Phys.* **16** 1971
[16] Brussaard P J and Glaudemans P W M 1977 *Shell-Model Applications in Nuclear Spectroscopy* (Amsterdam: North-Holland)

Angular correlation between proton and neutron rotors

N Tajima

Department of Applied Physics, University of Fukui, Fukui, 910-8507, Japan

E-mail: `tajima@quantum.apphy.u-fukui.ac.jp`

Abstract. A brief review is given on the controversy and its solution about the fact that the angular momentum vector of protons and that of neutrons in well-deformed nuclei at low total angular momenta have a strong correlation that they are oriented in opposite directions. In a simple two-rotor model in 2-dimensional space, this fact is explained as originating from the quantum mechanical uncertainty relation between the angle and the angular momentum for the relative rotation of the two rotors. As the second topic, a more realistic model consisting of two triaxial rotors in 3-dimensional space coupled with a QQ interaction is employed to investigate a possible shears-band-like collective rotation predicted by T. Otsuka, in which the angle at which the angular momentum of protons and that of neutrons intersect changes continuously from 180° at spin zero toward 0° at high spins within the same rotational band. The probability distributions of the angle between the two angular momenta and the angle between the longest principal axes of two rotors are calculated to examine the participation of the scissors mode in the evolution of the ground rotational band versus spin.

1. Introduction

We have recently given careful consideration to the meaning of the antiparallel proton and neutron angular momenta at low spins [1]. This fact that the total angular momenta of protons and that of neutrons are oriented in the opposite directions has first been demonstrated by Otsuka et al. for the angular-momentum projected Nilsson wave functions[2, 3, 4]. A question was raised immediately [5] that it might be incorrect because it would mean an unphysical free contra-rotation of proton and neutron ellipsoids as depicted in Fig. 1, while only small-amplitude oscillations in the scissors mode as depicted in Fig. 2 are possible from a physical point of view. Actually, Ref. [2, 3, 4] has never described this opposite rotation as a free contra rotation. However, it would be indeed a plausible picture of this fact from a classical point of view.

Ref.[5] argued that the removal of the spurious center-of-mass motion could make the angular momenta of protons and neutrons parallel. Afterward, two papers [6, 7] have cited Ref.[5] as solving the controversy. In Ref.[1], Otsuka and we have shown that the elimination of the spurious center-of-mass motion does not substantially change the situation in a classical treatment of independent nucleon motions. By using a simple two-rotor model in 2-dimensional space, we have also given a correct interpretation of the opposite-direction correlation of the two angular momenta, as is discussed in Sec. 2.

The principal intention of Ref.[2] was to suggest the possibility that collective rotational excitations are accompanied with a narrowing of the angle between the proton and neutron angular momenta. Namely, along a rotational band, the angle changes from 180° at spin zero

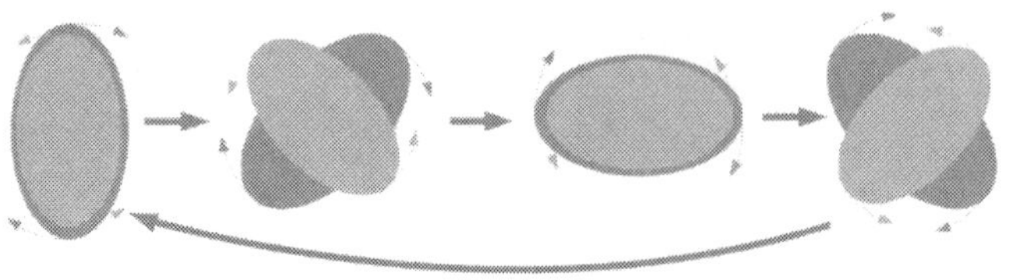

Figure 1. The free contra-rotation. The blue and red ovals represent neutrons and protons, respectively.

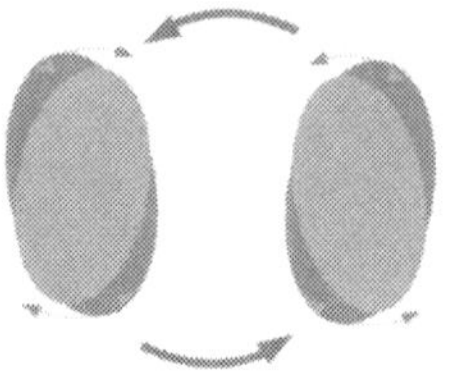

Figure 2. The scissors mode oscillation.

toward 0° at high spins. It was further suggested that this change is accompanied with a reduction of the overlap between the proton and neutron ellipsoids by means of the scissors-mode degree of freedom at intermediate spins. In order to investigate such a novel collective rotational excitation mechanism, we employ in sec. 3 a model in which two triaxial rotors in 3-dimensional space are coupled through an attractive quadrupole-quadrupole interaction. We calculate not only the expectation values but also the probability distribution of the angle between the angular momenta and the angle between the longest principal axes of rotors. These two angles have different kinds of information because of the quantum mechanical uncertainty in the directions of the angular momenta relative to the principal axes of the ellipsoids, and also owing to dynamically correlated rotational motions of the two rotors.

2. Analysis with a two-rotor model in 2-dimensional space

In Ref. [1] we have employed a model consisting of two rotors in 2-dimensional space to discuss the implication of the oppositely directed angular momenta of protons and neutrons. The Hamiltonian of the model is written as,

$$H = -\frac{\hbar^2}{2\mathcal{I}_\mathrm{p}}\frac{\partial^2}{\partial\varphi_\mathrm{p}^2} - \frac{\hbar^2}{2\mathcal{I}_\mathrm{n}}\frac{\partial^2}{\partial\varphi_\mathrm{n}^2} + V\left(\varphi_\mathrm{p} - \varphi_\mathrm{n}\right), \tag{1}$$

where φ_p and φ_n are the azimuths of proton and neutron rotors, respectively, as depicted in Fig. 3, $\mathcal{I}_\mathrm{p}$ and $\mathcal{I}_\mathrm{n}$ are the moment of inertia of protons and that of neutrons, respectively, and $V\left(\varphi_\mathrm{p} - \varphi_\mathrm{n}\right)$ is an attractive potential which favors alignments of the long axes of the proton and the neutron ellipses.

A wave-packet wavefunction of the form,

$$f(\varphi) = \sum_{m=-\infty}^{\infty} f_m e^{im\varphi}, \quad f_m = \mathcal{N}e^{-a^2m^2} \tag{2}$$

represents a state of each ellipse whose long axis is aligned to the space-fixed x-axis with a quantum mechanical fluctuation of $\sqrt{\langle\varphi^2\rangle} \approx a$. The products of such states, $f(\varphi_\mathrm{p})f(\varphi_\mathrm{n})$, can be viewed as an intrinsic state of a deformed nucleus. In order to gain energy from the interaction term $V\left(\varphi_\mathrm{p} - \varphi_\mathrm{n}\right)$, the size of the fluctuation (a) in the angles φ_p and φ_n should be decreased, which inevitably increases the width of the spreading of the angular momenta (m) of each rotor since $\sqrt{\langle m^2\rangle} \approx (2a)^{-1}$. One can construct a state of given total angular momentum M by means of the angular momentum projection method as,

$$\Psi = \hat{P}_M f(\varphi_\mathrm{p})f(\varphi_\mathrm{n}) = \sum_{m_\mathrm{p}=-\infty}^{\infty}\sum_{m_\mathrm{n}=-\infty}^{\infty} \delta_{m_\mathrm{p}+m_\mathrm{n},M} f_{m_\mathrm{p}} f_{m_\mathrm{n}} e^{im_\mathrm{p}\varphi_\mathrm{p}} e^{im_\mathrm{n}\varphi_\mathrm{n}}. \tag{3}$$

As depicted in Fig. 4, many combinations of angular momenta (m_p,m_n) in the line $m_\mathrm{p}+m_\mathrm{n} = M$ are mixed in the state of Eq. (3). When M is large, m_p, m_n, and M have the same sign.

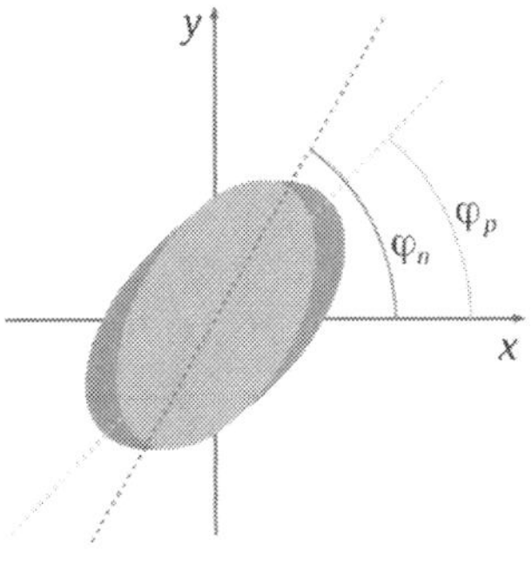

Figure 3. Two rotor model in 2-dimensional space.

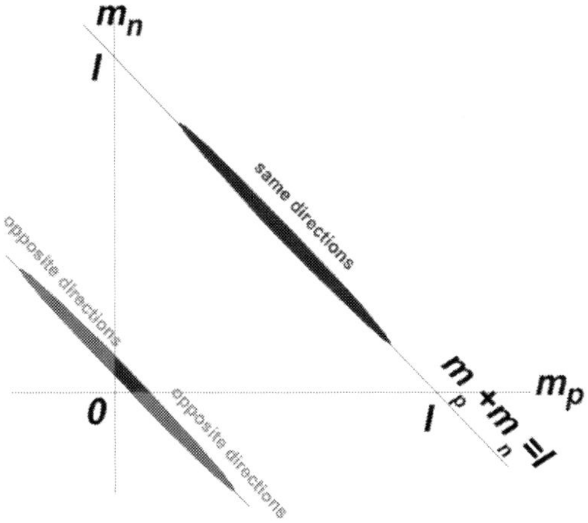

Figure 4. Mixture of $(m_{\rm p}, m_{\rm n})$ components in eigenstates of a system of two rotors in 2-dimensional space.

However, when M is smaller than the spreading $(2a)^{-1}$, $m_{\rm p}$ and $m_{\rm n}$ most often have different signs. Therefore, as far as this 2-dimensional model of deformed nuclei concerns, the opposite senses of the rotations of the proton and neutron rotors do not mean an unphysical free contra rotation, but a close binding of the two ellipses.

3. Analysis of a two-rotor model in 3-dimensional space

In this paper, we extend the two-rotor model to the 3-dimensional space. We assume the following Hamiltonian,

$$H = H_1 + H_2 - \kappa Q^1 \cdot Q^2, \quad H_{\rm q} = \sum_{i=1}^{3} \frac{\hbar^2}{2\mathcal{I}_i} J_{{\rm q},i}^2, \quad \mathcal{I}_i = 4B\beta^2 \sin^2\left(\gamma - \frac{2\pi}{3} i\right), \tag{4}$$

where $H_{\rm q}$ is the Hamiltonian for a triaxial rotor (the Davydov model[8]) with the irrotational-flow moment of inertia $\mathcal{I}_i$. The index q (=1,2) is introduced to distinguish the two rotors of this model. (It may be labeled as q=p,n as in the 2-dimensional model).

The two rotors are coupled through a quadrupole-quadrupole interaction. ($Q_\mu = \alpha_{2\mu}$. In the principal-axes frame, $\alpha_{2,0} = \beta\cos\gamma$, $\alpha_{2,\pm1} = 0$, $\alpha_{2,\pm2} = \frac{1}{\sqrt{2}}\beta\sin\gamma$, where β and γ are the quadrupole deformation parameters.)

We express the eigenstates of the coupled system in the laboratory frame as,

$$|IM\alpha\rangle = \sum_{J_1K_1J_2K_2} C^{I\alpha}_{J_1K_1J_2K_2} \sum_{M_1M_2} \langle J_1M_1J_2M_2|IM\rangle\, |J_1M_1K_1\rangle_1\, |J_2M_2K_2\rangle_2 . \tag{5}$$

Wavefunction of the Euler angles $\Omega = (\alpha, \beta, \gamma)$ of each rotor's orientation is given by

$$\langle\Omega|JMK\rangle = \sqrt{\frac{2J+1}{16\pi^2}} \cdot \frac{1}{\sqrt{1+\delta_{K0}}} \left[\mathcal{D}^J_{M,K}(\alpha,\beta,\gamma) + (-1)^I \mathcal{D}^J_{M,-K}(\alpha,\beta,\gamma)\right] \tag{6}$$

with $K = 0, 2, 4, \cdots \leq J$, because the signature quantum numbers are naturally restricted to $(r_1, r_2, r_3) = (+,+,+)$.

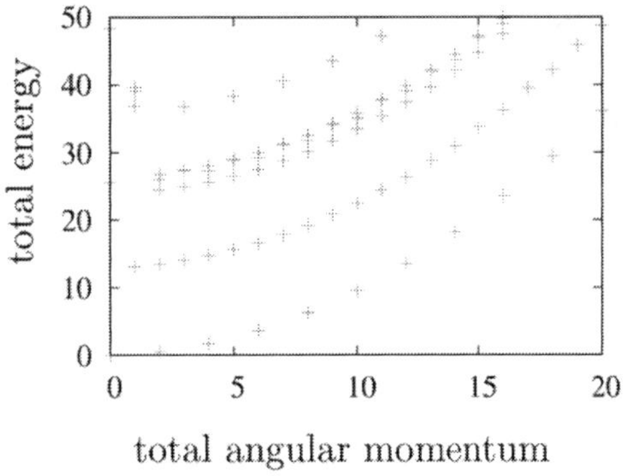

Figure 5. Energy spectrum of the 3-dimensional two-rotor model versus I in units of $\frac{\hbar^2}{B\beta^2}$.

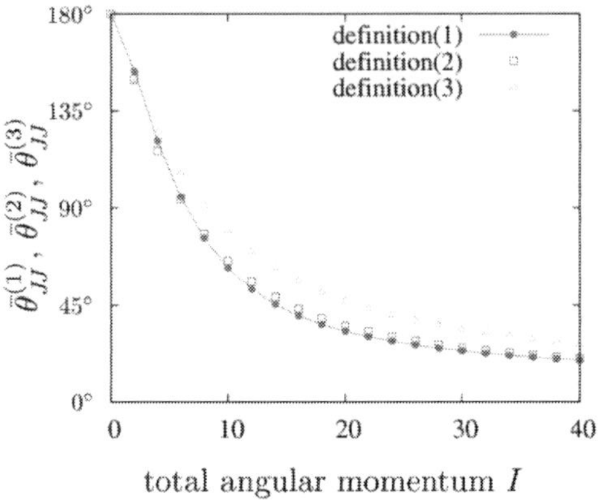

Figure 6. Three kinds of average angles between $\boldsymbol{J}_1$ and $\boldsymbol{J}_2$ versus I.

We use two identical rotors, each of which has $\gamma = 8°$, which determines two important eigenenergies of $\epsilon_{2^+_{\rm gr}} = 1.04$ and $\epsilon_{2^+_\gamma} = 26.16$ in units of $\frac{\hbar^2}{B\beta^2}$. The eigenstates of the total Hamiltonian are obtained by matrix diagonalizations with the strength of the QQ coupling of $\kappa\beta^2 = 1000 \times 0.3^2 = 90$ in units of $\frac{\hbar^2}{B\beta^2}$. The basis includes $J_i \leq 40$ and all (practically) possible values of K_i. Obtained eigenenergies of the coupled system are shown in Fig. 5. Two important excitation energies are $E_{2^+_{\rm gr}} = 0.52 \simeq \frac{1}{2}\epsilon^+_{\rm gr}$, and $E_{1^+_{K=1}} = 13.14$ (scissors mode) in units of $\frac{\hbar^2}{B\beta^2}$.

The angle between the angular momenta of two rotors, $\boldsymbol{J}_1$ and $\boldsymbol{J}_2$, is naturally defined by

$$\hat{\theta}_{JJ} = \arccos\frac{\hat{\boldsymbol{J}}_1 \cdot \hat{\boldsymbol{J}}_2}{\sqrt{\hat{\boldsymbol{J}}_1^2 \hat{\boldsymbol{J}}_2^2}}, \tag{7}$$

whose eigenvalues and eigenstates are given as

$$\hat{\theta}_{JJ} \left| J_1\alpha_1 J_2\alpha_2; IM \right\rangle = \arccos\frac{I(I+1) - J_1(J_1+1) - J_2(J_2+1)}{\sqrt{J_1(J_1+1)J_2(J_2+1)}} \left| J_1\alpha_1 J_2\alpha_2; IM \right\rangle, \tag{8}$$

for $J_1 \neq 0, J_2 \neq 0$. The average value of θ_{JJ} can be defined in a few different ways, e.g.,

$$\bar{\theta}^{(1)}_{JJ} = \left\langle \arccos\frac{\hat{\boldsymbol{J}}_1 \cdot \hat{\boldsymbol{J}}_2}{\sqrt{\hat{\boldsymbol{J}}_1^2 \hat{\boldsymbol{J}}_2^2}} \right\rangle', \quad \bar{\theta}^{(2)}_{JJ} = \arccos\left\langle \frac{\hat{\boldsymbol{J}}_1 \cdot \hat{\boldsymbol{J}}_2}{\sqrt{\hat{\boldsymbol{J}}_1^2 \hat{\boldsymbol{J}}_2^2}} \right\rangle', \quad \bar{\theta}^{(3)}_{JJ} = \arccos\frac{\left\langle \hat{\boldsymbol{J}}_1 \cdot \hat{\boldsymbol{J}}_2 \right\rangle}{\sqrt{\left\langle \hat{\boldsymbol{J}}_1^2 \right\rangle \left\langle \hat{\boldsymbol{J}}_2^2 \right\rangle}}, \tag{9}$$

where $\langle \cdots \rangle'$ are calculated after projecting out components having $J_1 = 0$ and/or $J_2 = 0$ and renormalizing the remaining part of the state vector. As shown in Fig.6, $\bar{\theta}^{(3)}_{JJ}$ [2] is larger than $\bar{\theta}^{(1)}_{JJ}$ ($\approx \bar{\theta}^{(2)}_{JJ}$) by ~ 0.3 radians. However, the decreasing behavior is common to all the definitions.

A way to understand the mechanism which gives rise to this behavior of $\bar{\theta}_{JJ}$ is to examine the probability distribution of the combination of angular momentum components (J_1, J_2) because θ_{JJ} is a simple function of (J_1, J_2) as shown in Fig. 7. The distributions at I=10 and 40 are shown in Fig. 8. One can observe the following points.

i) The spreading in $J_1 - J_2$ is only weakly dependent on I. (The tail is longer for larger I because of a trivial reason of the triangle condition.) By denoting the sum of two rotor's energies by $E(J_1, J_2)$, one can relate this approximate independence to the fact that

$$\Delta E = E\left(J_1 = \tfrac{1}{2}I + \Delta J, J_2 = \tfrac{1}{2}I - \Delta J\right) - E(J_1 = J_2 = \tfrac{I}{2}) = \tfrac{\hbar^2}{2\mathcal{I}}(\Delta J)^2 \tag{10}$$

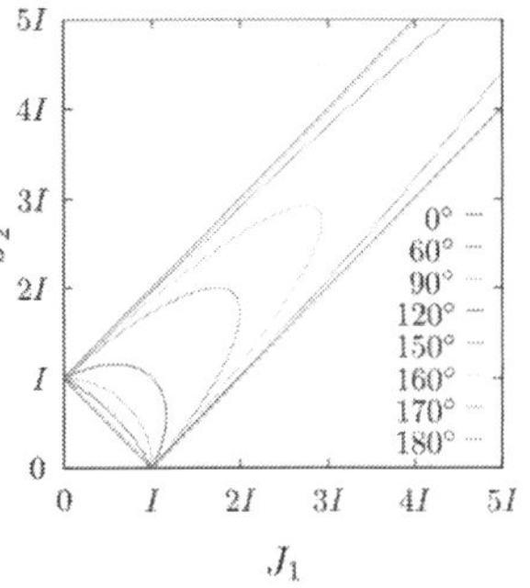

Figure 7. Lines in which θ_{JJ} takes on some constant values when J_1 and J_2 are coupled to I.

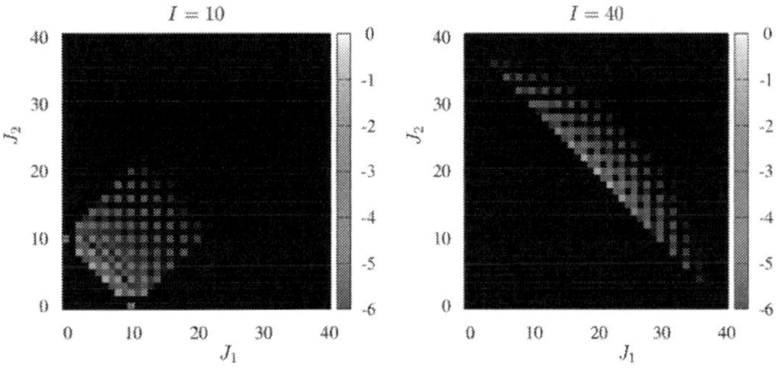

Figure 8. Distribution of the common logarithm of the probability of the combination of angular momentum components (J_1, J_2) for the yrast state of $I = 10$ and $I = 40$.

is independent of I.
ii) The width of the spreading in $J_1 + J_2$ is more squeezed for larger I, which is related to the fact that

$$\Delta E' = E\left(J_1 = J_2 = \tfrac{1}{2}I + \Delta J\right) - E(J_1 = J_2 = \tfrac{1}{2}I) = \frac{\hbar^2}{2\mathcal{I}}2\Delta J(I + \Delta J + 1) \tag{11}$$

is an increasing function of I.
The tight binding of the two rotors is created by mixing both $J_1 - J_2$ and $J_1 + J_2$ at low I and by mixing mainly $J_1 - J_2$ at high I. The pattern of mixture at low spins appears to be essentially different from that of the 2-dimensional case (Fig. 4). It may be related to a possible novel excitation modes such as the collective shears band[2].

Probability distributions of θ_{JJ} are shown in Fig.9. Instead of plotting original discrete-line spectra, we have replaced the delta functions representing the lines of the spectrum with a normalized Gaussian of width 3°. Around I=6-10 in the figure, there are sizeable components of $\theta_{JJ} \approx 90°$. We are interested in what takes place when the angular momentum vectors intersect at angles around 90°. Is it only for the angular momentum vectors? Or, do the principal axes of rotors also intersect at large angles in spite of the attractive QQ interaction?

The angle between the 3rd (longest) axis of rotor-1 with its orientation specified by the Euler angles $(\alpha_1, \beta_1, \gamma_1)$ and the 3rd axis of rotor-2 with $(\alpha_2, \beta_2, \gamma_2)$ is given by

$$\theta_{33} = \arccos\left[\sin\beta_1 \sin\beta_2 \cos(\alpha_1 - \alpha_2) + \cos\beta_1 \cos\beta_2\right]. \tag{12}$$

Its probability distribution, $\rho(\theta) = \langle IM\alpha|\delta(\theta_{33} - \theta)|IM\alpha\rangle$, can be calculated in principle via $\langle IM\alpha|P_l(\cos\theta_{33})|IM\alpha\rangle$ where P_l is the Legendre polynomial. Instead, we have roughly estimated it by the Monte Carlo integral in which two sets of Euler angles are sampled randomly and uniformly over $\alpha_i, \gamma_i \in [0, 2\pi]$ and $\cos\beta_i \in [-1, 1]$ $(i = 1, 2)$.

Fig.10 shows the probability distribution of θ_{33} for the yrast states at $I = 0$ and $I = 40$. One can see that the distribution of this angle is practically independent of I. Small discrepancy can be ascribed to the error due to the Monte-Carlo method.

The Hamiltonian given by Eq. (1) for 2-dimensional rotors is formally identical to the Hamiltonian for translational motions of two interacting objects. Hence, there holds the "Galilean" invariance so that the total and the relative rotational motions of two rotors are exactly decoupled and the probability distribution of the relative angle $\varphi_{\rm p} - \varphi_{\rm n}$ is independent of the total angular momentum M. From Fig.10, there probably also holds the "Galilean"

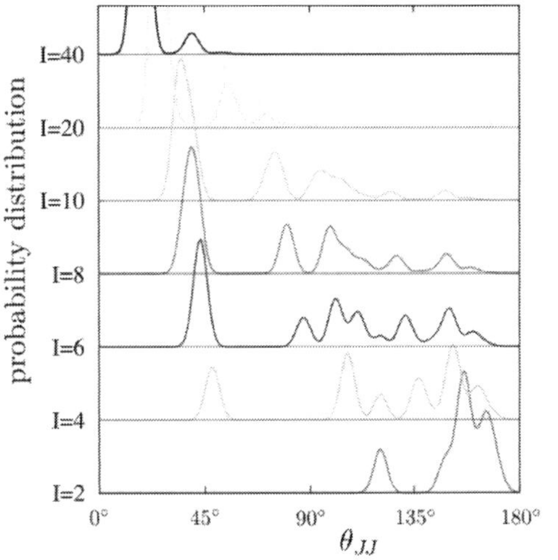

Figure 9. Probability distribution of θ_{JJ} for $I = 2, 4, \cdots, 40$ members of the ground rotational band.

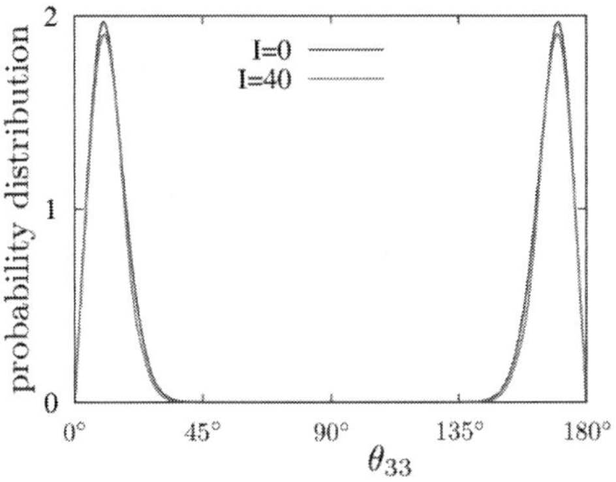

Figure 10. Probability distribution of θ_{33} for I=0 and 40 members of the ground rotational band.

invariance for two rotors in 3-dimensional space. It is a plausible conjecture within the classical mechanics, in which rotational motions of rigid bodies having a space fixed pivot are not affected by centrifugal and Coriolis forces because those forces are perpendicular to the velocity of the material points composing the rigid bodies.

However, this result does not exclude the possibility of the collective shears band [2] because in actual nuclei, i) the relative direction of angular momenta must affect the internal structure of the nucleus so that it cannot be represented by inert rotors, ii) the actual space is finite unlike the ideal rotor which has no band termination. We plan to include those effects in our model to obtain new insights into the dynamics of the relative orientations between partial angular momenta and the principal axes of deformations in nuclear collective rotations.

Acknowledgments

We are grateful to Professor Otsuka for useful discussions and continuous interest in this work. Sections 1 and 2 of this paper are reviews of a collaboration paper[1] with him.

References

[1] Tajima N and Otsuka T 2011 *Phys. Rev.* C **84** 064316
[2] Otsuka T 1993 *Phys. Rev. Lett.* **71** 1804
[3] Otsuka T, Mizusaki T and Honma M 1994 *Nucl. Phys.* A **570** 265c
[4] Otsuka T and Ginocchio J N 1994 *Nucl.Phys.* A **577** 197c
[5] Marshalek E R 1994 *Phys. Rev.* C **50** R5
[6] Sun Y, Wu C -L, Guidry K M and Feng D -H 1998 *Phys. Rev. Lett.* **80** 672
[7] Bes D R and Civitarese O 2001 *Phys. Rev.* C **63** 044323
[8] Davydov A S and Filippov G F 1958 *Nucl. Phys.* **8** 237

Interacting boson model with energy density functionals

K Nomura

Institut für Kernphysik, Universität zu Köln, D-50937 Köln, Germany

E-mail: `nomura@ikp.uni-koeln.de`

Abstract. A brief overview of the recent advancement in the microscopic study of the interacting boson model (IBM) is given. A new nucleon-boson mapping method has been proposed recently, that derives the IBM Hamiltonian based on the self-consistent mean- eld model with the microscopic energy density functional (EDF). The mean- eld total energy surface computed with a given EDF is mapped onto the analogous energy expectation value in the boson condensate, thereby the energy spectrum and the transition rates are generated. Since the EDF framework allows an accurate global description of nuclear intrinsic properties, the IBM is derived in a uni ed way, basically for any situation of low-energy quadrupole collective states of nuclei. The basic notion of the new mapping technique is sketched.

1. Introduction

Since its advent in the mid-1970s, the interacting boson model (IBM) of Arima and Iachello [1] has made tremendous successes in the study of low-lying collective states of medium-heavy and heavy nuclei. The ingredients of the IBM are monopole s and quadrupole d bosons, which correspond respectively to collective nucleon-pairs with angular momenta J $= 0^+$ and 2^+. The boson Hamiltonian is composed of only essential interactions acting among the bosons, and measurable quantities like energy spectrum and electromagnetic transition rates are obtained through the diagonalization of the Hamiltonian, either numerically or analytically when the Hamiltonian is written as a particular combinations of interaction terms (dynamical symmetry) [1]. Despite its success, however, the IBM has lacked in the microscopic foundation (and hence the predictive power) since the parameters of the boson Hamiltonian have been determined only from the phenomenological t to the known experimental data.

From a microscopic point of view, the IBM system represents a vast truncation of the full Hilbert space of nuclear shell model, and conventionally the parameters of the boson Hamiltonian have been calculated by associating the matrix element of a given fermion operator in the truncated shell-model space, comprising collective S and D (plus G, if needed) pairs, to the matrix element of the corresponding boson operator in the sd boson space [2–4]. Since a nucleus is consisting of protons and neutrons, the proton-neutron version of the IBM (IBM-2) has been introduced as a natural consequence, that is composed of neutron (proton) s_ν (s) and d_ν (d) bosons [2, 3]. The number of neutron (proton) bosons, denoted here as N_ν (N), equals the number of the collective pairs of valence neutrons (protons) outside of an inert core. Although the shell-model derivation of the IBM has worked out in many realistic cases of spherical vibrational and unstable systems [4, 5], since the underlying shell-model con guration becomes too complicated to be handled with the simple truncation scheme like the generalized seniority, the uni ed framework of deriving IBM Hamiltonian for general situations including rotational deformed system has long been missing.

Recently a novel robust way of deriving the IBM Hamiltonian has been developed, in conjunction with the self-consistent mean- eld model based on the microscopic energy density functional (EDF) [6, 7]. In this technique, a given boson Hamiltonian has been xed so that the energy expectation value of the nucleon many-body Hamiltonian in the mean- eld intrinsic state with varying deformation is mapped onto the analogous energy expectation value in the boson condensate. The self-consistent mean- eld method with a xed EDF parametrization allows an accurate global description of the nuclear bulk properties and collective excitations, e.g., mass, radii, deformation, shell structure, giant resonance, ... etc. [8], and can be thus exploited as a starting point from which the IBM Hamiltonian is constructed.

In this contribution, the basic notion of the new fermion-boson mapping technique is reviewed and we shall demonstrate how it works by taking a simple example. This contribution is based on Refs. [6, 7, 9] but discusses only a few essential results. For a complete review of the proposed methodology and its various extensions, the reader is referred to Ref. [10].

2. Formulation

To construct the boson Hamiltonian for general situations associated to various types of deformation, the energy surface of the collective deformation variables can be appropriate to start with. The energy surface is generated with a set of the constrained self-consistent mean- eld methods, such as Hartree-Fock, Hartree-Bogoliubov or Hartree-Fock-Bogoliubov models, etc., using a microscopic energy density functional. The mean- eld solution of the total energy surface at each collective coordinate (,), denoted here as $E(\ ,\ ) = \langle\Phi_F(\ ,\ )|\hat{H}_F|\Phi_F(\ ,\ )\rangle$ with $|\Phi_F\rangle$ being the corresponding intrinsic wave function, are obtained through the minimization of the total energy normally as a sum of kinetic energy, a given model of EDF, pairing functional and Coulomb energy with spurious center-of-mass motion being subtracted. The constraint means the one for the mass quadrupole moment associated to the deformation variables of the geometrical model (,) [11]. The e ective interaction for the particle-hole channel are typically of the Skyrme [12, 13] and the Gogny [14] types, and other parameterizations based on the relativistic mean- eld Lagrangian [15]. Note that, in this framework, only the total energy at the level of the mean- eld approximation is of relevance here. This means that neither collective potential energy surface nor mass parameters are considered explicitly and that any symmetry projection is not performed.

To associate the energy surface obtained through the EDF calculation to the analogous energy surface in the boson system, the so-called coherent-state framework [16] is used. The coherent state represents the intrinsic state of the boson system and is given as the following boson condensate $|\Phi_B\rangle$

$$|\Phi_B\rangle = \frac{1}{\sqrt{N_\nu! N_{\ }!}}(\lambda^\dagger_\nu)^{N_\nu}(\lambda^\dagger_{\ })^{N_\pi}|\ \rangle \tag{1}$$

with $|\ \rangle$ being inert core, and $\lambda^\dagger_\rho$ ($=\nu$ or) reads

$$\lambda^\dagger_\rho = \frac{1}{\sqrt{1+{}^{2}_{\rho}}}\Big[s^\dagger_\rho + {}_\rho\cos{}_\rho d^\dagger_{\rho 0} + \frac{1}{\sqrt{2}}{}_\rho\sin{}_\rho(d^\dagger_{\rho+2} + d^\dagger_{\rho\ 2})\Big]. \tag{2}$$

The IBM analogue of the mean- eld energy surface is obtained as an expectation value of the boson Hamiltonian $\langle\Phi_B|\hat{H}_B|\Phi_B\rangle$. The variables $_\rho$ and $_\rho$ in Eq. (2) are equivalent to geometrical quadrupole deformation (,). Since the following discussion is restricted to isoscalar motion and both neutrons and protons are supposed to have the same shape due to their strong coupling, $_\nu =$ $_B$ and $_\nu =$ $_B$ can be the rst good approximation. The variables ($_B$, $_B$) can be related to the geometrical variables (,) in the following way:

$$_B = C \quad \text{and} \quad _B = \ , \tag{3}$$

the former of which comes from the di erence in the model space [16], the fact that the IBM is comprised of only valence nucleons while the collective model entire nucleus. Thus, $_B$ is always larger than with $C \geq 1$. The latter relation in Eq. (3) implies that the angle should have the same physical meaning in both models. One can then map each point of the microscopic energy surface $E(\ , \ )$ onto the IBM analogue of the mean- eld energy surface. This requires that the following approximate equality be satis ed over a certain range of the deformation (,) corresponding to the low-energy excitation (typically up to several MeVs from the minimum).

$$\langle \Phi_F(\ , \ )|\hat{H}_F|\Phi_F(\ , \ )\rangle \quad \langle \Phi_B(\ _B, \ _B)|\hat{H}_B|\Phi_B(\ _B, \ _B)\rangle, \tag{4}$$

which means that the boson energy surface should be calculated to be identical in topology around the mean- eld minimum to the microscopic energy surface as much as possible [1]. By mimicking the basic topology of the microscopic energy surface most relevant to the low-energy excitation, e ects of the essential ingredients of nuclear many-body system, e.g., Pauli principle and nuclear force, are expected to be incorporated into the IBM system by renormalizing them in the values, and/or the boson-number dependence, of the strength parameters.

For the boson Hamiltonian $\hat{H}_B$, the possible simplest but the most essential form can be taken

$$\hat{H}_B = \ \hat{n}_d + \ \hat{Q}_\nu \ \hat{Q} \tag{5}$$

where $\hat{n}_d = \hat{n}_{d\nu} + \hat{n}_d$ stands for the d boson number operator, which drives nucleus into spherical shape, and the second term the quadrupole-quadrupole interaction between neutron and proton bosons, which makes dominant contribution to deformation, with quadrupole operator being $\hat{Q}_\rho = s_\rho^\dagger \tilde{d}_\rho + d_\rho^\dagger s_\rho + \chi_\chi [d_\rho^\dagger \ \tilde{d}_\rho]^{(2)}$. , , χ_ν and χ are parameters to be determined from the mapping. The boson Hamiltonian could be more general, but the simpli ed form Eq. (5) already embodies rich aspects of quadrupole collective states. The energy spectra and transition rates are obtained through the diagonalization of the mapped Hamiltonian in the laboratory frame.

3. Transition from spherical vibrational to deformed rotational states

The examples for the microscopic energy surfaces are shown on the left-hand side of Fig. 1 for axially deformed Sm isotopes, obtained through the HF+BCS method solved in a coordinate space [17] with the parametrization Skyrme SkM* [18] and the density-dependent zero-range pairing. One nds in the gure that the energy surfaces of the ^{146,148}Sm nuclei, which are in the vibrational regime, exhibit the minima close to the origin. Coming to the transitional nuclei ^{150,152}Sm, the minimum departs from the origin to 0.3 and also looks at in . Even larger deformation (0.35) is suggested in the ^{154,156}Sm nuclei, where the minimum becomes well isolated being very sharp with respect to both and . The mapped IBM energy surfaces on the right-hand side of Fig. 1 reproduce the topology of the microscopic energy surface around the minimum and the overall systematic trend with N.

If one looks at the resultant spectra in Fig. 2, the basic empirical features are reproduced: lowering of yrast levels with N, the phase-transitional rapid changes in the side-band levels at N 88 or 90. The present model well describes the vibrational spectra around N 86, being

[1] For technical details behind the mapping the reader is referred to Refs. [7, 10].

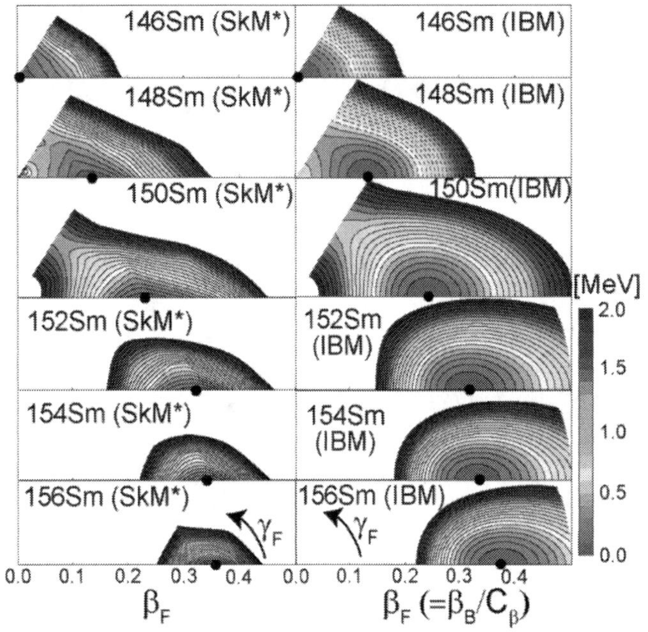

Figure 1. (Color online) Total (,) energy surface of ^{146} ^{156}Sm isotopes. On the left-hand side (SkM*) shown are microscopic energy surfaces with Skyrme SkM* interaction, while on the right-hand side the mapped IBM energy surfaces. The minimum is identi ed by the lled circle. $_F$ and $_F$ stand for the geometrical deformation variables.

close to the U(5) limit of IBM [1], and the typical rotational band around N 92, or the SU(3) limit of the IBM [1]. It is also worth noting that the result of the level energy systematics correlates with the variation of the deformation properties with neutron number indicated by the microscopic energy surface in Fig. 1.

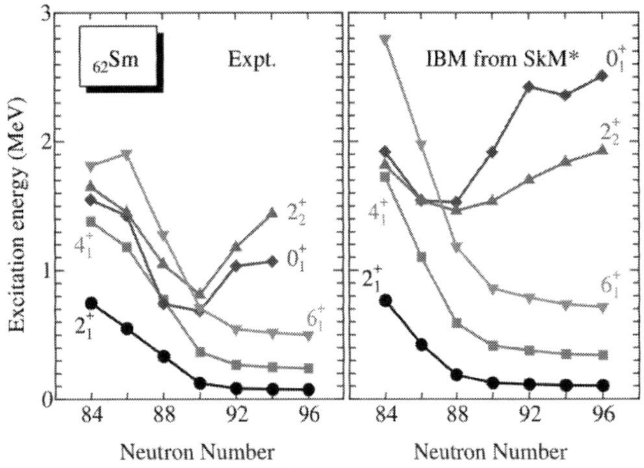

Figure 2. (Color online) Level energies of 2_1^+, 4_1^+, 6_1^+, 0_2^+ and 2_2^+ excited states with neutron number. Experimental data have been taken from Ref. [19].

At this point, one could notice in Fig. 2 that the rotational moment of inertia for deformed nuclei with $N \geq 92$, calculated in the IBM, underestimates to a large extent the experimental data. The problem is highlighted in Fig. 3, where excitation energies calculated in the microscopic IBM (energy bands on the left-hand side in each panel (a) and (b), denoted as "w/o") are much larger than the experimental [19] values. A reason for this deviation of the moment of inertia can be the limited degrees of freedom in the IBM consisting of only s and d bosons. Nevertheless, as far as the result of the energy surface mapping in Fig. 1 is concerned, the

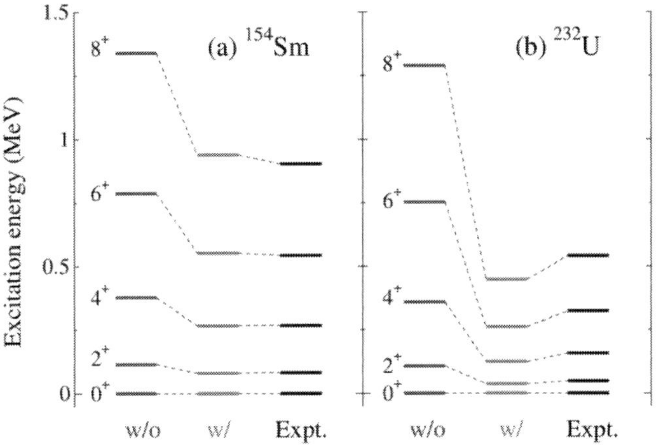

Figure 3. (Color online) Rotational bands of axially deformed (a) ^{154}Sm and (b) ^{232}U nuclei. The calculated spectra with (w/) and without (w/o) the $\hat{L} \quad \hat{L}$ term are compared with experiment [19].

con guration of *sd* bosons appears to be rather su cient. Thus, it would be more likely [9], that the problem occurs due to fact that the boson system di ers in its rotational property from the nucleon system (see [9] for details). To settle the problem, we have then proposed to introduce an extended mapping procedure, that includes a speci c rotational dynamics with non-zero angular frequency, in addition to the mapping of the energy surface with zero angular frequency (cf. Eq. (4)). The rotational dynamics could be incorporated through the standard cranking approximation [20]. Instead of the Hamiltonian $\hat{H}_X$ ($X = F, B$), the following Hamiltonian (with axial symmetry)

$$\hat{H}'_X = \hat{H}_X \quad \omega \hat{L}_y, \tag{6}$$

where $\hat{L}_y$ and ω are the y component of the total angular momentum and cranking frequency, respectively, should be minimized. Given the values of the strength parameters already derived through the mapping of Eq. (4), the response (or the energy shift of the system due) to rotational cranking should be approximately identical between fermion and boson systems at a particular point on the surface, such as the equilibrium ($^{(e)}$, $^{(e)}$):

$$\langle \Phi'_F(\ ^{(e)}, \ ^{(e)})|\hat{H}'_F|\Phi'_F(\ ^{(e)}, \ ^{(e)})\rangle \qquad \langle \Phi'_B(\ _B^{(e)}, \ _B^{(e)})|\hat{H}'_B|\Phi'_B(\ _B^{(e)}, \ _B^{(e)})\rangle, \tag{7}$$

where $|\Phi'_X\rangle$ represents the solution of $\hat{H}'_X$ in Eq. (6) at the equilibrium ($^{(e)}$, $^{(e)}$). To this end, we have further proposed to add the term to the boson Hamiltonian Eq. (5), that corrects the moment of inertia but does not alter the topology of the energy surface [9]. Such term should be of the form $\hat{L} \quad \hat{L}$, with $\hat{L} = \hat{L}_\nu + \hat{L}$ being the angular momentum operator. In practice the approximate equality of Eq. (7) can be ful lled by matching the cranking moment of inertia in the intrinsic state of IBM at the equilibrium ($^{(e)}$, $^{(e)}$) to the one resulting from the mean- eld calculation, leading to the coupling constant of the $\hat{L} \quad \hat{L}$ term [9].

To see the impact of the $\hat{L} \quad \hat{L}$ term on rotational band, we compare in Fig. 3 the energy levels calculated with the $\hat{L} \quad \hat{L}$ term with those without the term for strongly deformed ^{154}Sm and ^{232}U nuclei as examples. By the inclusion of the $\hat{L} \quad \hat{L}$ term, the agreement between theoretical and experimental rotational bands is remarkably improved for the nucleus ^{154}Sm so does for ^{232}U. It has to be stressed here that there is no adjustment to the experiment. It has been shown as well [9] that the e ect of the $\hat{L} \quad \hat{L}$ term is particularly important for axially deformed rotational nuclei but is normally much less for weakly deformed and -soft systems.

Given the results presented in Figs. 2 and 3, as well as in a number of applications [10] not shown here, we have come to the conclusion that the IBM can be derived in a uni ed way for various situations of the low-energy quadrupole collective states, ranging from spherical vibrational to strongly deformed rotational states, etc.

4. Summary

In this contribution, a novel robust way of deriving the IBM Hamiltonian from the self-consistent mean- eld model with microscopic EDF has been overviewed. By the proposed method, the basic collective excitation modes of nuclei are reproduced: not only vibrational and transitional states but the rotational deformed state. Particularly the latter point should more or less concern the question raised in the past by Bohr and Mottelson [21], as to whether the IBM is justi ed in the microscopic description of the rotational motion of deformed nuclei. While a conclusive answer to this question has still been missing, the present work sheds more light upon and should provide a new insight into this old problem. In practical applications, the IBM can be considered as an e ective theory to solve complex many-fermion problems for heavy-mass system, which would be computationally much more costly with a large-scale calculation, and is granted a predictive power for the spectroscopy of heavy exotic nuclei.

Acknowledgments

Many thanks to Taka Otsuka for inspiring me to set out a work in this eld and for pleasant discussions and collaborations for the past few years. I also acknowledge the nancial support through the JSPS Postdoctoral Fellowships for Research Abroad.

References

[1] Iachello F and Arima A 1987 *The interacting boson model* (Cambridge: Cambridge University Press)
[2] Arima A, Otsuka T, Iachello F and Talmi I 1977 *Phys. Lett.* B **66** 205
[3] Otsuka T, Arima A, Iachello F and Talmi I 1978 *Phys. Lett.* B **76** 139
[4] Otsuka T, Arima A and Iachello F 1978 *Nucl. Phys.* A **309** 1
[5] Mizusaki T and Otsuka T 1996 *Prog. Theor. Phys. Suppl.* **125** 97
[6] Nomura K, Shimizu N and Otsuka T 2008 *Phys. Rev. Lett.* **101** 142501
[7] Nomura K, Shimizu N and Otsuka T 2010 *Phys. Rev.* C **81** 044307
[8] Bender M, Heenen P H and Reinhard P G 2003 *Rev. Mod. Phys.* **75** 121
[9] Nomura K, Otsuka T, Shimizu N and Guo L 2011 *Phys. Rev.* C **83** 041302
[10] Nomura K 2012 Ph.D. thesis University of Tokyo to be published in Springer Theses Physics (2013)
[11] Bohr A and Mottelsson B R 1975 *Nuclear Structure* vol 2 (New York: Benjamin)
[12] Skyrme T H R 1958 *Nucl. Phys.* **9** 615
[13] Vautherin D and Brink D M 1972 *Phys. Rev.* C **5** 626
[14] Dechargé J, Girod M and Gogny D 1975 *Phys. Lett.* B **55** 361
[15] Nikšić T, Vretenar D and Ring P 2011 *Prog. Part. Nucl. Phys.* **66** 519
[16] Ginocchio J N and Kirson M W 1980 *Nucl. Phys.* A **350** 31
[17] Bonche P, Flocard H and Heenen P H 2005 *Compt. Phys. Commun.* **171** 49
[18] Bartel J, Quentin P, Brack M, Guet C and Hakansson H B 1982 *Nucl. Phys.* A **386** 79
[19] Brookhaven National Nuclear Data Center ENSDF http://www.nndc.bnl.gov
[20] Ring P and Schuck P 1980 *The nuclear many-body problem* (Berlin: Springer-Verlag)
[21] Bohr A and Mottelson B R 1980 *Physica Scripta* **22** 468

Impact on the r-process from the nuclear mass and lifetime in covariant density functional theory

J Meng[1,2,3], Y Chen[1], Z M Niu[4,1], B Sun[2,5] and P W Zhao[1]

[1]State Key Laboratory of Nuclear Physics and Technology, School of Physics, Peking University, Beijing 100871, China
[2]School of Physics and Nuclear Energy Engineering, Beihang University, Beijing 100191, China
[3]Department of Physics, University of Stellenbosch, Stellenbosch 7602, South Africa
[4]School of Physics and Material Science, Anhui University, Hefei 230039, China
[5]Justus-Liebig-Universit : at Giessen, Heinrich-Buff-Ring 14, Giessen 35392, Germany

E-mail: mengj@pku.edu.cn

Abstract. The covariant density functional theory with a few number of parameters allows a very successful description of the ground-state and excited-state properties for the nuclei all over the nuclear chart. The recent progresses on the application of the covariant density functional theory for the nuclear mass and β-decay half-life, as well as their influences on the astrophysical rapid neutron-capture process calculation are reviewed.

1. Introduction

In the past years, unstable nuclear beams have extended our knowledge of nuclear physics from the stable nuclei to the exotic nuclei far away from the stability. The study of exotic nuclei has attracted broad attention due to their large isospin values and interesting properties, such as the halo phenomenon [1, 2]. Furthermore, the properties of exotic nuclei, e.g. mass and β-decay half-life, are essential in understanding the nucleosynthesis via rapid neutron capture (r-process) which is responsible for roughly half of the enrichment of elements heavier than iron in the universe [3]. Nevertheless, although great efforts have been dedicated world-wide to experimental investigation, most of the neutron-rich nuclei of relevance to the r-process are still beyond the reach of experiments in the foreseeable future and therefore, can only rely on the nuclear theoretical models.

The density functional theory (DFT) with a small number of parameters is successful in describing the ground-state and excited-state properties of the nuclei all over the nuclear chart. In particular, the covariant density functional theory (CDFT) takes the Lorentz symmetry into account in a self-consistent way, and has received wide attention due to its successful description of lots of nuclear phenomenon [4, 5, 6, 7, 8].

There exist a number of attractive features in the CDFT. The most obvious one is the natural inclusion of the nucleon spin degree of freedom and resulting nuclear spin-orbit potential that emerges automatically with the empirical strength in a covariant way. Moreover, it can reproduce well the measurements of the isotopic shifts in the Pb region [9], and give more naturally the origin of the pseudospin symmetry [10, 11] as a relativistic symmetry [12, 13, 14, 15, 16, 17] and the spin symmetry in the anti-nucleon spectrum [18, 19]. It can also include the nuclear magnetism [20], i.e., a consistent description of currents and time-odd fields, which plays an important role in the nuclear magnetic moments [21, 22, 23, 24, 25] and nuclear rotations [26, 27, 28, 29]. Altogether, CDFT is a reliable and effective model for nuclear structure, and therefore, it is natural to describe the nuclei based on CDFT.

In this contribution, recent progress on the application of covariant density functional theory for nuclear masses and β-decay half-lives, as well as their influences on the astrophysical r-process calculation are reviewed.

2. Nuclear mass and β-decay half-life in CDFT

Taking the CDFT with the point-coupling interaction as an example, the starting point is an effective Lagrangian written as a power series in $\bar{\psi}\mathcal{O}\Gamma\psi$ and their derivatives,

$$\begin{aligned}\mathcal{L} &= \bar{\psi}(i\gamma_\mu\partial^\mu - m)\psi - \frac{1}{4}F^{\mu\nu}F_{\mu\nu} - e\frac{1-\tau_3}{2}\bar{\psi}\gamma^\mu\psi A_\mu - \frac{1}{2}\alpha_S(\bar{\psi}\psi)(\bar{\psi}\psi) - \frac{1}{2}\alpha_V(\bar{\psi}\gamma_\mu\psi)(\bar{\psi}\gamma^\mu\psi)\\ &\quad -\frac{1}{2}\alpha_{TV}(\bar{\psi}\vec{\tau}\gamma_\mu\psi)(\bar{\psi}\vec{\tau}\gamma^\mu\psi) - \frac{1}{3}\beta_S(\bar{\psi}\psi)^3 - \frac{1}{4}\gamma_S(\bar{\psi}\psi)^4 - \frac{1}{4}\gamma_V[(\bar{\psi}\gamma_\mu\psi)(\bar{\psi}\gamma^\mu\psi)]^2\\ &\quad -\frac{1}{2}\delta_S\partial_\nu(\bar{\psi}\psi)\partial^\nu(\bar{\psi}\psi) - \frac{1}{2}\delta_V\partial_\nu(\bar{\psi}\gamma_\mu\psi)\partial^\nu(\bar{\psi}\gamma^\mu\psi) - \frac{1}{2}\delta_{TV}\partial_\nu(\bar{\psi}\vec{\tau}\gamma_\mu\psi)\partial^\nu(\bar{\psi}\vec{\tau}\gamma_\mu\psi), \qquad (1)\end{aligned}$$

where all symbols have the usual meaning as in Ref. [8].

The mean-field approximation leads to the replacement of the operators $\bar{\psi}(\hat{\mathcal{O}}\Gamma)_i\psi$ by the expectation values which become bilinear forms of the nucleon Dirac spinor ψ_k,

$$\bar{\psi}(\hat{\mathcal{O}}\Gamma)_i\psi \rightarrow \langle\Phi|\bar{\psi}(\hat{\mathcal{O}}\Gamma)_i\psi|\Phi\rangle = \sum_k v_k^2\bar{\psi}_k(\hat{\mathcal{O}}\Gamma)_i\psi_k, \qquad (2)$$

where i indicates S, V, and TV. The sum $\sum$ runs over only positive energy states with the occupation probabilities v_k^2, i.e., the "no-sea" approximation. Based on these approximations, one finds the energy density functional for a nuclear system

$$E_{\rm DF}[\boldsymbol{\tau},\rho_S,j_i^\mu,A_\mu] = \int d^3r\ \mathcal{E}(\boldsymbol{r}) \qquad (3)$$

with the energy density

$$\mathcal{E}(\boldsymbol{r}) = \mathcal{E}^{\rm kin}(\boldsymbol{r}) + \mathcal{E}^{\rm int}(\boldsymbol{r}) + \mathcal{E}^{\rm em}(\boldsymbol{r}), \qquad (4)$$

which is composed of a kinetic part

$$\mathcal{E}^{\rm kin}(\boldsymbol{r}) = \sum_k v_k^2\ \psi_k^\dagger(\boldsymbol{r})\,(\boldsymbol{\alpha}\cdot\boldsymbol{p} + \beta m)\,\psi_k(\boldsymbol{r}), \qquad (5)$$

and an interaction part

$$\begin{aligned}\mathcal{E}^{\rm int}(\boldsymbol{r}) &= \frac{\alpha_S}{2}\rho_S^2 + \frac{\beta_S}{3}\rho_S^3 + \frac{\gamma_S}{4}\rho_S^4 + \frac{\delta_S}{2}\rho_S\triangle\rho_S + \frac{\alpha_V}{2}j_\mu j^\mu + \frac{\gamma_V}{4}(j_\mu j^\mu)^2 + \frac{\delta_V}{2}j_\mu\triangle j^\mu\\ &\quad + \frac{\alpha_{TV}}{2}\vec{j}_{TV}^{\,\mu}\cdot(\vec{j}_{TV})_\mu + \frac{\delta_{TV}}{2}\vec{j}_{TV}^{\,\mu}\cdot\triangle(\vec{j}_{TV})_\mu \qquad (6)\end{aligned}$$

with the local densities and currents

$$\rho_S(\boldsymbol{r}) = \sum_k v_k^2\bar{\psi}_k(\boldsymbol{r})\psi_k(\boldsymbol{r}), \qquad (7a)$$

$$j_V^\mu(\boldsymbol{r}) = \sum_k v_k^2\bar{\psi}_k(\boldsymbol{r})\gamma^\mu\psi_k(\boldsymbol{r}), \qquad (7b)$$

$$\vec{j}_{TV}^{\,\mu}(\boldsymbol{r}) = \sum_k v_k^2\bar{\psi}_k(\boldsymbol{r})\vec{\tau}\gamma^\mu\psi_k(\boldsymbol{r}), \qquad (7c)$$

and an electromagnetic part

$$\mathcal{E}^{\rm em}(\boldsymbol{r}) = \frac{1}{4}F_{\mu\nu}F^{\mu\nu} - F^{0\mu}\partial_0 A_\mu + eA_\mu j_p^\mu. \tag{8}$$

Minimizing the energy density functional Eq. (3) with respect to $\bar{\psi}_k$, one obtains the Dirac equation for the single nucleon

$$[\gamma_\mu(i\partial^\mu - V^\mu) - (m+S)]\psi_k = 0. \tag{9}$$

The single-particle effective Hamiltonian contains local scalar $S(\boldsymbol{r})$ and vector $V^\mu(\boldsymbol{r})$ potentials,

$$S(\boldsymbol{r}) = \Sigma_S, \quad V^\mu(\boldsymbol{r}) = \Sigma^\mu + \vec{\tau}\cdot\vec{\Sigma}^\mu_{TV}, \tag{10}$$

where

$$\begin{aligned}
\Sigma_S &= \alpha_S\rho_S + \beta_S\rho_S^2 + \gamma_S\rho_S^3 + \delta_S\triangle\rho_S, && \text{(11a)}\\
\Sigma^\mu &= \alpha_V j_V^\mu + \gamma_V(j_V^\mu)^3 + \delta_V\triangle j_V^\mu + eA^\mu, && \text{(11b)}\\
\vec{\Sigma}^\mu_{TV} &= \alpha_{TV}\vec{j}^\mu_{TV} + \delta_{TV}\triangle\vec{j}^\mu_{TV}. && \text{(11c)}
\end{aligned}$$

As the translation symmetry is broken in the mean-field approximation, proper treatment of center-of-mass (c.m.) motion is very important. The c.m. correction energy can be taken into account in the oscillator approximation or in a microscopic way [30, 31, 32].

Based on the nuclear ground state, one can perform the random phase approximation (RPA) calculation for the low-lying excited states and giant resonances, as furthermore, for the β-decay rates of neutron-rich nuclei. With the individual transition strengths B_m, the β-decay half-life of an even-even nucleus can be calculated in the allowed Gamow-Teller approximation,

$$T_{1/2} = \frac{D}{g_A^2\sum_m B_m f(Z,E_m)}, \tag{12}$$

where $D = 6163.4 \pm 3.8$ s and $g_A = 1$. The sum runs over all final states with an excitation energy smaller than the Q_β value. The intergrated phase volume $f(Z,E_m)$ is

$$f(Z,E_m) = \int_{m_e}^{E_m} p_e E_e (E_m - E_e)^2 F_0(Z,E_e) dE_e, \tag{13}$$

where p_e, E_e, and $F_0(Z,E_e)$ denote the emitted electron momentum, energy, and Fermi function, respectively [33]. The β-decay transition energy E_m corresponds to the energy difference between the initial and final state calculated using the RPA

$$E_m = \Delta_{np} - E_{\rm RPA}, \tag{14}$$

where $E_{\rm RPA}$ is the RPA energy with respect to the ground-state of the parent nucleus and corrected by the difference of the neutron and proton Fermi energies in the parent nucleus [34], and Δ_{np} is the neutron-proton mass difference.

3. Influence of nuclear mass on astrophysical r-process calculation

Nuclear mass is one of the key nuclear inputs for astrophysical r-process calculation, from which one can directly determine the neutron separation energy, shell gap and also the β-decay energy. However, the majority of neutron-rich nuclei related to the astrophysical r-process are still out of the reach of present experimental capabilities, and therefore reliable predictions of theoretical

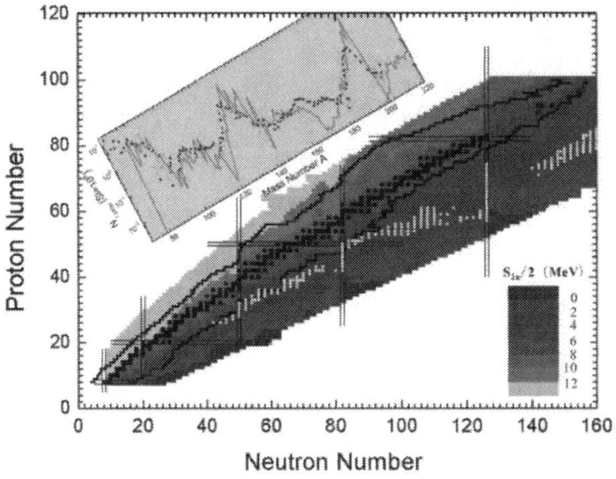

Figure 1. Features of the r-process calculated using the CDFT mass table and available data. The solid line denotes the border of nuclei with known masses in both neutron-deficient and neutron-rich sides. The yellow squares show the r-process path. The observed and calculated solar r-process abundance curves are plotted versus the mass number A in the inset, whose x-axis is curved slightly to follow the r-process path. Taken from Ref. [36].

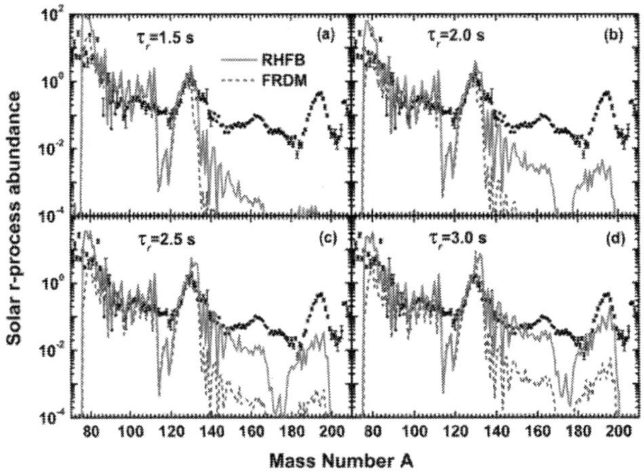

Figure 2. The impact of nuclear β-decay half-lives on the calculated r-process abundance. The solid (dashed) curves correspond to r-process abundance calculated with the RHFB+QRPA (FRDM+QRPA) β-decay half-lives in comparison with the data denoted by the points. Panels (a)-(d) correspond to the process times $\tau_r = 1.5, 2.0, 2.5$ and 3.0 s, respectively. Taken from Ref.[48].

models have to be used [35, 36, 37, 38, 39, 40, 41]. By employing the state-dependent BCS method with a δ force, the first systematic study of the ground-state properties for over 7000 nuclei was performed by using CDFT [42], where the rms deviation of neutron separation energy S_n with respect to the known experimental data is 0.7 MeV.

Initiated by the development of nuclear masses, the r-process calculation based on the classical r-process approach is performed [43, 36]. The best simulation using the CDFT mass table and the available data [44] is presented in Figure 1. The β-decay properties are taken from the calculation results of the finite-droplet model (FRDM) plus quasiparticle RPA [45]. In general, the solar r-process abundance peaks can be well reproduced. However, there are still large troughs before the second and third abundance peaks for the calculated r-process abundance distribution. These deficiencies in reproducing the solar r-process abundances can be mainly traced back to two aspects from the nuclear physics point of view: the shell effect and the location of nuclear shape transition. More detailed investigations on the nuclei with neutron numbers around magic numbers are demanded in microscopic nuclear physics models, which indicate that serious effort is needed to improve the formulation of the CDFT.

Recently, a new parametrization PC-PK1 for CDFT with nonlinear point-coupling interaction is proposed, which can provide very good descriptions for the nuclear masses as well as its isospin dependence [46, 47]. Therefore, a new CDFT mass table with PC-PK1 is expected, and it will be interesting to investigate its influence on the astrophysical r-process calculation.

4. Influence of nuclear lifetime on astrophysical r-process calculation

Another key nuclear input for astrophysical r-process calculation is β-decay half-life, which can set the timescale of the astrophysical r-process and hence determine the production of heavy elements in the universe. Recently, fully self-consistent proton-neutron QRPA based on the relativistic Hartree-Fock-Bogoliubov (RHFB) theory is developed [48]. With an isospin-dependent proton-neutron pairing interaction in the isoscalar channel, the systematical calculation of β-decay half-lives for neutron-rich nuclei with $20 \geq Z \geq 50$ is performed. The results well reproduce the experimental β-decay half-lives, especially for nuclei with half-lives less than one second.

Based on the predicted β-decay half-lives, a classical r-process calculation is performed [48]. The snapshots of r-process abundance for different process time τ_r is presented Figure 2. The available nuclear masses are taken from Ref. [49], otherwise the predictions of CDFT [42] are employed. Compared with the FRDM+QRPA calculation, the half-lives calculated with RHFB+QRPA model produce a faster r-matter flow at the $N = 82$ region, and thus yield higher r-process abundances of elements with $A \gtrsim 140$, which is an important result for the estimate of the duration of the r-process, and hence the origin of heavy elements in the universe. In addition, the abundance at $A \sim 80$ is higher using the RHFB+QRPA result can be easily understood, because it takes more time to pass the r-path nuclei ^{78}Ni and ^{80}Zn and, as a result, higher abundances are accumulated.

5. Summary

Recent progress on the nuclear masses and β-decay half-lives in CDFT, as well as their influences on the r-process calculation are reviewed. The solar r-process abundance distribution can be generally well reproduced by simulations using the CDFT, where the large troughs before the second and third abundance peaks for the calculated r-process abundance distribution can be mainly traced back to the shell effect and the location of nuclear shape transition. Based on the β-decay half-lives of neutron-rich nuclei with $20 \geq Z \geq 50$ calculated by RHFB+QRPA model, a remarkable speeding up of r-matter flow is predicted which leads to enhanced r-process abundances of elements with $A \gtrsim 140$, an important result for the understanding of the origin of heavy elements in the universe. In the future, with the new parametrization PC-PK1, the systematic calculation for nuclear masses and β-decay half-lives based on CDFT as well as astrophysical r-process calculation with these nuclear inputs will be performed.

Acknowledgments

We would like to thank H. Z. Liang, W. H. Long, T. Nikšić, Y. F. Niu, D. Vretenar for their contributions to the investigations presented in this proceeding. This work was supported in part by the Major State 973 Program of China (Grant No. 2013CB834400), the Natural Science Foundation of China (Grants No. 10975007, No. 10975008, No. 11175002, No. 11105005, No. 11105010, No. 11035007, No. 11128510, No. 11235002, and No. 11205004), the 211 Project of Anhui University under Grant No. 02303319-33190135, and the Research Fund for the Doctoral Program of Higher Education under Grant No. 20110001110087 and the China Postdoctoral Science Foundation Grant No. 2012M520101. B.S. is also partially supported by NCET-09-0031.

References

[1] Tanihata I, Hamagaki H, Hashimoto O, Shida Y, Yoshikawa N, Sugimoto K, Yamakawa O, Kobayashi T and Takahashi N 1985 *Phys. Rev. Lett.* **55** 2676
[2] Meng J and Ring P 1996 *Phys. Rev. Lett.* **77** 3963
[3] Burbidge E M, Burbidge G R, Fowler W A and Hoyle F 1957 *Rev. Mod. Phys.* **29** 547
[4] Serot B and Walecka J 1986 *Adv. Nucl. Phys.* **16** 1
[5] Ring P 1996 *Prog. Part. Nucl. Phys.* **37** 193
[6] Vretenar D, Afanasjev A V, Lalazissis G A and Ring P 2005 *Phys. Rep.* **409** 101
[7] Meng J, Toki H, Zhou S G, Zhang S Q, Long W H and Geng L S 2006 *Prog. Part. Nucl. Phys.* **57** 470
[8] Nikšić T, Vretenar D and Ring P 2011 *Prog. Part. Nucl. Phys.* **66** 519
[9] Sharma M M, Lalazissis G A and Ring P 1993 *Phys. Lett.* B **317** 9
[10] Arima A, Harvey M and Shimizu K 1969 *Phys. Lett.* B **30** 517
[11] Hecht K T and Adler A 1969 *Nucl. Phys.* A **137** 129
[12] Ginocchio J 1997 *Phys. Rev. Lett.* **78** 436
[13] Meng J, Sugawara-Tanabe K, Yamaji S, Ring P and Arima A 1998 *Phys. Rev.* C **58** 628
[14] Meng J, Sugawara-Tanabe K, Yamaji S and Arima A 1999 *Phys. Rev.* C **59** 154
[15] Long W H, Sagawa H, Meng J and Giai N V 2006 *Phys. Lett.* B **639** 242
[16] Liang H Z, Zhao P W, Zhang Y, Meng J and Van Giai N 2011 *Phys. Rev.* C **83** 041301
[17] Lu B N, Zhao E G and Zhou S G 2012 *Phys. Rev. Lett.* **109** 072501
[18] Zhou S G, Meng J and Ring P 2003 *Phys. Rev. Lett.* **91** 262501
[19] Liang H Z, Long W H, Meng J and Van Giai N 2010 *Eur. Phys. J.* A **44** 119
[20] Koepf W and Ring P 1989 *Nucl. Phys.* A **493** 61
[21] Yao J M, Chen H and Meng J 2006 *Phys. Rev.* C **74** 024307
[22] Arima A 2011 *Sci. China Ser. G-Phys. Mech. Astron.* **54** 188
[23] Li J, Meng J, Ring P, Yao J M and Arima A 2011 *Sci. China Ser. G-Phys. Mech. Astron.* **54** 204
[24] Li J, Yao J M, Meng J and Arima A 2011 *Prog. Theor. Phys.* **125** 1185
[25] Wei J, Li J and Meng J 2012 *Prog. Theor. Phys.* **196** 400
[26] Afanasjev A V, Ring P and König J 2000 *Nucl. Phys.* A **676** 196
[27] Zhao P W, Peng J, Liang H Z, Ring P and Meng J 2012 *Phys. Rev.* C **85** 054310
[28] Zhao P W, Peng J, Liang H Z, Ring P and Meng J 2011 *Phys. Rev. Lett.* **107** 122501
[29] Zhao P W, Zhang S Q, Peng J, Liang H Z, Ring P and Meng J 2011 *Phys. Lett.* B **699** 181
[30] Bender M, Rutz K, Reinhard P G and Maruhn J A 2000 *Eur. Phys. J.* A **7** 467
[31] Long W H, Meng J, Van Giai N and Zhou S G 2004 *Phys. Rev.* C **69** 034319
[32] Zhao P W, Sun B Y and Meng J 2009 *Chin. Phys. Lett.* **26** 112102
[33] Langanke K and Martínez-Pinedo G 2003 *Rev. Mod. Phys.* **75** 819
[34] Engel J, Bender M, Dobaczewski J, Nazarewicz W and Surman R 1999 *Phys. Rev.* C **60** 014302
[35] Casten R F and Zamfir N V 2000 *Phys. Rev. Lett.* **85** 3584
[36] Sun B and Meng J 2008 *Chin. Phys. Lett.* **25** 2429
[37] Li Z, Niu Z M, Sun B H, Wang N and Meng J 2012 *Acta Physica Sinica* **61** 72601
[38] Hua X M, Heng T H, Niu Z M, Sun B H and Guo J Y 2012 *Sci. China-Phys. Mech. Astron.* **55** 2414
[39] Niu Z M, Sun B H and Meng J 2009 *Phys. Rev.* C **80** 065806
[40] Meng J, Niu Z, Liang H and Sun B 2011 *Sci. China-Phys. Mech. Astron.* **54** 119
[41] Zhang W H, Niu Z M, Wang F, Gong X B and Sun B H 2012 *Acta Physica Sinica* **61** 112601
[42] Geng L S, Toki H and Meng J 2005 *Prog. Theor. Phys.* **113** 785
[43] Sun B, Montes F, Geng L S, Geissel H, Litvinov Y A and Meng J 2008 *Phys. Rev.* C **78** 025806
[44] Audi G, Bersillon O, Blachot J and Wapstra A H 2003 *Nucl. Phys.* A **729** 3
[45] Möller P, Pfeiffer B and Kratz K L 2003 *Phys. Rev.* C **67** 055802
[46] Zhao P W, Li Z P, Yao J M and Meng J 2010 *Phys. Rev.* C **82** 054319
[47] Zhao P W, Song L S, Sun B, Geissel H and Meng J 2012 *Phys. Rev.* C **86** 064324
[48] Niu Z M, Niu Y F, Liang H Z, Long W H, Nikšić T, Vretenar D and Meng J 2012 *Preprint* arXiv:1210.0680
[49] Audi G, Wapstra A and Thibault C 2003 *Nucl. Phys.* A **729** 337

International Symposium on Exotic Nuclear Structure From Nucleons (ENSFN 2012) IOP Publishing
Journal of Physics: Conference Series **445** (2013) 012017 doi:10.1088/1742-6596/445/1/012017

Time-scaled scenario of low-energy heavy-ion collisions

Yoritaka Iwata
Graduate School of Science, The University of Tokyo
E-mail: y.iwata@gsi.de

Abstract. The underlying scenario of low-energy heavy-ion collisions is presented based on time-dependent density-functional calculations. A classification of several types of reaction dynamics is given with respect to their time-scales.

1. Introduction

Heavy-ion reactions play an essential role in the evolution of our universe. Indeed, our universe consists of nuclei, electrons and so on. In particular most of the chemical elements (nuclei) are realised as final products of low-energy heavy-ion reactions such as fusion reactions. Therefore studies of the low-energy heavy-ion collision provide fundamental information about the chemical evolution of our universe.

In this paper, the reaction dynamics of heavy-ion reactions is discussed based on the time-dependent density-functional theory (TDDFT), with special attention to the time scales. It was quite recently that we had a certain understanding of the underlying scenario of low-energy heavy-ion reactions [1, 2], even though it is rather limited to the early and the intermediate stages of heavy-ion reactions. This issue is not only related to the answer to the fundamental question "how do heavy-ion reactions take place", but also yields some constraints on both the dynamics and the final products of heavy-ion reactions.

2. Time scales

2.1. Some energies decisive for heavy-ion reactions

We consider collisions between two heavy ions (two nuclei),

$$^{A_1}Z_1 \; + \; ^{A_2}Z_2,$$

where Z_i and A_i denote the proton and mass numbers of the two colliding nuclei, respectively. Let us denote heavy-ion collisions with an incident energy of a few MeV per nucleon as low-energy heavy-ion collisions. It corresponds to energies around or lower than the Fermi energies of each colliding nucleus, so that low-energy heavy-ion reactions mostly include fusion, fission, and fragmentation.

Let us begin with rough energy estimates. The nucleus is a many-nucleon system in which there are two kinds of nucleons: protons and neutrons. Based on the mass of the nucleons, the typical energy scale for the nucleon is the order of 1 GeV. The nucleon is rather stable during

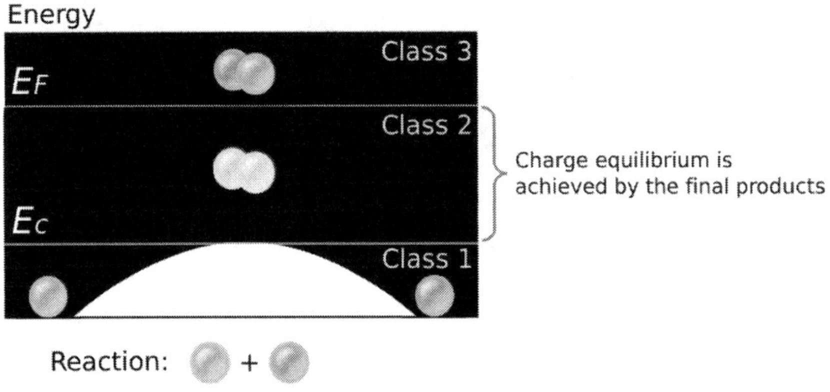

Figure 1. (Colour Online) Classification of heavy-ion collisions according to the incident energy. There are three classes in low-energy heavy-ion collisions, where the colliding two heavy-ions are represented by red and blue spheres. The colour of each sphere indicates the proton-to-neutron ratio. Two nuclei with different proton-to-neutron ratios collide (in terms of the most probable time evolution) if the incident energy is higher than the Coulomb energy E_C. For the energy region between E_C and E_F, charge equilibrium is achieved by the final products (indicated by the green colour of the spheres), i. e., an equilibrium with respect to proton-to-neutron ratio. Another class is determined by the Fermi energy E_F. For the definition of E_C and E_F, see Eqs. (2) and (3), respectively.

the reaction process if the incident energies are sufficiently smaller than 1 GeV per nucleon. It guarantees the stable existence of nucleons, as well as the validity of treating low-energy heavy-ion collisions in the theoretical framework based on the nucleonic degrees of freedom. Meanwhile, there is another criterion energy; namely, the Fermi energy (denoted by E_F). The Fermi energy is related to the stability and existence of the nucleus. Considering the application to heavy-ion collisions, we derive estimates for the Fermi energy in the centre-of-mass frame. The kinetic energy of the projectile nucleus moving at the speed of x % of the speed of light ($x < 50$) in the laboratory frame is calculated to be

$$E_{\rm lab}/A_1 = \frac{1}{2} m_N \left(\frac{x}{100} c \right)^2 = \frac{1}{2} m_N c^2 \left(\frac{x}{100} \right)^2 = 469 \left(\frac{x}{100} \right)^2 \ ({\rm MeV}) \tag{1}$$

per nucleon, where A_1 denotes the mass of the projectile and the mass of the nucleon was inserted as $m_N = 938\,{\rm MeV}/c^2$. Meanwhile, the centre-of-mass energy per nucleon is

$$E_x = E_{\rm cm}/(A_1 + A_2) = (E_{\rm lab}/A_1) \frac{A_1 A_2}{(A_1 + A_2)^2} = 469 \left(\frac{x}{100} \right)^2 \frac{A_1 A_2}{(A_1 + A_2)^2} \ ({\rm MeV}).$$

Consequently the Fermi energy per nucleon in the centre-of-mass frame is estimated as

$$E_F/(A_1 + A_2) = E_{x=25} \ ({\rm MeV}), \tag{2}$$

where $x = 25$ % (of the speed of light) has been known as the typical value for the Fermi velocity of nuclei (for more accurate estimates, see Eq. (1) of [1]). Here the Fermi energy per nucleon means the incident energy per nucleon which reaches the Fermi energy in collision situations; cf. the typical value of Fermi energy is equal to 40 MeV. For example, in mass symmetric collision $A_1 = A_2$ with $E_{\rm lab}/A_1$ and $E_F/(A_1 + A_2)$ are calculated to be 29.3 MeV and 7.3 MeV

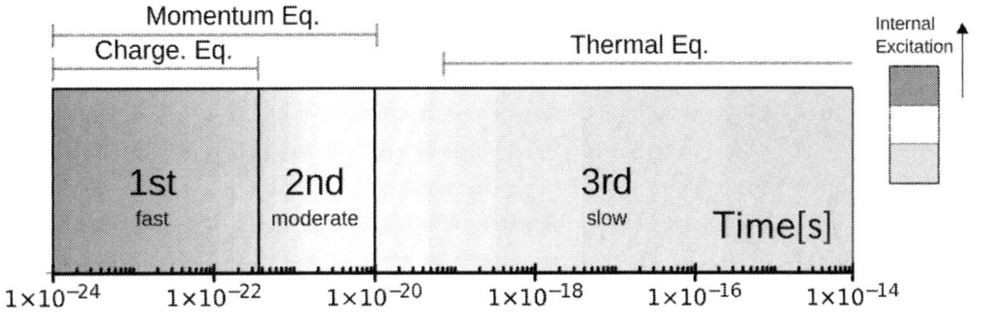

Figure 2. (Colour Online) Time evolution of low-energy heavy-ion collisions. Contact between the two ions takes place at 0 s. Three typical time scales of low-energy heavy-ion collisions are illustrated in which the first, second, and third period correspond to fast charge, momentum and thermal equilibration, respectively.

respectively. In particular the latter value (7.3 MeV) is reasonable in comparison with the results shown in Fig. 2 of [1].

In addition, the Coulomb energy (E_C) of the binary system, which is calculated by the Coulomb energy between the two colliding nuclei, can be another criterion-energy in determining the type of reaction dynamics. Two nuclei cannot touch each other if the total incident energies are less than the Coulomb energy of the binary system, since the proton is a charged particle. Using a simple approximation of the touching distance, the Coulomb energy per nucleon is estimated by

$$E_C/(A_1 + A_2) = \frac{1.43987\ Z_1\ Z_2}{R\ (\text{fm})} \frac{1}{A_1 + A_2} \sim \frac{1.43987\ Z_1\ Z_2}{1.2\ (A_1^{1/3} + A_2^{1/3})\ (\text{fm})} \frac{1}{A_1 + A_2}\ (\text{MeV}). \qquad (3)$$

This energy is 1.7 MeV (per nucleon) for collisions between two identical ^{238}U ($Z_1 = Z_2 = 92$ and $A_1 = A_2 = 238$), and 0.4 MeV (per nucleon) for collisions between two identical ^{4}He ($Z_1 = Z_2 = 2$ and $A_1 = A_2 = 4$). Therefore we see that the Coulomb energy is a few MeV per nucleon at most.

In low-energy heavy-ion reactions, there are two different criterion-energies. The first one is the Fermi energy (E_F) determining whether the fast charge equilibration appears (lower) or not (higher) (for the definition of the fast charge equilibration, see [1]). The second one is the Coulomb energy (E_C) determining whether fusion appears (higher) or not (lower). Depending on the energy, the reaction dynamics is classified into three types (Fig. 1). For energy less than the Coulomb energy, two ions cannot have a contact. For energy larger than the Coulomb energy and less than the Fermi energy, this energy region is characterised by the fast charge equilibration phenomena, in which the proton-to-neutron ratio of each final product is almost similar to that of the total system. For higher energies, the fast charge equilibration does not appear, and the proton-to-neutron ratio of each final product is not similar to that of the total system. In these transitions of reaction types, the fusion is the concept completely included in the charge equilibration. Note that, although it has been believed that the Fermi energy is decisive to whether multi-fragmentation appears or not, multi-fragmentation cannot generally take place at energies 5 to 10 MeV (per nucleon) higher than the Fermi energy. On the other hand, fragmentation into a few pieces can appear even at energies sufficiently less than the Fermi energy.

2.2. Time-scales in heavy-ion reactions

The criterion-energies lead to the concept of time-scales of heavy-ion reactions.

$$\hbar\omega = 0.65820 \times 10^{-21}\ \omega\ (\text{MeV}),$$

where ω [s^{-1}] denotes the angular frequency, and the typical duration time is estimated (as the time period) using $\omega = 2\pi/T$. Accordingly the typical time interval arising from the Fermi and the Coulomb energies are T_F =1.0×10^{-22} s (the representative energy is 40 MeV) and T_C =4.1×10^{-21} s (the representative energy is 1 MeV), respectively. Note that the typical duration time of low-energy heavy-ion collisions is 1000 fm/c (3.3×10^{-21} s). In particular the Fermi energy is related to the motion of nucleons inside nuclei, and more precisely to the fastest motion inside nuclei. Such a motion is also related to the collective aspect of many nucleon systems (cf. zero sound propagation [2]). Meanwhile, the Coulomb energy is related to the relative motion of the two colliding nuclei. For reference, the energy 1 GeV, which has been referred to as the typical energy scale for the nucleon, corresponds to the time interval 4.1×10^{-24} s.

From a different point of view there are three equilibria in classifying heavy-ion collisions: charge, momentum and thermal equilibria. The typical equilibration times of those three equilibria are different from each other, and three different time-scales follow. Let the three equilibration processes in low-energy heavy-ion reactions be denoted by the fast, moderate, and slow processes (Fig. 2). Typically the first period is associated with the fast charge equilibration, which is the process at a time scale of 10^{-22} s. The fast processes arise from the motion of nucleons inside a composite nucleus at the Fermi velocity. The second period is associated with the momentum equilibration, which is the process at a time scale of 10^{-20} s. This time interval corresponds to the typical duration time of heavy-ion reactions. The moderate processes arise from the motion of nucleons inside a composite nucleus, where the composite nuclei formed in the first period do not correspond to stationary states, therefore the nuclear and the Coulomb forces play a role in reducing the corresponding imbalance. The third period is associated with the thermal equilibration, which is the process taking more than 10^{-20} s. The slow processes arise from the diffusion of nucleons. This kind of motion gives rise to non-stationary time evolution even under charge and momentum equilibrium situations.

2.3. Time scales of TDDFT calculations

TDDFT calculations are usually utilised to simulate events whose time scale is from 10^{-23} s to 10^{-20} s. This utility of TDDFT calculations originally comes from the unit time step of calculations. Indeed, by taking into account the accumulation of errors, we usually have 10000 steps of calculations with a unit step of the order of 10^{-24} s. Note that the accumulation of errors should be more carefully treated for time-dependent calculations compared to the calculations for stationary states. Consequently the present TDDFT calculations can only be applied to clarify the reaction processes whose time scale is from 10^{-23} s to 10^{-20} s. In this sense, based on the TDDFT calculations, it is difficult to investigate reaction processes located on the third period in Fig. 2 without any special treatments. In addition, the direct binary nucleon-nucleon collisions will play a role in the long-term evolution, while they are mostly Pauli-blocked in the initial phase.

The thermal effect cannot be fully treated in the TDDFT calculations, where the total energy of the system is conserved during the calculation. The iso-energetic time evolution is ultimately reduced to the unitary time evolution in quantum mechanics, and the thermal effect treated in the TDDFT is only the energy transfer between kinetic and potential energies, where the energy deposit in the potential energy is often called the internal excitation energy in the context of the TDDFT calculations. Such energy transfers are only a part of possible energy transfers; indeed, even gamma-ray emission is not included. It is worth mentioning here that the energy density functionals used in TDDFT are described only by a small number of densities (for example, see [3]), so that special time evolution such as heat flow, emissions and so on due to the localised distribution of internal excitation is highly suppressed in the TDDFT calculations. Note that efficient cooling of composite nuclei is realized by particle emission.

3. Events in low-energy heavy-ion reactions

In the first period in Fig. 2, fast processes appear as consequences of the fast charge equilibration. It provides a strict limitation for the synthesis of exotic nuclei. Indeed, in case of fast charge equilibration, the proton-to-neutron ratios of the final products are not far from the proton-to-neutron ratios of the the total system: $(Z_1 + Z_2)/(N_1 + N_2)$. To avoid the fast charge equilibration, it is necessary to make the total reaction time be shorter than 10^{-22} s at the least, and the incident energy is required to be higher than E_F. If the incident energy is higher than E_F, the reaction is already completed at the time 10^{-22} s, and several fragments with different proton-to-neutron ratios appear (absence of fast charge equilibration).

In the second period in Fig. 2, moderate processes appear as consequences of momentum equilibration. For moderate processes, fusion, deep-inelastic reactions, or fragmentation into a few pieces takes place depending on the incident energy and the impact parameter. All the moderate processes are affected by the fast process, so that most of the fragments in low-energy heavy-ion collisions (whose incident energy is equal to a few MeV per nucleon) are in charge equilibrium.

In the third period in Fig. 2, slow processes appear as consequences of thermal equilibration. For slow processes, fission, decay, and emission of particles are known. Note that thermal transfer is related to fission and decay, while not essentially to fragmentation into a few pieces. All the slow processes follow after the fast and moderate processes.

4. Conclusion

TDDFT can reveal the scenario of low-energy heavy-ion reactions, where TDDFT simulates only the most probable time evolutions. Low energy heavy-ion collisions proceed step by step;

- in the early stage, fast charge equilibration inevitably appears mostly independent of the incident energy and the impact parameter;
- in the intermediate stage, fusion or fragmentation appears depending on the incident energy and the impact parameter;
- in the late stage, fission, decay, or particle-emission appears.

Each time period corresponds to its specific equilibration dynamics: the fast charge, momentum and thermal equilibrations respectively. In particular, charge and momentum equilibria are already achieved in the first and second periods, respectively.

Acknowledgments

This work was partially supported by the Helmholtz alliance HA216/EMMI. Valuable comments from Prof. J. A. Maruhn are appreciated. The author would like to express his sincere gratitude to Prof. T. Otsuka for perpetual encouragement.

References

[1] Iwata Y, Otsuka T, Maruhn J A and Itagaki N 2010 *Phys. Rev. Lett.* **104** 252501
[2] Iwata Y 2012 *J. Mod. Phys.* **3** 476
[3] Iwata Y and Maruhn J A 2012 *Energy Density Functional in Nuclear Physics* (New York: Nova publishers) in press (*Preprint* arXiv:1211.2355)

Journal of Physics: Conference Series **445** (2013) 012018 doi:10.1088/1742-6596/445/1/012018

Spin-isospin responses studied with RI-beams

Tomohiro Uesaka
RIKNE Nishina Center, 2-1 Hirosawa, Wako, Saitama 351-0198, Japan
E-mail: uesaka@riken.jp

Abstract. In light of availabilities of a variety of radioactive isotope (RI) beams and progress in physics of unstable nuclei, new possibilities in experimental studies of spin-isospin responses are foreseen. They include: 1) studies of spin-isospin responses of unstable nuclei, 2) studies of higher multipoles, in particular spin-dipole responses, 3) studies of higher tones, such as isovector monopole resonances, and 4) studies of many phonon states. In this article, we discuss how the study of spin-isospin responses will be proceeded, taking examples from experiments with the SHARAQ spectrometer at RI Beam Factory and with the Grand Raiden spectrometer at Research Center for Nuclear Phyics, Osaka University.

1. Introduction

Charge exchange reactions at intermediate energies can extend studies of nuclear spin-isospin responses to higher excitation energies that are out of the reach of β-decay experiments[1, 2, 3]. Rich information on nuclear structure, correlations in nuclei[4], nuclear matter properties, and quantities needed to understand astrophysical phenomena[5] has been obtained through experimental studies of spin-isospin responses. In particular, experiments with light-ion induced reactions, such as (p, n), $(d, {}^2\mathrm{He})$, and $({}^3\mathrm{He}, t)$ reactions, have been carried out intensively, with some emphases on Gamow-Teller (GT) transitions that are characterized by $\Delta n = 0$, $\Delta L = 0$, $\Delta S = 1$, and $\Delta T = 1$. Here Δn, ΔL, ΔS, and ΔT are changes of a principal quantum number, orbital and spin angular momenta, and isospin. The experimental data bring about reasonable understanding of GT transitions in stable nuclei.

On the other hand, availability of RI-beams has increased rapidly since 1990's and has brought intense attention of nuclear scientists to physics of unstable nuclei. RI Beam Factory (RIBF) at RIKEN[6, 7, 8] is a new-generation RI-beam facility that can provide highly-intense RI beams by means of fragmentation reactions of heavy-ion beams and by in-beam fission of an uranium beam.

In light of availabilities of a variety of RI-beams and progress in physics of unstable nuclei, new possibilities in experimental studies of spin-isospin responses are foreseen. They include:

(i) Spin-isospin responses of unstable nuclei
(ii) Studies of higher multipoles ($\Delta L = 1, 2, ...$)
(iii) Studies of higher tones ($\Delta n \geq 1$)
(iv) Studies of many phonon states

Case 1 is a straightforward extension of studies with the light-ion beams on stable targets, while cases 2–4 are not. In cases 2–4, RI-beam induced charge exchange reactions with unique capabilities which are missing in stable-beam induced reactions are used to reveal new aspects of spin-isospin responses in nuclei. The capabilities of RI-beam induced charge exchange reactions are based on

- A variety of selectivities in transferred quantum numbers, ΔS, ΔT, ΔT_z, ΔL etc, and
- Kinematical conditions which can not be reached via the stable-beam induced reactions.

We have constructed the high-resolution SHARAQ spectrometer [9] and the dedicated beam-line [10] at RI Beam Factory to proceed studies of spin-isospin responses with RI-beams. The ion-optical system is designed to achieve a momentum resolution of Δp/p=1/14700 and an angular resolution of $\Delta\theta \sim$1 mrad for RI-beams with a momentum spread of $\pm$0.3%. The performances facilitate missing-mass spectroscopy experiments with RI-beams.

In this article, we discuss how the study of spin-isospin responses will be proceeded in light of the availability of RI-beams and/or of physics of unstable nuclei, putting emphasis on experiments with the SHARAQ spectrometer.

2. Spin-isospin responses of unstable nuclei

The (p, n) reaction study of unstable nuclei was initiated in late 1990's at RIKEN. In the first experiment with a ^{11}Li beam, Teranishi and his collaborators succeeded in observing an isobaric analog state of the ^{11}Li ground state in an invariant mass spectrum of neutrons and a heavy residue[11]. The invariant mass method were applied to (p, n) reactions studies of other drip-line nuclei[12, 13].

The invariant mass method is attractive in that it enables a high excitation energy resolution even in experiment with a thick target. On the other hand, it has weak points that it is not sensitive to bound excited states and an experimental acceptance at high excitation energies can not be so large. The latter would prevent one from investigating giant resonances locating at excitation energies higher than 10 MeV. Thus a missing mass approach is definitely needed.

Recently, new neutron detection setups for the missing mass spectroscopy with an inverse-kinematics (p, n) reaction have been developed by the Michigan State University group and by the Tokyo-RIKEN group. The Michigan group has applied the setup called LENDA[14] to extract Gamow-Teller strengths in ^{56}Ni[15, 16] which is of astrophysical importance.

At RI Beam Factory, the ^{12}Be$(p, n)^{12}$B experiment at 200 MeV/nucleon has been performed[17] with the newly-developed neutron detector array WINDS[18]. The GT transitions to the ground and excited states at around 10 MeV in ^{12}B are clearly observed. The experimental technique established here will furnish excellent opportunities to investigate GT and higher multipole strengths in unstable nuclei.

3. Higher multipoles: search for collective spin-dipole 0^- resonances

Study of spin-isospin responses with higher multipolarities, $\Delta L \geq 1$, is another interesting subject. Spin-dipole (SD) states excited with $\Delta L = 1$, $\Delta S = 1$, and $\Delta J = 0^-, 1^-, 2^-$, attract recent theoretical attention due to its strong relevances to tensor-force effects. For example, self-consistent HF+RPA calculations in Ref. [19] predict that the tensor correlations cause hardening of the collective 0^- resonance, resulting in a shift of the strength toward higher excitation energy. It is also predicted that the effect is sensitive to the magnitude of tensor strength. Thus experimental data of the SD 0^- distribution enables us to quantitatively examine the tensor correlation effects in nuclei.

In spite of its interesting properties described above, experimental information on 0^- states is limited. This is because, as a recent polarization transfer experiment for the ^{208}Pb(p, n) reaction shows[20], 0^- states are populated only weakly compared with 1^- and 2^- states and they are strongly fragmented over a wide range of excitation energy. A well-suited experimental tool to probe the 0^- states is definitely desired.

Dozono and his collaborators propose a new experiment with the parity-transfer (^{16}O,^{16}F(0^-)) reaction[21]. The parity-transfer reaction uses $0^+ \longrightarrow 0^-$ transition in the projectile to probe 0^- states in a target nucleus. This reaction has a uniques sensitivity to unnatural parity states, which is an advantage of the parity-transfer reaction over other reactions used so far. The first experiment for a ^{12}C target is planned with the SHARAQ spectrometer in 2014.

4. Higher tones: isovector monopole resonances

It is well known that an isoscalar giant monopole resonance (ISGMR) is an oscillation mode of (isoscalar) density and can be related to the incompressibility of nuclear matter[22]. Microscopically, on the other hand, ISGMR comprises coherent particle-hole excitations with $\Delta n = 1$, $\Delta L = \Delta S = 0$.

Isovector analogies of ISGMR are an isovector monopole resonance (IVGMR) and an isovector spin monopole resonance (IVSGMR). By the analogy, IVGMR (IVSGMR) is considered to be an oscillation mode of isovector (isovector spin) density (see Fig. 1) and can be related to (spin-) isospin properties of nuclear matter. IVGMR and IVSGMR are regarded as higher tones ($\Delta n = 1$) of an isobaric analogue resonance and a giant Gamow-Teller resonance with $\Delta n = 0$.

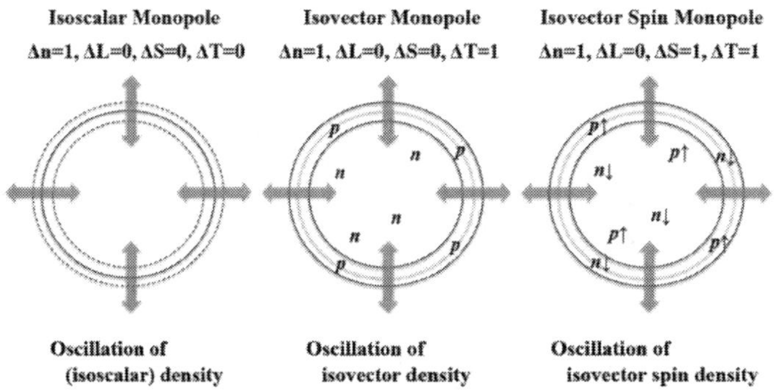

Figure 1. Isoscalar and isovector monopole resonances.

Although IVGMR and IVSGMR were observed in several previous experiments[23, 24], our knowledge on the resonances were still poor when compared with that on IAS and GT resonances. The first series of experiments with the SHARAQ spectrometer were performed to extract new information of IVSGMR and IVGMR.

4.1. (t, ^{3}He) *reaction*

The (t, ^{3}He) reaction is the most etablished RI-beam induced charge exchange reaction. It has been used to investigate β^+ strengths in nuclei at 115 MeV/nucleon[25] and at lower energies. Effectiveness of the reaction in investigating isovector spin-flip strengths in nuclei should be more emphasized at the RIBF energy of ~300 MeV/nucleon, because $\sigma\tau$-modes are most strongly populated and nuclear distortion effects are the weakest at the energies.

The first (t, ^{3}He) experiment with the SHARAQ spectrometer was performed to search for β^+-type IVSGMR in ^{90}Zr and ^{208}Pb[26]. A 300-MeV/nucleon triton beam with an intensity of 10^7 sec^{-1} was produced by the projectile fragmentation of a primary 320-MeV/nucleon ^{4}He beam on a 4-cm thick beryllium target and was irradiated onto the secondary targets. The scattered ^{3}He ions were momentum-analyzed by the SHARAQ spectrometer.

Red histograms in the upper panels of Fig. 2 show the monopole ($\Delta L = 0$) cross section for the ^{208}Pb, ^{90}Zr(t, ^{3}He) reactions. The monopole cross section is deduced by the multipole

Journal of Physics: Conference Series **445** (2013) 012018 doi:10.1088/1742-6596/445/1/012018

decomposition analysis. Yellow histograms overlaid on the $(t,{}^3\mathrm{He})$ cross section present cross section for the (p,n) reaction. Since a nucleon can penetrate more deeply in the target nucleus than a triton, strong cancellation of the transition amplitude takes place in the case of IVSGMR. Thus the (p,n) cross section is considered to be dominated by GT strengths. Thus, enhancement of the $(t,{}^3\mathrm{He})$ cross section over (p,n) observed in Fig. 2 is considered to be a clear signature of IVSGMR. The monopole strength is compared with HF-RPA calculations in the lower panels of Fig. 2. The theoretical calculations reasonably reproduce the experimental data.

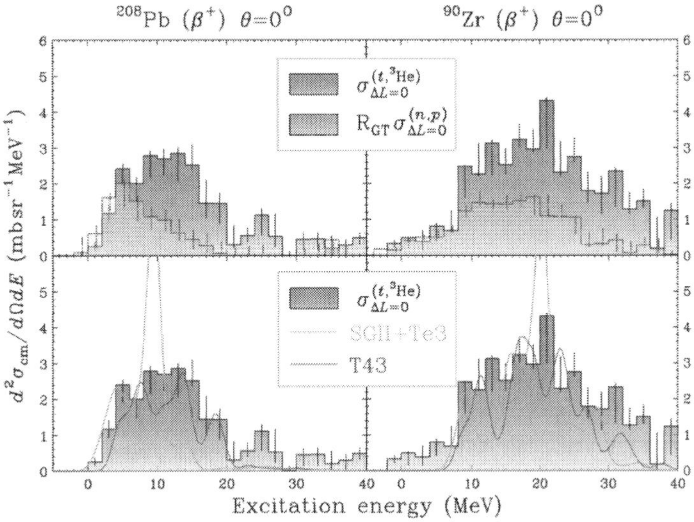

Figure 2. Monopole cross sections of the $^{90}\mathrm{Zr}, {}^{208}\mathrm{Pb}(\mathrm{t},{}^3\mathrm{He})$ reaction (red), together with those of the (n,p) reactions (yellow). Reprinted from Ref. [26]. Copyright (2012) by the American Physical Society.

In 2010, two experiments were carried out to reveal different aspects of IVGMR and IVSGMR. One was the $^7\mathrm{Li}, {}^{90}\mathrm{Zr}({}^{10}\mathrm{C}, {}^{10}\mathrm{B}(0^+;\mathrm{IAS}))$ experiment to search for IVGMR[27]. The $({}^{10}\mathrm{C}, {}^{10}\mathrm{B}(0^+;\mathrm{IAS}))$ reaction with a unique selectivity to isovector ($\Delta T = 1$) non-spin-flip ($\Delta S = 0$) modes was employed in the experiment. In the Doppler-corrected gamma-ray spectrum, the 1.022-MeV peak that is a signature of $\Delta S = 0$ transitions is clearly observed[27]. The other was the $^{90}\mathrm{Zr}({}^{12}\mathrm{N}, {}^{12}\mathrm{C})$ experiment aiming at populating IVSGMR with a small momentum transfer. In the experiment, the RI-beam induced $({}^{12}\mathrm{N}, {}^{12}\mathrm{C})$ reaction with a large positive Q value ($Q = +17$ MeV), which we call the *exothermic charge exchange reaction*, was used to populate the high-lying β^--IVSGMR in a recoilless manner. Since IVSGMR is strongly excited when an accompanied momentum transfer is small, prominent strength is expected to be observed by the exothermic reaction. Preliminary results show that the cross section ratio of IVSMR to GT is largely enhanced in the $({}^{12}\mathrm{N}, {}^{12}\mathrm{C})$ compared to the (p,n) reaction[28].

5. Many phonon states: search for double GT resonances and spectroscopy of unstable nuclei

Double charge exchange (DCX) reactions can be unique spectroscopic tools through which we can transfer quantum numbers by an amount of two to a target nucleus. Their obvious applications are population of multi-phonon excitation in nuclei and spectroscopy of unstable nuclei.

Among possible multi-phonon excitations, double Gamow-Teller resonances (DGTR)[29] are of particular interest, partly because of its relevance to double beta decay processes and partly because of the fact that it is remained to be discovered. Attempts to find DGTR via the heavy ion double charge exchange reactions were made at GANIL/MSU and at RCNP by using the (^{18}O, ^{18}Ne)[30] and (^{11}B, ^{11}Li)[31] reactions, respectively, but no clear evidence was found so far.

Another application of double charge exchange reaction is a spectroscopic study of light unstable nuclei. Through DCX, we can easily access to neutron/proton-rich nuclei even with stable-beam induced reactions. Prior to RI-beam experiments, we carried out an experimental program with DCX reactions at Research Center for Nuclear Physics, Osaka University, aiming at establishing DCX reactions as spectroscopic tools. In the first experiment, the $(\beta^+)^2$-type DCX ^{12}C(^{18}O, ^{18}Ne) reaction at 80 MeV/nucleon were used to investigate structure of the neutron-rich ^{12}Be nucleus[32]. In the missing mass spectrum, three peaks corresponding to the ground and excited states of ^{12}Be were clearly observed.

In 2012, an experiment to search for tetraneutron states via the DCX ^{4}He(^{8}He, ^{8}Be) reaction has been conducted with the SHARAQ spectrometer[33]. The exothermicity of the (^{8}He, ^{8}Be) reaction is advantageous in populating the fragile tetraneutron states from the ^{4}He target. The experiment at SHARAQ is expected to present decisive information on the nature of tetraneutron states. Another RI-beam induced $(\beta^-)^2$-type DCX reaction, the (^{10}C,^{10}Be) reaction, which accompanies momentum transfer as small as 10–30 MeV/c at around 0-degrees will be an innovative probe to DGTR in future.

6. Summary

Possibilities of new experimental studies of spin-isospin responses are discussed. In studies of spin-isospin responses of unstable nuclei, the missing mass spectroscopy with an inverse-kinematics (p, n) reaction will play a central role in future experiments. New neutron detection setups, LENDA at MSU and WINDS at Tokyo-RIKEN will be widely used to investigate GT and SD strengths in unstable nuclei. One of the main topic in studies of higher multipoles should be detailed study of the 0^- strength. The parity-transfer (^{16}O,^{16}F(0^-)) reaction will serve as a unique probe to 0^-. It has been shown that the RI-beam induced charge exchange reactions, $(t, {}^3\text{He})$, (^{10}C, ^{10}B(0^+)), (^{12}N, ^{12}C), are quite useful in observing IVSGMR and IVGMR in nuclei. Use of DCX reaction will be the next step to proceed.

Acknowledgements

The present work was partly supported by the Grant-in-Aid for Scientific Research (Nos. 17002003 and 19204024) by the Ministry of Education, Culture, Sports, Science and Technology of Japan. The authors also thank the JSPS Core-to-Core program.

References

[1] Taddeucci T, Goulding C, Carey T, Byrd R, Goodman C, Gaarde C, Larsen J, Horen D, Rapaporrt J and Sugarbaker E 1987 *Nuclear Physics* A **469** 125–72
[2] Osterfeld F 1992 *Reviews of Modern Physics* **64** 491–557
[3] Spicer W and Spicer B 1998 *Advances in Nuclear Physics* **24** 1–82
[4] Ichimura M, Sakai H and Wakasa T 2006 *Progress in Particle and Nuclear Physics* **56** 446–531
[5] Langanke K and Martínez-Pinedo G 2003 *Rev. Mod. Phys.* **75** 819–62
[6] Motobayashi T and Sakurai H 2012 *Progress of Theoretical and Experimental Physics* **2012** 03C001
[7] Okuno H, Fukunishi N and Kamigaito O 2012 *Progress of Theoretical and Experimental Physics* **2012** 03C002
[8] Kubo T, Kameda D, Suzuki H, Fukuda N, Takeda H, Yanagisawa Y, Ohtake M, Kusaka K, Yoshida K, Inabe N, Ohnishi T, Yoshida A, Tanaka K and Mizoi Y 2012 *Progress of Theoretical and Experimental Physics* **2012** 03C003
[9] Uesaka T, Shimoura S and Sakai H 2012 *Progress of Theoretical and Experimental Physics* **2012** 03C007
[10] Kawabata T, Berg G P A, Kubo T, Sakai H, Shimoura S and Uesaka T 2008 *Nuclear Instruments and Methods in Scientific Research* B **266** 4201–4
[11] Teranishi T, Shimoura S, Ando Y, Hirai M, Iwasa N, Kikuchi T, Moriya S, Motobayashi T, Murakami H, Nakamura T, Nishio T, Sakurai H, Uchibori T, Watanabe Y, Yanagisawa Y and Ishihara M 1997 *Physics Letters* B **407** 110 – 4

[12] Takeuchi S, Shimoura S, Motobayashi T, Akiyoshi H, Ando Y, Aoi N, Fü Zs, Gomi T, Higurashi Y, Hirai M, Iwasa N, Iwasaki H, Iwata Y, Kobayashi H, Kurokawa M, Liu Z, Minemura T, Ozawa S, Sakurai H, Serata M, Teranishi T, Yamada K, Yanagisawa Y and Ishihara M 2001 *Physics Letters* B **515** 255 – 60

[13] Satou Y, Nakamura T, Kondo Y, Matsui N, Hashimoto Y, Nakabayashi T, Okumura T, Shinohara M, Fukuda N, Sugimoto T, Otsu H, Togano Y, Motobayashi T, Sakurai H, Yanagisawa Y, Aoi N, Takeuchi S, Gomi T, Ishihara M, Kawai S, Ong H, Onishi T, Shimoura S, Tamaki M, Kobayashi T, Matsuda Y, Endo N and Kitayama M 2011 *Physics Letters* B **697** 459 – 62

[14] Perdikakis G, Austin S M, Bazin D, Caesar C, Deaven J M, Guess C J, Hitt G W, Meharchand R, Nguyen D, Shimbara Y, Thorne K and Zegers R G T 2009 *IEEE Transaction on Nuclear Science* **56** 1174–8

[15] Sasano M, Perdikakis G, Zegers R G T, Austin S M, Bazin D, Brown B A, Caesar C, Cole A L, Deaven J M, Ferrante N, Guess C J, Hitt G W, Meharchand R, Montes F, Palardy J, Prinke A, Riley L A, Sakai H, Scott M, Stolz A, Valdez L and Yako K 2011 *Phys. Rev. Lett.* **107** 202501

[16] Sasano M, Perdikakis G, Zegers R G T, Austin S M, Bazin D, Brown B A, Caesar C, Cole A L, Deaven J M, Ferrante N, Guess C J, Hitt G W, Honma M, Meharchand R, Montes F, Palardy J, Prinke A, Riley L A, Sakai H, Scott M, Stolz A, Suzuki T, Valdez L and Yako K 2012 *Phys. Rev.* C **86** 034324

[17] Yako K, Sakai H, Dozono M, Fujii T, Kawase S, Kikuchi Y, Kisamori K, Kubota Y, Matsubara H, Michimasa S, Miki K, Miya H, Noji S, Lee C, Ota S, Sasamoto Y, Sasano M, Shimoura S, Takaki M, Tang L, Tokieda H, Uesaka T, Xu Z and Yokoyama R 2012 *RIKEN Accel. Prog. Rep.* **45** v

[18] Yako K, Sakai H, Dozono M, Fujii T, Kawase S, Kikuchi Y, Kisamori K, Kubota Y, Matsubara H, Michimasa S, Miki K, Miya H, Noji S, Lee C, Ota S, Sasamoto Y, Sasano M, Shimoura S, Takaki M, Tang L, Tokieda H, Uesaka T, Xu Z and Yokoyama R 2012 *RIKEN Accel. Prog. Rep.* **45** 137

[19] Bai C L, Zhang H Q, Sagawa H, Zhang X Z, Colò G and Xu F R 2010 *Phys. Rev. Lett.* **105** 072501

[20] Wakasa T, Okamoto M, Dozono M, Hatanaka K, Ichimura M, Kuroita S, Maeda Y, Miyasako H, Noro T, Saito T, Sakemi Y, Yabe T and Yako K 2012 *Phys. Rev.* C **85** 064606

[21] Dozono M *in International symposium on frontiers in nuclear physics*, Beijing, China, November 2011.

[22] Blaizot J 1980 *Physics Reports* **64** 171–248

[23] Erell A, Alster J, Lichtenstadt J, Moinester M A, Bowman J D, Cooper M D, Irom F, Matis H S, Piasetzky E and Sennhauser U 1984 *Physical Review Letters* **52** 2134

[24] Zegers R G T, Abend H, Akimune H, van den Berg A M, Fujimura H, Fujita H, Fujita Y, Fujiwara M, Galès S, Hara K, Harakeh M N, Ishikawa T, Kawabata T, Kawase K, Mibe T, Nakanishi K, Nakayama S, Toyokawa H, Uchida M, Yamagata T, Yamasaki K and Yosoi M 2003 *Phys. Rev. Lett.* **90** 202501

[25] Zegers R G T, Akimune H, Austin S M, Bazin D, van den Berg A M, Berg G P A, Brown B A, Brown J, Cole A L, Daito I, Fujita Y, Fujiwara M, Galès S, Harakeh M N, Hashimoto H, Hayami R, Hitt G W, Howard M E, Itoh M, Jänecke J, Kawabata T, Kawase K, Kinoshita M, Nakamura T, Nakanishi K, Nakayama S, Okumura S, Richter W A, Roberts D A, Sherrill B M, Shimbara Y, Steiner M, Uchida M, Ueno H, Yamagata T and Yosoi M 2006 *Phys. Rev.* C **74** 024309

[26] Miki K, Sakai H, Uesaka T, Baba H, Bai C L, Berg G P A, Fukuda N, Kameda D, Kawabata T, Kawase S, Kubo T, Michimasa S, Miya H, Noji S, Ohnishi T, Ota S, Saito A, Sasamoto Y, Sagawa H, Sasano M, Shimoura S, Takeda H, Tokieda H, Yako K, Yanagisawa Y and Zegers R G T 2012 *Phys. Rev. Lett.* **108** 262503

[27] Sasamoto Y, Uesaka T, Shimoura S, Michimasa S, Ota S, Noji S, Miki K, Yako K, Tokieda H, Miya H, Kawase S, Kikuchi Y, Kisamori K, Takaki M, Dozono M, Matsubara H, Sakai H, Kubo T and Yanagisawa Y 2012 *Proceedings of the French-Japanese Symposium on Nuclear Structure Problems* (World Scientific) p 148

[28] Noji S 2012 development of a new spectroscopic tool for spin-isospin responses by the exothermic charge-exchange reaction *Doctoral dissertation* (University of Tokyo)

[29] Auerbach N, Zamick L and Zheng D Z 1989 *Annals of Physics* **192** 77

[30] Blomgren J, Lindh K, Anantaraman N, Austin S M, Berg G, Brown B, Casandjian J M, Chartier M, Cortina-Gil M, Fortier S, Hellström M, Jongman J, Kelley J, Lepine-Szily A, Lhenry I, Cormicke M M, Mittig W, Nilsson J, Olsson N, Orr N A, Ramakrishnan E, Roussel-Chomaz P, Sherrill B M, Tegntr R E, Winfield J S and Winger J 1995 *Physics Letters* B **362** 34–8

[31] Takahisa K, *private communication*

[32] Matsubara H, Takaki M, Uesaka T, Shimoura S, Aoi N, Dozono M, Fujii T, Hatanaka K, Hashimoto T, Kawabata T, Kawase S, Kisamori K, Kikuch Y, Kubota Y, Lee C S, Lee H C, Maeda Y, Michimasa S, Miki K, Miya H, Noji S, Ota S, Sakaguchi S, Sasamoto Y, Suzuki T, Tang L T, Takahisa K, Tokieda H, Tamii A, Yako K, Yasuda Y, Yokota N, Yokoyama R and Zenihiro J 2013 *Few-Body Systems, in print*

[33] Shimoura S 2010 Tetra-neutron system produced by exothermic double-charge exchange reaction *Proposal for Nuclear Physics Experiment at RI Beam Factory*

Continuum dipole response near the threshold and the direct neutron capture cross section at astrophysical energies

Masayuki Matsuo

Department of Physics, Faculty of Science, Niigata University, Niigata 950-2181, Japan

E-mail: matsuo@nt.sc.niigata-u.ac.jp

Abstract. We apply the continuum quasiparticle random phase approximation to calculate the direct neutron capture cross section relevant to the r-process nucleo-synthesis. The electric dipole strength function in an even-even n-rich nucleus is decomposed with respect to the channels of direct neutron decays. Using the detailed balance relation, the partial dipole strengths are converted to obtain the direct neutron-capture cross sections. Numerical examples are given for ^{142}Sn.

1. Introduction

The rapid neutron capture process is physics that plays a central role in understanding the origin of heavy elements in our universe[1,2]. The r-process passes through neutron-rich nuclei in essentially all the mass regions of the nuclear chart via various reactions and decays, such as the neutron capture, the photo-absorption, the beta-decay and the fission. Besides experimental efforts to produce and measure the relevant neutron-rich nuclei in the advanced RI beam facilities, a large number of theoretical studies have been performed aiming at quantitative description of the relevant nuclear structures and reactions. The nuclear density functional theories are, in our opinion, one of the most promising nuclear models for this purpose since they provide not only accurate description of the nuclear masses but also the nuclear responses for the weak and electromagnetic interactions. In the present work, I show that the nuclear density functional theory also has a ability to describe the key nuclear reaction in the r-process, i.e., the direct neutron capture.

The low-energy neutron capture reaction may be modeled using two different mechanisms[2]. One is the statistical mechanism: it is assumed that the capture of a neutron leads to formation of compounds states which then decay by statistical E1-gamma transitions. The Hauser-Feshbach statistical reaction theory is usually adopted, and the gamma-ray strength function associated with the compounds states is one of the key quantity. Recently microscopic calculations of the strengh functions based on the quasiparticle random phase approximation to the density functional theories or their extensions [3,4] are applied in order to improve predictabilities. The other is the direct neutron capture mechanism where the gamma-decay occurs immediately after the neutron capture. The relative importance of the two mechanisms depends on the neutron separation energy S_n[2,5]. For nuclei near the stability line, relevant to the s-process, the separation energy $S_n \sim 8$ MeV is so large that the Bohr's compounds states are formed because of the high level density. In this situation the statistical mechanism dominates. The neutron-rich nuclei involved in the r-process, however, have much smaller separation energy $S_n \sim O(1)$ MeV, and it is suggested that the direct neutron capture dominates[5]. The neutron single-particle model is often adopted in calculating the direct neutron capture cross section[6-8]. However, the neutron-rich nuclei is known to exhibit anomalous features in the electromagnetic responses, for instance, the enhanced E1 response arising from the pygmy dipole resonance. The enhanced E1 response and the pygmy resonance may imply correlation or collectivity of a new kind although their properties are under current investigates. We need to go beyond the single-particle model in order to include these effects. It is the purpose of our study to provide such a description on the basis of the recent advances in the nuclear density functional theories.

2. The continuum QRPA and the E1 strength function

The starting point of our formulation is the continuum quasiparticle random phase approximation (QRPA) [9] applied to the Skyrme-Hartree-Fock-Bogoliubov model. It is a linear response theory describing excitations of a nucleus caused by a perturbing external field V_{ext}, the electric dipole field in our case. Adopting the time-dependent Hartree-Fock-Bogoliubov theory as a fundamental framework to describe the time-evolution of the system, the perturbation causes fluctuations in the particle density and the pair density, which in turn induce fluctuations in the selfconsistent Hartree-Fock-Bogoliubov mean fields. The selfconsistent relation between these quantities is written as

$$\begin{pmatrix} \delta\rho(r,\omega) \\ \delta\tilde{\rho}(r,\omega) \\ \delta\tilde{\rho}^*(r,\omega) \end{pmatrix} = \int dr' \left(R_0^{\alpha\beta}(r,r',\omega)\right) \begin{pmatrix} \delta\Gamma(r',\omega) + v_{ext}(r') \\ \delta\tilde{\Delta}(r',\omega) \\ \delta\tilde{\Delta}^*(r',\omega) \end{pmatrix} \tag{1}$$

where $\delta\rho(r,\omega)$ and $\delta\tilde{\rho}(r,\omega)$ are the fluctuations in the density and the pair density, respectively, and $\delta\Gamma(r,\omega) = (\partial^2 E_{edf}/\partial^2\rho)\,\delta\rho(r,\omega)$ and $\delta\tilde{\Delta}(r,\omega) = (\partial^2 E_{edf}/\partial^2\tilde{\rho})\,\delta\tilde{\rho}(r,\omega)$ are the induced Hartree-Fock and pair fields derived from the energy density functional E_{edf}. Propagations of the quasiparticles excited by the external field and the induced selfconsistent fields bring about the fluctuation in the densities. This is described by the response function $R_0^{\alpha\beta}(r,r',\omega)$, which is expressed in terms of the quasiparticle Green's function $G(r,r',E)$. It is then straightforward to obtain the strength function for the E1 operator: we solve Eq.(1) for the dipole field $v_{ext}(r) = (Z/A\delta_{\tau p} - N/A\delta_{\tau n})r$, and then evaluate the E1 strength function

$$S_{E1}(\omega) = \sum_k |\langle 1^- k|v_{ext}|0^+\mathrm{gs}\rangle|^2\delta(\Omega_k - \omega) = -\frac{1}{\pi}\mathrm{Im}\int dr v_{ext}(r)\delta\rho(r,\omega) \tag{2}$$

where $|k\rangle$ is the QRPA excited states with excitation energy Ω_k. The quantum numbers are omitted in the above equations. The details of the continuum QRPA are given in Refs.[9-11], and in the present work we follow Ref.[10]. The photo-absorption cross section via the E1 gamma's is given by

$$\sigma_\gamma(E_\gamma) = \frac{16\pi^3 e^2}{9\hbar c} E_\gamma S_{E1}(E_\gamma). \tag{3}$$

3. The direct neutron capture cross section

Let us outline how we use the continuum QRPA to calculate the direct neutron capture cross section. It is noted here that the excited states populated by the photo-absorption have a decay branch of direct one-neutron emission if the excitation energy exceeds the neutron separation energy. The continuum QRPA has an advantage, compared to the standard QRPA frameworks, that it enables to describe the direct neutron decays. This is because the neutron wave functions in the continuum energy region (i.e. the quasineutron states above the separation threshold) are connected to the scattering waves in the external region. We wish to decompose the total photo-absorption cross section into partial cross sections corresponding to different decay channels. Once this is achieved, we can convert the partial photo-absorption cross sections to the direct neutron capture cross sections by using the detailed balance relation since the direct neutron capture with the E1 gamma decay is the inverse process of the E1 photo-absorption followed by one-neutron emission.

We apply and extend the prescription of Zangwill and Soven [12,13] for this decomposition. In the preceding works, the method is formulated in the framework of the continuum RPA which have been applied to the photo-absorption and associated electron emission of molecules and atomic clusters[12,13]. In the present case, however, the nucleus is in the superfluid phase caused by the neutron pair correlation. A clear difference is that once the pairing is switched on both one and two neutron emissions are allowed. Concerning the one-neutron decays, the channels are specified by the partial wave of the emitted neutron (i.e. the quasineutron states in the continuum) and the states of the residual $A-1$ nucleus, which correspond to the bound one-quasineutron states.

The partial E1 strength function specific to a one-neutron decay channel with the residual quasiparticle configuration i is calculated as

$$S_{E1,1c,i}(\omega) = -\frac{1}{\pi}\mathrm{Im}\int\int drdr'\bar{\phi_{\tilde{i}}}(r)^{\dagger}(\mathcal{V}_{scf}(r,\omega))^{\dagger}G_{>}(r,r',\omega-i0-E_i)\mathcal{V}_{scf}(r',\omega)\bar{\phi_{\tilde{i}}}(r')+(\text{backward}) \tag{4}$$

where $\phi_i(r)$ is the two-component wave function of the bound quasiparticle state i with energy E_i, and $\bar{\phi_{\tilde{i}}}(r)$ being its negative energy counter part. $\mathcal{V}_{scf}(r,\omega)$ is the 2×2 matrix representation of the selfconsistent fields $(\delta\Gamma(r,\omega)+v_{ext}(r'), \delta\tilde{\Delta}(r,\omega), \delta\tilde{\Delta}^{*}(r,\omega))$. $G_{>}(r,r',E)$ is the quasiparticle Green's function which describe propagation of the quasiparticle with continuum spectrum, i.e., the neutron escaping from the nucleus under the influence of the Hartree-Fock-Bogoliubov mean-fields. One can show that the partial strength $S_{E1,1c,i}(\omega)$ is equivalent to

$$S_{E1,1c,i}(\omega) = \sum_{p\in\text{cont.}} |\langle ip|\mathcal{V}_{scf}(\omega)|0\rangle|^2\delta(E_i+E_p-\omega) \tag{5}$$

where $|\langle ip|\mathcal{V}_{scf}(\omega)|0\rangle|^2$ is the partial E1 strength in which the final state is the one-quasineutron state i (the residual $A-1$ nucleus) plus the continuum neutron state p. The crucial point is that the operator $\mathcal{V}_{scf}(\omega)$ appearing in these expressions includes, in addition to the bare E1 operator $v_{ext}(r)$, the induced fields $\delta\Gamma(r,\omega)$ and $\delta\tilde{\Delta}(r,\omega)$ representing effects of the RPA correlations or the collectivity of the excited states at energy $E_x=\omega$.

4. Results

We have performed numerical calculations for a very neutron-rich tin isotope ^{142}Sn. This is an isotope whose one-neutron separation energy is calculated to be around 2 MeV, and it may be located around the r-process path. We adopt the parameter set SLy4 for the Skyrme functional and the density-dependent delta interaction model of the mix type for the effective pairing interaction. We use the spherical HFB + continuum QRPA code[10] adopting the mesh representation in the radial coordinate, and the space size is chosen as $R_{max}=20$ fm. The Landau-Migdal approximation is introduced for the residual interaction to obtain the induced Hartree-Fock field in the continuum QRPA calculation. The solid curve in Fig.1 is the total E1 strength function obtained with Eqs.(1) and (2). Here the strength function is smoothed with the smearing width of 1 MeV.

The large strengths distributed around $E_x\sim 10-18$ MeV correspond to the giant dipole resonance. In addition, a significant amount of strength is seen at low excitation energies

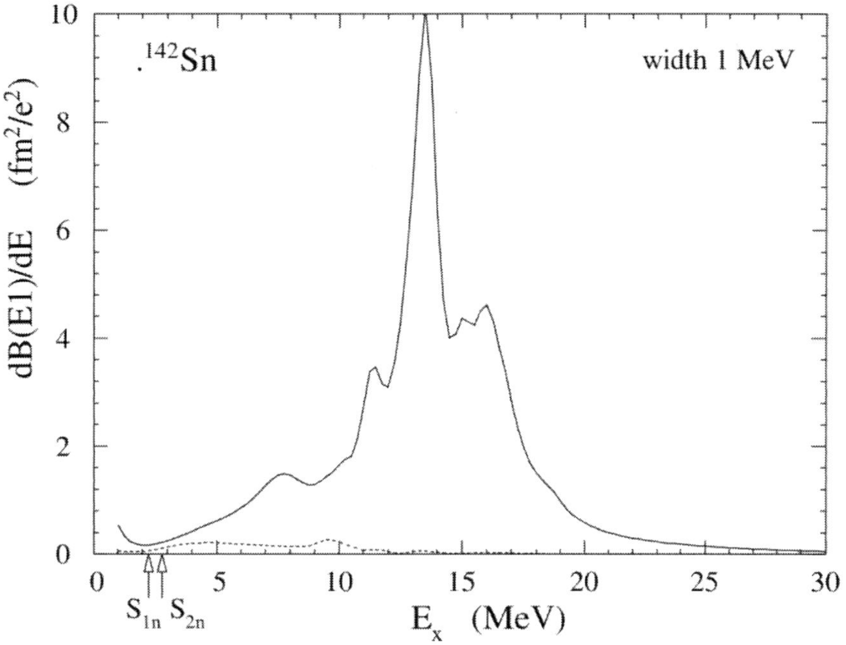

Figure 1. The E1 strength function of ^{142}Sn as a function of the excitation energy, calculated with the Skyrme-HFB plus continuum QRPA method (the solid curve). The dotted curve is the partial E1 strength function for the decay channels with direct one-neutron emission. The smoothing width is chosen to 1 MeV. The arrows indicate the one- and two-neutron separation energies.

$E_x \sim 2 - 8$ MeV located just above the one-neutron separation energy $S_{1n} = 2.25$ MeV. Our model predicts that the low-lying E1 strength appears in neutron-rich isotopes heavier than ^{132}Sn, and the strength increases with the neutron excess, suggesting larger photo-absorption cross section and larger direct neutron capture cross section.

There are only two bound quasineutron states in the present HFB calculation. They correspond to the $3p_{3/2}$ and $3p_{1/2}$ orbits located above the $N = 82$ shell gap. Correspondingly the $A-1$ system ^{141}Sn, described as one-quasineutron states in the present model, has only two bound states $3/2^-$ and $1/2^-$ having the $3p_{3/2}$ and $3p_{1/2}$ quasineutron configurations, respectively. The 1^- excited states of ^{142}Sn populated by the E1 gamma can decay via one-neutron emission leaving the $3p_{3/2}$ or $3p_{1/2}$ quasineutron state of ^{141}Sn. The dotted curve in Fig.1 is the partial E1 strength, Eq.(4), associated with these decay channels. It is noted here that the partial strength with one-neutron decay is only a small fraction of the total E1 strength at excitation energies $E_x \gtrsim 5$ MeV. We found that the direct two-neutron decay, i.e. the decay mode emitting two-neutrons and populating directly the ground state of the $A-2$ system ^{140}Sn dominates among the remaining E1 strength. This channel has a considerable fraction since the calculated two-neutron separation energy is small $S_{2n} = 2.80$ MeV, and the presence of the pair correlation makes many of the hole-like quasiparticle states unbound.

Let us now focus on the strength near the neutron separation energy $E_x \sim S_{1n}$. Figure 2 shows the partial E1 strengths for the direct one-neutron decay channels. Here the total strength is decomposed according to both the partial waves of the escaping neutron ($s_{1/2}$, $d_{3/2}$ or $d_{5/2}$) and the configurations ($3p_{1/2}$ or $3p_{3/2}$) of the populated state of the residual ^{141}Sn. Here we adopt a negligibly small smoothing width 10^{-6} MeV in order to guarantee a sufficiently fine

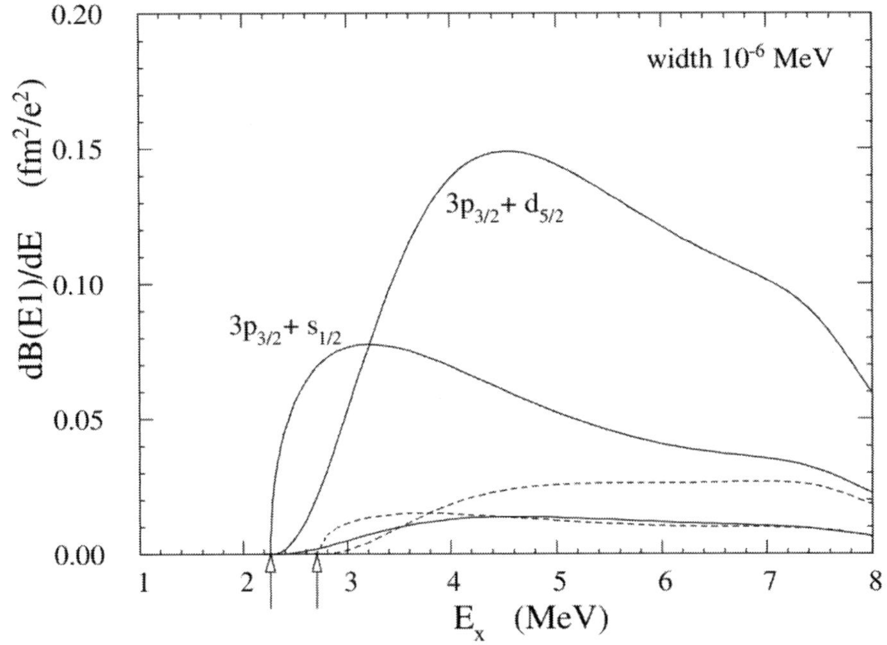

Figure 2. The partial E1 strength functions in ^{142}Sn for different channels of the direct one-neutron decay. The solid curves are for the decays populating the $3p_{3/2}$ one-quasineutron state in ^{141}Sn with one neutron emission in the partial waves $s_{1/2}$, $d_{3/2}$, and $d_{5/2}$. The dotted curves are for the decays to the $3p_{1/2}$ one-quasineutron state in ^{141}Sn and the neutron partial waves $s_{1/2}$ and $d_{3/2}$. The smoothing width is 10^{-6}fm.

energy resolution. At energies just above the one-neutron separation energy, only the channels populating the $3p_{3/2}$ configuration are open since the $3p_{3/2}$ configuration has lower energy than $3p_{1/2}$ (i.e. it is the ground state of ^{141}Sn). It is seen in Fig.2 that the partial strengths for decays with the s-wave neutron steeply rise at the threshold energy. This is consistent with the energy dependence $\propto (E - E_{th})^{l+1}$ which is expected from the low-energy behavior of the scattering neutron waves[6-8].

Using the partial E1 strength functions thus obtained and Eq.(3), we calculate the partial photo-absorption cross sections $\sigma(\gamma + A \rightarrow (A-1)_i + n_p)$. We then relate this to the direct neutron capture cross section $\sigma((A-1)_i + n_p \rightarrow \gamma + A)$ using the detailed balance:

$$\sigma((A-1)_i + n_p \rightarrow \gamma + A) = S_{J_A J_{A-1}} \frac{k_\gamma^2}{k_n^2} \sigma(\gamma + A \rightarrow (A-1)_i + n_p) \quad (6)$$

with k_γ and k_n being the photon and neutron momenta, and $S_{J_A J_{A-1}}$ is a spin factor. Provided that the ground state of ^{141}Sn is the one-quasineutron state $3p_{3/2}$, the direct neutron capture of ^{141}Sn followed by the E1 transition to the ground state of ^{142}Sn takes place for neutron partial waves $s_{1/2}$, $d_{3/2}$ and $d_{5/2}$. The calculated direct neutron capture cross sections are plotted in Fig.3 for an interval of neutron kinetic energy from $E_n = 0.01$ MeV (10 keV) to 6 MeV, which is relevant to the r-process nucleo-synthesis. Note that the typical temperature of the astronomical environment is order of $T \sim 10^9$ K, corresponding to $E_n \sim k_B T \sim 0.1$ MeV.

It is seen that the s-wave capture is dominant, and the d-wave capture cross section decreases as the energy becomes low. The energy dependence is consistent with $\propto E_n^{l-1/2}$ (the dashed

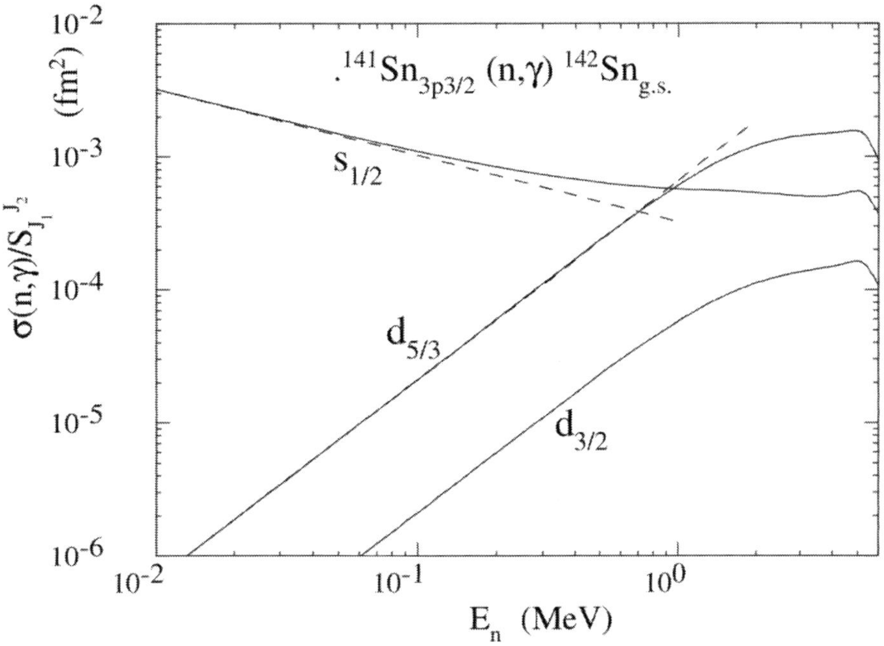

Figure 3. The calculated direct neutron capture cross sections of ^{141}Sn, plotted as a function of the captured neutron energy E_n. Plotted is the cross section divided by the spin factor $S_{J_A J_{A-1}}$ in Eq.(6).

lines), which comes from the low-energy behavior of the partial E1 strength $S_{E1} \propto E_n^{l+1/2}$ seen in Fig. 3. We found that the calculated cross sections are influenced by the RPA correlations. Such effects are taken into account via the induced fields $\delta\Gamma$ and $\delta\Delta$ in the selfconsistent field $\mathcal{V}_{scf}$ used in Eqs.(4) and (5). Consequently the calculated cross sections differ from the values evaluated in the simple single-particle models.

In conclusion, the density functional theory plus the continuum QRPA provides a new scheme to calculate the direct neutron capture cross section relevant to the r-process nucleo-synthesis.

This work is supported by the Grant-in-Aid for Scientific Research (No. 21340073, No. 23540294 and No. 24105008) from the Japan Society for the Promotion of Science.

References

[1] Burbidge E M, Burbidge G R, Fowler W A and Hoyle F 1957 *Rev. Mod. Phys.* **29** 547
[2] Arnould M, Goriely S and Takahashi K 2007 *Phys. Rep.* **450** 97
[3] Goriely S, Khan E and Shamyn M 2004 *Nucl. Phys.* A **739** 331
[4] Litvinova E *et al* 2009 *Nucl. Phys.* A **823** 26
[5] Mathews G J *et al* 1983 *Astrophys. J.* **270** 740
[6] Raman S *et al* 1985 *Phys. Rev.* C **32** 18
[7] Typel S and Bauer G 2005 *Nucl. Phys.* A **759** 247
[8] Nagarajan M A, Lenzi S and Vitturi A 2005 Eur. Phys. J. A **24** 63
[9] Matsuo M 2001 *Nucl. Phys.* A **696** 371
Matsuo M 2002 *Prog. Theor. Phys. Suppl.* **146** 110
[10] Serizawa Y and Matsuo M 2009 *Prog. Theor. Phys.* **121** 97
[11] Mizuyama K, Matsuo M and Serizawa Y 2009 *Phys. Rev.* C **79** 024313
[12] Zangwill A and Soven P 1980 *Phys. Rev.* A **21** 1561
[13] Nakatsukasa T and Yabana K 2001 *J. Chem. Phys.* **114** 2550

Journal of Physics: Conference Series **445** (2013) 012020 doi:10.1088/1742-6596/445/1/012020

High-lying Gamow-Teller excited states in the deformed nuclei, ^{76}Ge, ^{82}Se and N = 20 nuclei in the island of inversion by the Deformed QRPA (DQRPA)

Myung-Ki Cheoun and Eunja Ha

Department of Physics, Soongsil University, Seoul 156-743, Korea

E-mail: cheoun@ssu.ac.kr

Abstract. With the advent of high analysis technology in detecting the Gamow-Teller (GT) excited states beyond one nucleon emission threshold, the quenching of the GT strength to the Ikeda sum rule (ISR) seems to be recovered by the high-lying (HL) GT states. We address that these HL GT excited states result from the smearing of the Fermi surface by the increase of the chemical potential owing to the deformation within a framework of the deformed quasi-particle random phase approximation (DQRPA). Detailed mechanism leading to the smearing is discussed, and comparisons to the available experimental data on ^{76}Ge, ^{82}Se and N = 20 nuclei are shown to explain the strong peaks on the HL GT excited states.

1. Introduction

The deformation in nuclei becomes more important than last decades with the recent development of rare isotope (RI) accelerator facilities, from which one may perform lots of challenging experiments related to the RI nuclei. They are usually produced in the successive nuclear capture reactions in the cosmos, *i.e.* slow- and rapid- process, presumed to be occurred, respectively, at the initial and explosive stage of stellar evolution. Although they decay fractions of a second, their existence are imprinted on the nuclear abundances of stars [1].

On the other hand, recent experimental GT data on the high-lying (HL) states deduced by the more energetic projectiles shed a new light on the GT states above one nucleon threshold, whose contributions are thought to enable us to explain the quenching problem with the HL GT excited states through the multi-particle and multi-hole configuration mixing [2]. Of course, the contributions from the Δ excitation and the two-body current may contribute more or less to the quenching problem.

Conventional approach to understand the nuclear structure is based on the spherical symmetry. But, in order to describe neutron-rich nuclei and their relevant nuclear reactions occurred in the nuclear processes, one needs to develop theoretical formalism including explicitly the deformation [3, 4]. Ref. [3] exploited the Nilsson basis to the DQRPA. But the two-body interaction was derived from the effective separable force. Realistic two-body interaction is firstly considered at Ref. [4] only with neutron-neutron (nn) and proton-proton (pp) pairing correlations which have only isospin T = 1 and J = 0 interaction. But, to properly describe the deformed nuclei, the T= 0 and J = 1 pairing should be also taken into account because it may easily lead to the deformation by the J = 1 pairing through the $^3S_0$ tensor force. In this work, we extend our previous QRPA based on the spherical symmetry [5], which has been exploited as a useful framework for describing the nuclear reactions sensitive on the nuclear structure of medium-heavy and heavy nuclei [6].

2. Theoretical frameworks

Our DQRPA formalism is fully discussed at Ref. [7]. We start from a deformed Wood Saxon potential [8] and transform a physical state given by the diagonalization of total Hamiltonian in the Nilsson basis into the spherical basis, in which one may perform more easily theoretical calculations. In a cylindrical coordinate, eigenfunctions of a single particle state and its time-reversed state denoted as α and $\bar{\alpha}$ in the deformed Woods-Saxon potential are given as

$$|\alpha\rho_\alpha> = \sum_{Nn_z}[b^{(+)}_{Nn_z\Omega_\alpha}|N, n_z, \Lambda_\alpha, \Omega_\alpha = \Lambda_\alpha + 1/2 > + b^{(-)}_{Nn_z\Omega_\alpha}|N, n_z, \Lambda_\alpha + 1, \Omega_\alpha = \Lambda_\alpha + 1 - 1/2 >], \tag{1}$$

where $N = n_\perp + n_z$ ($n_\perp = 2n_\rho + \Lambda_\alpha$) is a major shell number, and n_z and n_ρ are numbers of nodes of the physical state on the deformed harmonic oscillator wave functions in z and ρ directions, respectively. $\Lambda_\alpha(\Omega_\alpha)$ is the projection of the orbital (total) angular momentum onto the nuclear symmetric axis z. The coefficients $b^{(+/-)}_{Nn_z\Omega_\alpha}$ are obtained by the eigenvalue equation of the total Hamiltonian in the Nilsson basis. The deformed wave function, $|Nn_z\Lambda_\alpha\Sigma\ (= \pm 1/2)>$ in Eq. (1) can be expanded in terms of the spherical harmonic oscillator wave function $|N_0 l\Lambda\Sigma>$

$$|Nn_z\Lambda_\alpha\ \Sigma> = \sum_{N_0,l} A^{N_0 l,\ n_r = \frac{N_0 - l}{2}}_{Nn_z\Lambda} \sum_j C^{j\Omega_\alpha}_{l\Lambda_\alpha\frac{1}{2}\Sigma}|N_0 lj\ \Omega_\alpha>, \tag{2}$$

with the Clebsch-Gordan coefficient $C^{j\Omega_\alpha}_{l\Lambda_\alpha\frac{1}{2}\Sigma}$ and the spatial overlap integral $A^{N_0 l}_{Nn_z\Lambda} =< N_0 l\Lambda|Nn_z\Lambda>$ numerically calculated in the spherical coordinate system.

The Deformed BCS equation is solved by using *ab initio* Brueckner G-matrix calculated from the realistic Bonn CD potential for the nucleon-nucleon interaction. The β^+ transition amplitudes from the ground state of an initial nucleus to the excited state are expressed by

$$< 1(K), m|\hat{\beta}^-_K|\ QRPA > = \sum_{\alpha\alpha''\rho_\alpha\beta\beta''\rho_\beta} \mathcal{N}_{\alpha\alpha''\rho_\alpha\beta\beta''\rho_\beta} \times \tag{3}$$
$$< \alpha\alpha''p\rho_\alpha|\sigma_K|\beta\beta''n\rho_\beta > [u_{p\alpha\alpha''}v_{n\beta\beta''}X_{\alpha\alpha''\beta\beta'',K} + v_{p\alpha\alpha''}u_{n\beta\beta''}Y_{\alpha\alpha''\beta\beta'',K}]\ ,$$

where $|\ QRPA>$ denotes the correlated QRPA ground state in the intrinsic frame with the nomalization factor $\mathcal{N}_{\alpha\alpha''\beta\beta''}(J)$. To compare to the experimental data, the GT($\pm$) strength functions

$$B^{\pm}_{GT}(m) = \sum_{K=0,\pm1} |< 1(K), m||\hat{\beta}^{\pm}_K||\ QRPA > |^2 \tag{4}$$

for ^{76}Ge and ^{82}Se are presented in this report. The single particle states are used up to $4\hbar\omega$ for two nuclei, in the spherical limit. Since the GT strength distribution turns out to rely on the deformation parameter β_2 [8], we exploited the deformation parameter, $|\beta_2| \leq 0.35$, as default values.

An interesting and important point to be noticed is the wide smearing of the Fermi surface in deformed nuclei, which makes the uv coefficients for the paring gaps small around the Fermi surface and enlarges the pairing strength to fit the empirical pairing gap. This tendency is explicitly revealed in the occupation probabilities of ^{82}Se and ^{76}Ge in Fig.1, which shows a wide smearing, *i.e.* significant change of the occupation probabilities of the particles around the Fermi surface by the increase of the deformation parameter β_2.

Journal of Physics: Conference Series **445** (2013) 012020 doi:10.1088/1742-6596/445/1/012020

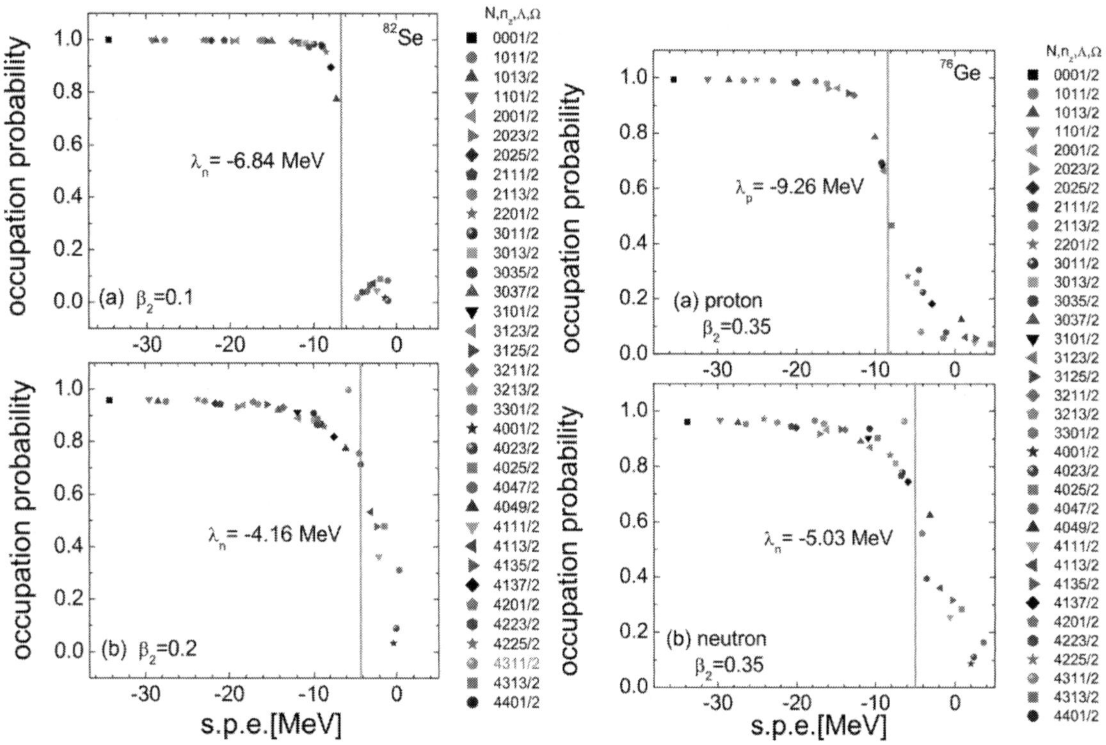

Figure 1. (Color online) Occupation probabilities of the nucleons in ^{82}Se and ^{76}Ge as a function of the single particle energy given by Nilsson basis for two deformation parameters β_2. In left panel, with the increase of β_2 from 0.1 to 0.2, the Fermi energy λ_n is increased to the wider smearing of the Fermi surface.

3. Results

By the β_2 deformation leading to the smearing, some states above the Fermi surface are reallocated below or around the surface, because the single particle state energies adopted from the deformed Woods Saxon potential depend on the parameter β_2 [7]. The deformation of nuclei may be conjectured to come from macroscopic phenomena, for example, the core polarization, the high spin states and so on. Microscopic reasons may be traced to the tensor force in the nucleon-nucleon interaction, which is known to account for the shell evolution according to the recent shell model calculations [9, 10]. Therefore, the deformation parameter adopted in this work may include implicitly and effectively such effects, because the single particle states from the deformed Wood Saxon potential show strong dependence on the β_2.

In Fig.2, we show the GT strength distributions on ^{82}Se and ^{76}Ge in terms of the β_2 parameter as a function of excited energy of parent nuclei, so that experimental data are presented by adding the empirical Q values from the measured data. The GT strength distributions are widely scattered due to deformation. With the redistribution of the GT strengths, the GT data around 12MeV are well reproduced at $\beta_2 = 0.2$ and 0.35, respectively. For a reference, the β_2 values from the relativistic mean field (RMF) are 0.133 and 0.157[12]. Ikeda sum rules are also satisfied within a few %. It means that these HL GT states are intimately associated with the smearing of some physical states around the Fermi surface in deformed nuclei. Similar results are for N=20 nuclei will be appeared elsewhere.

In summary, by taking the deformed axially symmetric Woods-Saxon potential, we performed

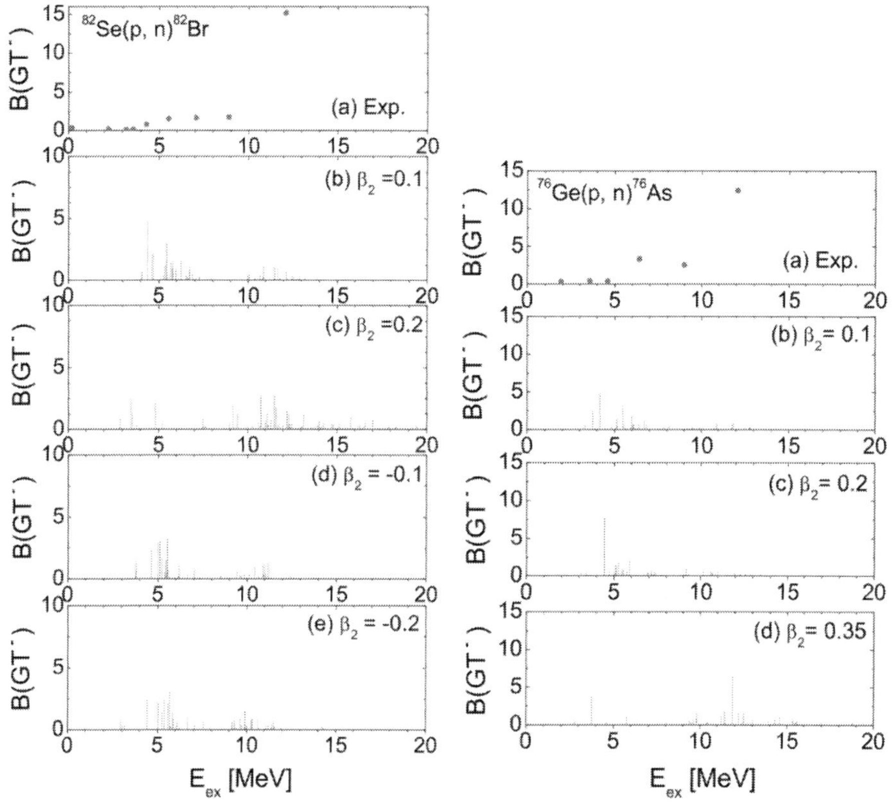

Figure 2. (Color online) GT strength distributions B(GT$^-$) on ^{82}Se and ^{76}Ge as a function of the excitation energy E_{ex} w.r.t. the ground state of ^{82}Se. Experimental data denoted as filled (red) points in uppermost panels, which show strong GT state peaks around 12 MeV, are deduced from the (p,n) reaction at 134.4MeV [11]. In each panel, we indicate β_2.

the deformed BCS and deformed QRPA with a realistic two-body interaction calculated by *ab initio* G-matrix based on Bonn potential. Results of the GT strength, B(GT), for ^{76}Ge and ^{82}Se show that the deformation effect leads to a fragmentation of the GT strength and reproduces the HL GT excitation deduced by higher energy projectiles, which turns out to result from the wide smearing by the increase of the Fermi surface energy due to the deformation. This work was supported by the National Research Foundation of Korea (2012R1A1A3009733, 2011-0015467).

References

[1] Hayakawa T, Iwamoto N, Shimizu T, Kajino T, Umeda H and Nomoto K 2004 *Phys. Rev. Lett.* **93** 161102
[2] Yako K *et al.* 2005 *Phys. Lett.* B **615** 193
[3] Simkovic F, Pacearescu L and Faessler A 2004 *Nucl. Phys.* A **733** 321
[4] Yousef M, Rodin V, Faessler A and Simkovic F 2009 *Phys. Rev.* C **79** 014314
[5] Cheoun M K, Bobyk A, Faessler A, Simcovic F and Teneva G 1993 *Nucl. Phys.* A **561** 74
[6] Cheoun M K, Ha E, Kim K S and Kajino T 2010 *J. Phys.* G **37** 055101
[7] Ha E and Cheoun M K 2012 *Preprint* arXiv:1205.4561 [nucl-th] and arXiv:1206.2156[nucl-th]
[8] Hamamoto I 2004 *Phys. Rev.* C **69** 041306(R); 2007 *Phys. Rev.* C **76** 054319
[9] Otsuka T *et al.* 2005 *Phys. Rev. Lett.* **95** 232502
[10] Otsuka T *et al.* 2010 *Phys. Rev. Lett.* **104** 012501
[11] Madey R *et al.* 1989 *Phys. Rev.* C **40** 540
[12] Lalazissis G A, Raman S and Ring P 1999 *At. Data and Nucl. Data tables* **71** 1

Journal of Physics: Conference Series **445** (2013) 012021 doi:10.1088/1742-6596/445/1/012021

Systematic investigation of $E1$ strength for the isotopes from $Z = 28$ to 50

Shuichiro Ebata[1,2], Takashi Nakatsukasa[2,3] and Tsunenori Inakura[2]

[1] Center for Nuclear Study, University of Tokyo, Wako-shi 351-0198, Japan

[2] RIKEN Nishina Center, Wako-shi 351-0198, Japan

[3] Center for Computational Sciences, University of Tsukuba, Tsukuba 305-8571, Japan

E-mail: ebata@cns.s.u-tokyo.ac.jp

Abstract. We carry out a systematic study of electric dipole mode ($E1$) for neutron-rich isotopes from nickel (Z=28) to tin (Z=50) using a time-dependent mean-field theory. Our time-dependent scheme is the canonical-basis time-dependent Hartree-Fock-Bogoliubov theory which can self-consistently describe nuclear dynamics with pairing correlation. We focus our discussion on the pygmy dipole resonance (PDR) and $E1$ polarizability (α_{D}). The correlation between neutron-skin thickness and PDR strongly depends on the neutron number, but the correlation between the skin thickness and α_{D} is much more stable.

1. Introduction

Recent progress in radioactive isotope facilities and experimental techniques allows us to generate and to study neutron-rich nuclei. In the neutron-rich region far from stability line, the neutron-skin and halo structures may appear in the ground state [1]. Due to their exotic structures in the ground state, new elementary modes are expected. The pygmy dipole resonance (PDR) is known as a candidate of such a characteristic excited state in the neutron-rich nuclei, which appears in low energy region, sometimes interpreted as a neutron-skin mode. Previous studies indicate that there is the linear relation between PDR and neutron-skin thickness [2, 3]. On the other hand, Ref.[4] indicates that the correlation between PDR and the skin is small from the covariance analysis, and that the electric dipole ($E1$) polarizability is much more correlated to the neutron skin than PDR [4]. Recently, the dipole polarizability has been measured for ^{206}Pb at RCNP, Osaka University [5].

In the present paper, the relation among neutron-skin, PDR and $E1$ polarizability, will be shown for over 350 isotopes, which is obtained by the linear response calculation using the time-dependent scheme.

2. Formulation

In order to study the dynamics of many nuclei including superfluid deformed nuclei, we proposed the canonical-basis time-dependent Hartree-Fock-Bogoliubov (Cb-TDHFB) theory. The Cb-TDHFB can be derived from the full TDHFB equation represented in the canonical basis and on the assumption that the pair potential can be approximated to be diagonal in the canonical basis [6].

Introducing the time-dependent canonical states $|\phi_k(t)\rangle$ and $|\phi_{\bar{k}}(t)\rangle$, we express the TDHFB state in the canonical (BCS) form as

$$|\Psi(t)\rangle = \prod_{k>0}\left\{u_k(t) + v_k(t)c_k^\dagger(t)c_{\bar{k}}^\dagger(t)\right\}|0\rangle, \tag{1}$$

where $u_k(t), v_k(t)$ are time-dependent BCS factors. It should be noted that the pair of states, k and $\bar{k}$, are not necessarily related to each other by the time-reversal, $|\phi_{\bar{k}}(t)\rangle \neq T|\phi_k(t)\rangle$. The diagonal approximation of the pair potential leads to the following equations:

$$i\frac{\partial}{\partial t}|\phi_k(t)\rangle = (h(t) - \eta_k(t))|\phi_k(t)\rangle, \qquad i\frac{\partial}{\partial t}|\phi_{\bar{k}}(t)\rangle = (h(t) - \eta_{\bar{k}}(t))|\phi_{\bar{k}}(t)\rangle, \tag{2a}$$

$$i\frac{d}{dt}\rho_k(t) = \kappa_k(t)\Delta_k^*(t) - \kappa_k^*(t)\Delta_k(t), \tag{2b}$$

$$i\frac{d}{dt}\kappa_k(t) = (\eta_k(t) + \eta_{\bar{k}}(t))\,\kappa_k(t) + \Delta_k(t)\,(2\rho_k(t) - 1)\,. \tag{2c}$$

These basic equations determine the time evolution of the canonical states, $|\phi_k(t)\rangle$ and $|\phi_{\bar{k}}(t)\rangle$, their occupation $\rho_k(t) = |v_k(t)|^2$, and pair probabilities $\kappa_k(t) = u_k(t)v_k(t)$. $\Delta_k(t)$ is the gap energy,

$$\Delta_k(t) = \sum_{l>0} Gf(\varepsilon_k^0)f(\varepsilon_l^0)\kappa_l(t), \tag{3}$$

where G is a the pairing strength determined by the smoothed pairing method [7], and $f(\varepsilon)$ is a cutoff function which defines the number of canonical basis [6]. The cutoff function is time-independent with the single-particle energies ε_k^0 at the ground state.

We solve the Cb-TDHFB equations in real time and calculate the linear response of the nucleus. The numerical procedure of linear response is same as that of Ref.[6], we add an external field $\hat{V}_{\text{ext}}(\boldsymbol{r}, t) = -k\hat{F}_{E1}\delta(t)$ to the ground state of the nucleus, where $\hat{F}_{E1} = (N/A)\sum_p \hat{r}_p - (Z/A)\sum_n \hat{r}_n$, here $r = (x, y, z)$ and k is an arbitrary small parameter. Then, we compute the time-evolution of the nuclear density using the time-dependent scheme. We obtain the strength function $S(E1; E)$ through the Fourier transformation of the time-dependent expectation value of $\hat{F}_{E1}$[6]. To quantify the PDR, we use the ratio:

$$\frac{m_1(E_c)}{m_1} \equiv \frac{\int^{E_c} E \times S(E1; E)dE}{\int^{E_T} E \times S(E1; E)dE}, \tag{4}$$

where we adopt the cutoff energy $E_c = 10$ MeV and the total energy $E_T = 100$ MeV in the present calculation. The $E1$ polarizability α_D is defined using $E1$ strength function as,

$$\alpha_D \equiv 2\int^{E_T} \frac{S(E1; E)}{E} dE. \tag{5}$$

In this work, we use the three-dimensional Cartesian coordinate for the canonical states, $\phi_k(\boldsymbol{r}, \sigma; t) = \langle \boldsymbol{r}, \sigma|\phi_k(t)\rangle$ with $\sigma = \pm 1/2$. The coordinate space is discretized in the square mesh of $\Delta x = \Delta y = \Delta z = 1.0$ fm in a sphere with radius of 15 fm.

3. Results

Figure 1 shows the PDR ratio as a function of neutron-skin thickness which is defined by the difference between root mean square radii of neutrons and protons, from Ge to Sn isotopes. In both panels, open circles and triangles indicate the isotopes with N=50 and 82, respectively. We can see the linear relation between the PDR ratio and the neutron-skin thickness for each isotopic chain with $N=50 \rightarrow 58$. The slope of the linear correlation becomes small as the proton number approaches Z=50. The slope for Sn isotope from N=50 to 58 is about a half of that in Ge isotope. The PDR ratio strength jumps up again over N=82. The correlation between PDR and N-skin strongly depends on the neutron number.

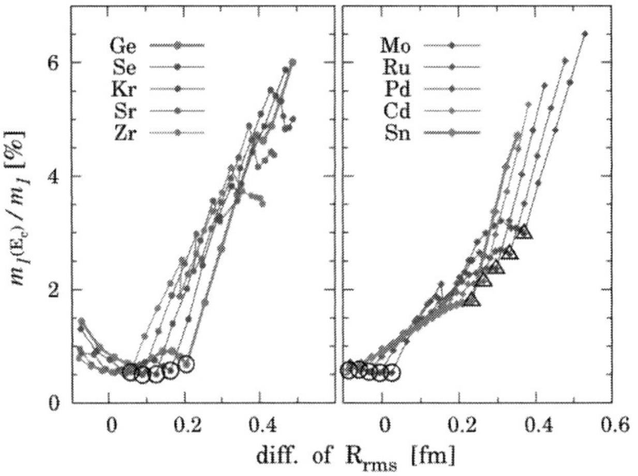

Figure 1. PDR ratio defined by Eq.(4), as a function of N-skin thickness from Ge to Sn isotopes. Open circle and triangle symbols indicate the isotope with N=50 and 82, respectively.

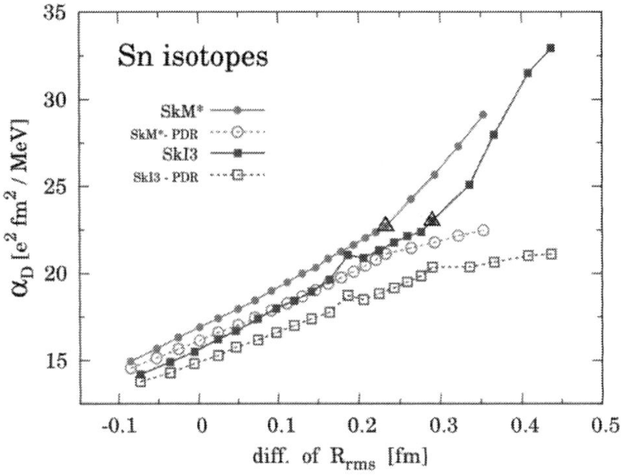

Figure 2. $E1$ polarizability α_D defined by Eq.(5), as a function of neutron-skin thickness for Sn isotopes with $N = 50 - 90$. Circles (squares) with solid lines indicate results with SkM* (SkI3) parameter set. Open symbols with dashed lines are defined by Eq.(6)

Figure 2 shows the $E1$ polarizability α_D as a function of N-skin thickness for Sn isotopes. Circles (squares) are obtained with the SkM* (SkI3) Skyrme energy functional. Open symbols in Fig.2 are computed from the $E1$ strength excluding the PDR part;

$$\tilde{\alpha}_D \equiv 2 \int_{E_c}^{E_T} \frac{S(E1;E)}{E} dE. \tag{6}$$

This can be regarded as the dipole polarizability associated with the giant dipole resonance. The SkI3 parameters are made so as to reproduce the ordering of single-particle levels in Pb isotopes obtained by the relativistic mean-field model. Especially, the part of the spin-orbit force is different from that of SkM*. The polarizability $\tilde{\alpha}_D$ is strongly correlated with the skin-thickness, although the contribution of the PDR becomes large in both results over N=82. In contrast to the relation between PDR and the skin, the correlation between α_D and the skin thickness is more stable with respect to change of the neutron number and the energy functional.

In summary, we found the linear correlation between the PDR and the neutron-skin thickness, however, the correlation rather strongly depends on the neutron number. The correlation between α_D and the neutron-skin thickness is more robust, especially if we remove the contribution from the PDR. This may suggest a better way to constrain the neutron-skin thickness by the observation of $E1$ strength function.

Acknowledgments

This work is supported by HPCI Strategic Program Field 5 " The origin of matter and the universe" . The computational resources were provided by the RIKEN Integrated Cluster of Clusters (RICC), by the SR16000 at YITP in Kyoto University, and by the Joint Research Program at Center for Computational Sciences, University of Tsukuba.

References

[1] Tanihata I *et al.* 1985 *Phys. Rev. Lett.* **55** 2676
[2] Piekarewicz J 2006 *Phys. Rev.* C**73** 044325
[3] Inakura T Nakatsukasa T and Yabana K 2011 *Phys. Rev.* C**84** 021302
[4] Reinhard P-G and Nazarewicz W 2010 *Phys. Rev.* C**81** 051303(R)
[5] Tamii A *et al.* 2011 *Phys. Rev. Lett.* **107** 062502
[6] Ebata S *et al.* 2010 *Phys. Rev.* C**82** 034306
[7] Strutinsky V M 1967 *Nucl. Phys.* A**95** 420

International Symposium on Exotic Nuclear Structure From Nucleons (ENSFN 2012) IOP Publishing
Journal of Physics: Conference Series **445** (2013) 012022 doi:10.1088/1742-6596/445/1/012022

Neutrinos and rare isotopes

A B Balantekin

Physics Department, University of Wisconsin, Madison WI 53706 USA

E-mail: baha@physics.wisc.edu

Abstract. The close connection between neutrino physics and the physics explored at rare isotope facilities is explored. The duality between the Hamiltonian describing the self-interacting neutrino gas near the proto-neutron star in a core-collapse supernova and the BCS theory of pairing is elucidated. This many neutrino system is unique as it is the only many-body system driven by weak interactions. Its symmetries are discussed.

As you read the title of this contribution to the proceedings of the conference celebrating contributions of Taka Otsuka to nuclear physics on the occasion of his 60th birthday, you may wonder what neutrinos have to do with exotic nuclei. The answer follows from an examination of the scientific motivations of the physics enterprise with rare isotopes. Exploring physics with rare isotopes would impact many areas of inquiry. It would contribute to our understanding of nuclear structure by exploring the limits of nuclear existence, new shapes and new collective behaviors. Studies of nuclear astrophysics with rare isotope facilities encompass a broad range of subjects including searches for the origin of elements, investigating rare isotopes present in explosive phenomena in astrophysical settings, and understanding neutron star crusts. Rare isotopes can be viewed as laboratories to study fundamental symmetries where symmetry violations are sometimes amplified. Last, but not the least, research at rare isotope facilities would lead to societal applications. Neutrinos connect all these venues, forming an intellectual bridge between them.

One topic where the connection between rare isotope physics and neutrino physics is manifest is the nucleosynthesis of various elements. A complete understanding of element nucleosynthesis requires input from both areas. Not only the knowledge of neutrino properties such as masses and mixing angles, but also the spin-isospin response of a broad range of nuclei from stable isotopes to rare ions are the crucial components of a successful description of the element nucleosynthesis. This synergy is perhaps most obvious for the ν-process, the nucleosynthesis via neutrino-induced nucleon emission [1]. The most quoted example of the ν-process nucleosynthesis is the production of ^{19}F by neutrino capture on ^{20}Ne in the outer shells of the supernovae, which accounts for the entire observed abundance of ^{19}F. Of course the ν-process in more broadly operational. For example, recent work nicely ties the ν-process nucleosynthesis yields of ^{11}B and ^{7}Li in the He shells of supernovae to the neutrino properties [2]. Matter-enhanced neutrino oscillations governed by the mixing between the first and the third neutrino generations is operative at matter densities that exist in those outer shells of a supernova. This boosts the ν-process nucleosynthesis yields of ^{11}B and ^{7}Li for the inverted neutrino mass hierarchy, but not for the normal one.

The cosmic site of the r-process nucleosynthesis, driven by a succession of rapid neutron captures on heavy seed nuclei, is still unknown [3], but the neutrino-driven wind in the core-collapse supernovae is among the possible sites. Recent hydrodynamical simulations of the neutrino-driven wind do not seem to always result in the necessary extreme conditions [4]. However, since our theoretical understanding of core-collapse supernovae is still evolving, it is not possible to rule out the neutrino-driven wind as a possible site. Numerical modeling of core-collapse supernovae has made significant recent progress, in particular two- and three-dimensional models were successfully implemented unveiling a complex interplay between neutrino physics, thermonuclear reactions and turbulence [5, 6, 7]. Core-collapse supernovae are neutrino-dominated dynamical systems [8] and one crucial ingredient to understand the r-process nucleosynthesis they may host is a better knowledge of neutrino physics. A second crucial input is a better working understanding of the spin-isospin response of a broad range of nuclei.

There is another, more formal, connection between neutrino physics and the nuclear many-body problem. The sheer number of neutrinos ($\sim 10^{58}$) emitted from the cooling proto-neutron star following the core collapse necessitates inclusion of neutrino-neutrino interactions in the description of the neutrino transport in supernovae. Unlike the one-body Hamiltonian of the matter-enhanced neutrino oscillations where neutrinos interact with a mean-field (generated by the background particles other than neutrinos), the Hamiltonian describing the many-neutrino gas in a core-collapse supernova contains both one- and two-body terms, making it technically much more challenging. Inclusion of the neutrino-neutrino interaction terms leads to very interesting collective effects [9, 10]. In contrast to the many-body systems studied in condensed-matter physics where the systems are primarily driven by electromagnetic interactions, and in nuclear physics, where the systems are primarily driven by strong interactions, the many-neutrino system in the core-collapse supernovae is the only many-body system in nature driven by the weak interactions. In addition, the latter system contains many more particles than the other two. Partially motivated by these features many-body aspects of the supernova neutrinos recently started to receive more attention [11, 12, 13].

There is a further analogy between supernova neutrino many-body system and the pairing problem in nuclear physics. To demonstrate this analogy let us consider only two flavors of neutrinos: electron neutrino, ν_e, and another flavor, ν_x which mixes with the electron neutrinos. For simplicity, we will also ignore the antineutrinos, which are of course present in supernovae[1]. Introducing the creation and annihilation operators for a neutrino with three momentum $\mathbf{p}$, we can write down the generators of the neutrino flavor isospin algebras [11]:

$$\begin{aligned} J_+(\mathbf{p}) &= a_x^\dagger(\mathbf{p})a_e(\mathbf{p}), \quad J_-(\mathbf{p}) = a_e^\dagger(\mathbf{p})a_x(\mathbf{p}), \\ J_0(\mathbf{p}) &= \frac{1}{2}\left(a_x^\dagger(\mathbf{p})a_x(\mathbf{p}) - a_e^\dagger(\mathbf{p})a_e(\mathbf{p})\right). \end{aligned} \tag{1}$$

The integrals of these operators over all possible values of momenta generate the global flavor isospin algebra. Using the operators in Eq. (1) the Hamiltonian for a neutrino propagating through matter takes the form

$$H = H_\nu + H_{\nu\nu} = \left(\sum_p \frac{\delta m^2}{2p}\hat{B}\cdot\vec{J}_p - \sqrt{2}G_F N_e J_p^0\right) + \frac{\sqrt{2}G_F}{V}\sum_{\mathbf{p},\mathbf{q}}(1-\cos\vartheta_{\mathbf{pq}})\,\vec{J}_{\mathbf{p}}\cdot\vec{J}_{\mathbf{q}} \tag{2}$$

where the auxiliary vector quantity $\hat{B}$ is given by

$$\hat{B} = (\sin 2\theta, 0, -\cos 2\theta), \tag{3}$$

N_e is the background electron density and δm^2 is the difference between squares of the masses associated with mass eigenstates. In the above equations θ is the mixing angle between electron

[1] Inclusion of antineutrinos requires a second set of SU(2) algebras. For three flavors two sets of SU(3) algebras are needed.

neutrino and the other neutrino flavor and $\vartheta_{\mathbf{pq}}$ is the angle between neutrino momenta $\mathbf{p}$ and $\mathbf{q}$. In writing this equation a term proportional to identity is omitted as such terms do not contribute to the neutrino oscillations. Note that the presence of the $(1 - \cos\vartheta_{\mathbf{pq}})$ term in the integral above is crucial to recover the effects of the Standard Model weak interaction physics in the most general situation. In the idealized case of an isotropic neutrino distribution and a very large number of neutrinos, this term may average to a constant and the neutrino-neutrino interaction term in the Hamiltonian simply reduces to the Casimir operator of the global SU(2) algebra. If the background electron neutrino density is negligible (a good approximation near the proto-neutron star where neutrino-neutrino interactions are dominant) and a single angle dominates the second term, the Hamiltonian of Eq. (2) takes the form

$$H = \sum_p \frac{\delta m^2}{2p} \hat{B} \cdot \vec{J}_p + \frac{\sqrt{2} G_F}{V} \vec{J} \cdot \vec{J}, \tag{4}$$

where $\vec{J} = \sum_p \vec{J}_p$. The Hamiltonian in Eq. (4) is mathematically very similar to the BCS Hamiltonian in the quasi-spin basis[2]

$$\hat{H}_{\mathrm{BCS}} = \sum_k 2\epsilon_k \hat{t}^0_k - |G| \hat{T}^+ \hat{T}^-, \tag{5}$$

where the quasi-spin operators also form an SU(2) algebra:

$$[\hat{t}^+_k, \hat{t}^-_l] = 2\delta_{kl} \hat{t}^0_k \ , \qquad [\hat{t}^0_k, \hat{t}^\pm_l] = \pm \delta_{kl} \hat{t}^\pm_k \tag{6}$$

with $T^+ = \sum_i \hat{t}^+_i$ and so on. An exact solution of this problem was given by Richardson some time ago [14]. This solution was later generalized by Gaudin [15]. A similar solution also exists for the neutrino Hamiltonian of Eq. (4) [12]. It is manifest that the Hamiltonian of Eq. (4) possesses an $SU(N)_f$ rotation symmetry in the neutrino flavor space with N flavors [11, 17, 18]. Since the BCS Hamiltonian is shown to be integrable, there must be constants of motion associated with it [16]. One can write down the constants of motion of the collective neutrino Hamiltonian in Eq. (4) as [12]

$$\hat{h}_p = \hat{B} \cdot \vec{J}_p + 2 \sum_{q(\neq p)} \frac{\vec{J}_p \cdot \vec{J}_q}{\omega_p - \omega_q}. \tag{7}$$

where we defined $\omega_p = \frac{1}{\mu} \frac{\delta m^2}{2p}$ with $\mu = \frac{\sqrt{2} G_F}{V}$. The Hamiltonian of Eq. (4) itself can be written to include a linear combination of these invariants:

$$H = \sum_p w_p \hat{h}_p + \sum_p \vec{J}_p \cdot \vec{J}_p \tag{8}$$

It was shown that existence of such constants of motion could lead to collective neutrino oscillations [19].

To solve the full Hamiltonian of Eq. (2) for the large number of neutrinos present in the supernovae is a numerically very challenging, if not almost impossible, task. The full Hamiltonian is usually simplified using an RPA-like linearization procedure by writing the two-body terms

[2] Note that neutrino Hamiltonian in the single-angle approximation is *not* identical to the BCS Hamiltonian as the sign of the two-body term is different. Also the Hamiltonian in Eq. (4) does not include the antineutrino terms.

as a one-body operator times its expectation value in a suitably chosen basis. For example applying this procedure to the Hamiltonian of Ref. (4) one gets

$$H \to \hat{H}^{\text{RPA}} = \sum_p \omega_p \hat{B} \cdot \vec{J}_p + \vec{P} \cdot \vec{J} \tag{9}$$

where the polarization vector $\vec{P}$ ensues from the linearization of the flavor isospin operators:

$$\vec{P}_{\mathbf{p},s} = 2\langle \vec{J}_{\mathbf{p},s} \rangle. \tag{10}$$

Using the SU(2) coherent states associated with the flavor isospin to calculate the operator averages in the above equations yields a reduced collective neutrino Hamiltonian, widely used in the literature [11]. This introduces an approximation to the description of the neutrino gas, however abandoning this approximation seems to be unlikely in the near future for practical reasons.

Collective neutrino oscillations produce an interesting effect, called spectral swappings or splits, on the final neutrino energy spectra: at a particular energy these spectra are almost completely divided into parts of different flavors [20, 21]. The nature of the such swaps can be elucidated by referring to the Hamiltonian in Eq. (9). As in the BCS theory, this Hamiltonian does not conserve particle number. Particle number conservation can be enforced by introducing a Lagrange multiplier:

$$\hat{H}^{\text{RPA}} \to \hat{H}^{\text{RPA}} + \omega_c \hat{J}^0. \tag{11}$$

Diagonalization of this Hamiltonian using an appropriately chosen Bogoliubov transformation gives rise to spectral splits or swaps in the neutrino energy spectra with the Lagrange multiplier playing the role of the swap frequency [12].

It should be emphasized that the collective neutrino oscillations could significantly impact r-process nucleosynthesis yields [22, 23]. To understand the r-process is clearly one of the drivers of the experiments with rare ion beams. This physics program requires measuring the beta-decay rates of nuclei both at and far from stability, in particular half-lifes at the r-process ladders as well as the initial and final state energies. A complete knowledge of the spin-isospin response of a broad range of nuclei to a various probes is crucial for not only for the r-process nucleosynthesis, but also for many other applications. Recently much progress was made in understanding the nuclear spin-isospin response, both experimentally and theoretically. On the experimental side the matrix elements of the Gamow-Teller operator $\vec{\sigma}\,\vec{\tau}$ between the initial and final states were successfully measured using inverse kinematics [24]. A major theoretical development is the proper inclusion of the tensor force in the shell model Hamiltonians. Monopole component of the nucleon-nucleon force is the same in nuclear medium and free space, however the monopole effect of the tensor force alters the shell structure in a significant way [25]. This is because the monopole component of the tensor interaction changes depending on whether the nucleon spin is parallel or antiparallel to its orbital angular momentum. In most cases the monopole component is an average over all possible spin orientations, so the tensor component does not contribute for the filled orbits. However, near the Fermi surfaces where the spin-orbit force splits the orbits, the $j = \ell + s$ orbit fills first altering the mean field. Indeed residual effective force between the valence nucleons, beyond that represented by the mean field, is very well described by the tensor force [26]. A new p-sd shell model Hamiltonian including up to 2-3 $\hbar\Omega$ excitations can describe the magnetic moments and Gamow-Teller (GT) transitions in p-shell nuclei well with a small quenching for the spin g-factor and the axial-vector coupling constant [27]. These new developments lead to a description of the spin properties of such nuclei better than the conventional shell model Hamiltonians resulting in a better description of the weak interactions for astrophysical applications. For example, this new Hamiltonian significantly improves the description of the cross sections for the reactions ν_e+^{12}C [28] and ν_e+^{13}C [29], potentially important for scintillator-based neutrino experiments. These new Hamiltonians can successfully describe not only the Gamow-Teller, but also the first-forbidden transitions. Inclusion of the latter terms can significantly change the lifetimes [30]. These recent advances in the shell model studies of stable and exotic rare nuclei as well as their use in the description of the spin-dependent nuclear weak processes are reviewed in references [32] and [31].

Neutrinos form the bridge between many astrophysical phenomena and laboratory nuclear physics, investigating stable as well as unstable, exotic, and rare elements. Exploiting this connection in an intellectually beneficial way necessitates a multitude of experimental and theoretical efforts. On the theoretical side, these include a better understanding of neutrino properties and improving our knowledge of nuclear structure to calculate neutrino interactions with nuclei more accurately. On the experimental side, these include measurements of the spin-isospin response of both stable and exotic nuclei to various probes as well as measuring salient properties of such nuclei. On the observational side, these include better determinations of cosmic abundances. One of the prizes of this quest is understanding the origin of the elements. At least to some of us, another prize will be the full appreciation of the considerable role neutrinos play in the cosmos.

Acknowledgments

I was lucky to be associated with Taka Otsuka during the span of both of our careers and I am grateful to him for many physics discussions over many years. I also would like to thank T. Kajino, Y. Pehlivan and T. Suzuki with whom most of the work presented here was carried out. This work was supported in part by the U.S. National Science Foundation Grant No. PHY-1205024, in part by the University of Wisconsin Research Committee with funds granted by the Wisconsin Alumni Research Foundation and in part through JUSTIPEN (Japan-U.S. Theory Institute for Physics with Exotic Nuclei) under grant number DEFG02- 06ER41407 (U. Tennessee).

References

[1] Woosley S E, Hartmann D H, Hoffman R D and Haxton W C 1990 *Astrophys. J.* **356** 272

[2] Yoshida T, Kajino T and Hartmann D H 2005 *Phys. Rev. Lett.* **94** 231101 (*Preprint* astro-ph/0505043); Mathews G J, Kajino T, Aok Wi, Fujiya W and Pitts J B 2012 *Phys. Rev* D **85** 105023 (*Preprint* arXiv:1108.0725 [astro-ph.HE])

[3] Qian Y Z and Wasserburg G J 2007 *Phys. Rept.* **442** 237 (*Preprint*: arXiv:0708.1767 [astro-ph])

[4] Roberts L F, Woosley S E and Hoffman R D 2010 *Astrophys. J.* **722** 954; Arcones A and Thielemann F K 2013 *J. Phys.* G **40** 013201 (*Preprint* arXiv:1207.2527 [astro-ph.SR])

[5] Lentz E J, Mezzacappa A, Bronson Messer O E, Hix W R and Bruenn S W 2012 *Astrophys. J.* **760** 94 (*Preprint* arXiv:1206.1086 [astro-ph.SR])

[6] Muller B, Janka H-T and Marek A 2012 *Astrophys. J.* **756** 84 (*Preprint* arXiv:1202.0815 [astro-ph.SR])

[7] Brandt T D, Burrows A, Ott C D and Livne E 2011 *Astrophys. J.* **728** 8 (*Preprint* arXiv:1009.4654 [astro-ph.HE])

[8] Balantekin A B and Fuller G M 2013 *Preprint* arXiv:1303.3874 [nucl-th]

[9] Duan H, Fuller G M and Qian Y-Z 2010 *Ann. Rev. Nucl. Part. Sci.* **60** 569 (*Preprint* arXiv:1001.2799 [hep-ph])

[10] Raffelt G 2012 *Preprint* arXiv:1201.1637 [astro-ph.SR]

[11] Balantekin A B and Pehlivan Y 2007 *J. Phys.* G **34** 47 (*Preprint* astro-ph/0607527)

[12] Pehlivan Y, Balantekin A B, Kajino T and Yoshida T 2011 *Phys. Rev.* D **84** 065008 (*Preprint* arXiv:1105.1182 [astro-ph.CO])

[13] Volpe C, Vaananen D and Espinoza C 2013 *Preprint* arXiv:1302.2374 [hep-ph]

[14] Richardson R W 1967 *Phys. Rev.* **159** 792

[15] Gaudin M 1976 *J. Physique* **37** 1087

[16] Yuzbashyan A A, Altshuler B L, Kuznetsov V B and Enolskii V E 2005 *J. Phys. A: Math. Gen.* **38** 7831

[17] Duan H, Fuller G M and Qian Y Z 2009 *J. Phys.* G **36** 105003 (*Preprint* arXiv:0808.2046 [astro-ph])

[18] Balantekin A B 2010 *Nucl. Phys.* A **844**, 14c (*Preprint* arXiv:0910.1814 [nucl-th])

[19] Raffelt G G 2011 *Phys. Rev.* D **83** 105022 (*Preprint* arXiv:1103.2891 [hep-ph])
[20] Raffelt G G and Smirnov A Y 2007 *Phys. Rev* D **76** 081301 [Erratum-ibid. 2008 **77** 029903] (*Preprint* arXiv:0705.1830 [hep-ph])
[21] Duan H, Fuller G M and Qian Y-Z 2008 *Phys. Rev.* D **77** 085016 (*Preprint* arXiv:0801.1363 [hep-ph])
[22] Balantekin A B and Yuksel H 2005 *New J. Phys.* **7** 51 (*Preprint* astro-ph/0411159)
[23] Duan H, Friedland A, McLaughlin G C and Surman R 2011 *J. Phys.* G **38** 035201 (*Preprint* arXiv:1012.0532 [astro-ph.SR]); Chakraborty S, Choubey S, Goswami S and Kar K 2010 *J. Cosmol. Astropart. Phys.* JCAP06(2010)007 (*Preprint* arXiv:0911.1218 [hep-ph])
[24] Thies J H *et al.* 2012 *Phys. Rev.* C **86** 054323; Thies J H *et al.* 2012 *Phys. Rev.* C **86** 044309
[25] Otsuka T, Suzuki T, Fujimoto R, Grawe H and Akaishi Y 2005 *Phys. Rev. Lett.* **95** 232502
[26] Otsuka T, Suzuki T, Honma M, Utsuno Y, Tsunoda N, Tsukiyama K and Hjorth-Jensen M 2010 *Phys. Rev. Lett.* **104** 012501 (*Preprint* arXiv:0910.0132 [nucl-th])
[27] Suzuki T, Fujimoto R and Otsuka T 2003 *Phys. Rev.* C **67** 044302
[28] Suzuki T, Chiba S, Yoshida T, Kajino T and Otsuka T 2006 *Phys. Rev.* C **74** 034307 (*Preprint* nucl-th/0608056)
[29] Suzuki T, Balantekin A B and Kajino T 2012 *Phys. Rev.* C **86** 015502 (*Preprint* arXiv:1204.4231 [nucl-th])
[30] Suzuki T, Yoshida T, Kajino T and Otsuka T 2012 *Phys. Rev.* C **85** 015802 (*Preprint* arXiv:1110.3886 [nucl-th])
[31] Suzuki T, Honma M, Yoshida T, Mao H, Kajino T and Otsuka T 2011 *Prog. Part. Nucl. Phys.* **66** 385
[32] Suzuki T, Chiba S, Yoshida T, Higashiyama K, Honma M, Kajino T and Otsuka T 2007 *Prog. Part. Nucl. Phys.* **59** 486

Updates of the nuclear equation of state for core-collapse supernovae and neutron stars: effects of 3-body forces, QCD, and magnetic fields

G J Mathews[1], M Meixner[1], J P Olson[1], I-S Suh[1], T Kajino[2], T Maruyama[3], J Hidaka[2], C-Y Ryu[4], M-K Cheoun[5] and N Q Lan[6]

[1]Department of Physics/Center for Astrophysics, University of Notre Dame, Notre Dame, United States
[2]Theory Division, National Astronomical Observatory of Japan, Mitaka, Tokyo, Japan
[3]College of Bioresource Sciences, Nihon University, Fujisawa 252-8510, Japan
[4]General Education Curriculum Center, Hanyang University, Seoul 133-791, Korea
[5]Department of Physics, Soongsil University, Seoul, South Korea
[6]Hanoi National University of Education, 136 Xuan Thuy, Hanoi, Vietnam

E-mail: `gmathews@nd.edu`

Abstract. We summarize several new developments in the nuclear equation of state for supernova simulations and neutron stars. We discuss an updated and improved Notre-Dame-Livermore Equation of State (NDL EoS) for use in supernovae simulations. This Eos contains many updates. Among them are the effects of 3- body nuclear forces at high densities and the possible transition to a QCD chiral and/or super-conducting color phase at densities. We also consider the neutron star equation of state and neutrino transport in the presence of strong magnetic fields. We study a new quantum hadrodynamic (QHD) equation of state for neutron stars (with and without hyperons) in the presence of strong magnetic fields. The parameters are constrained by deduced masses and radii. The calculated adiabatic index for these magnetized neutron stars exhibit rapid changes with density. This may provide a mechanism for star-quakes and flares in magnetars. We also investigate the strong magnetic field effects on the moments of inertia and spin down of neutron stars. The change of the moment of inertia associated with emitted magnetic flares is shown to match well with observed glitches in some magnetars. We also discuss a perturbative calculation of neutrino scattering and absorption in hot and dense hyperonic neutron-star matter in the presence of a strong magnetic field. The absorption cross-sections show a remarkable angular dependence in that the neutrino absorption strength is reduced in a direction parallel to the magnetic field and enhanced in the opposite direction. The pulsar kick velocities associated with this asymmetry comparable to observed pulsar velocities and may affect the early spin down rate of proto-neutron star magnetars with a toroidal field configuration.

1. Introduction

To describe the structure and hydrodynamics of compact matter; an equation of state (EoS) is needed to relate the physics of the state variables [1]. In supernovae the EoS determines the dynamics of the collapse and the outgoing shock, and determines whether the remnant ends up as a neutron star or a black hole. In a neutron star, it determines the mass-radius relationship, stellar composition, cool-down time and dynamics of neutron star spin down and mergers. In this paper we summarize some progress in the development of Equations of state for supernova and neutron star simulations. In particular we highlight the role of 3-body forces, QCD, magnetic fields and neutrino transport.

2. Three body forces and the EoS

At present, only a few hadronic EoSs are commonly employed that cover large enough ranges in density, temperature and electron fraction to be of use in core-collapse supernova simulations. The two most employed in astrophysical simulations are the EoS of Lattimer & Swesty (LS91) [2] and that of H. Shen et. al. (Shen98) [3, 4]. The former utilizes a non relativistic parameterization of nuclear interactions in which nuclei are treated as a compressible liquid drop including surface effects. The latter is based upon a Relativistic Mean Field (RMF) model using the TM1 parameter set in which nuclei are calculated in a Thomas-Fermi approximation. Baryonic matter was parameterized with a new RMF model that treated nuclei and non-uniform matter with the statistical model of Hempel et. al. [5].

Here, we discuss a new Notre Dame-Livermore (NDL) EoS [6]. This EoS evolves from the original Livermore formulation [7, 8], but unlike the previous version this NDL EoS, is consistent with known experimental nuclear matter constraints and recent [14] mass and radii measurements of neutron stars. Below nuclear matter density, the conditions for nuclear statistical equilibrium (NSE) are achieved at a temperature of $T \approx 0.5$ MeV. Below this temperature the nuclear matter is a approximated by a nine element reaction network which must be evolved dynamically. Above this temperature, the nuclear constituents are represented by free nucleons, alphas and a single "representative" heavy nucleus. Among the new features in the NDL EoS is that high density phase of the EoS is treated with a parameterized Skyrme energy density functional that utilizes a modified zero range 3-body interaction. The effects of pions on the state variables at high densities is also included as well as the consequences of a phase transition to a QGP.

Above nuclear matter saturation density we include both 2-body ($v_{ij}^{(2)}$) and 3-body ($v_{ijk}^{(3)}$) interactions in the many-nucleon system. The Skyrme two-body potential is given in the standard form [9].

$$v_{12}^{(2)} = t_0\left(1 + x_0\hat{P}_s\right)\delta\left(\mathbf{r_1} - \mathbf{r_2}\right) + \tfrac{1}{2}t_1\left(\delta\left(\mathbf{r_1} - \mathbf{r_2}\right)\hat{k}^2 + \hat{k}'^2\delta\left(\mathbf{r_1} - \mathbf{r_2}\right)\right) \quad (1)$$

$$+t_2\hat{\mathbf{k}}^{\mathbf{2}}\cdot\delta\left(\mathbf{r_1} - \mathbf{r_2}\right)\hat{\mathbf{k}} + iW_0\left(\hat{\sigma}_1 + \hat{\sigma}_2\right)\cdot\hat{\mathbf{k}}'\times\delta\left(\mathbf{r_1} - \mathbf{r_2}\right)\hat{\mathbf{k}} \quad (2)$$

Here [6] we consider the possibility that the Skyrme potential can be dominated by a 3-body repulsive interaction at high density. This term is taken to be a zero range force of the form $v_{123} = t_3\delta\left(\mathbf{r_1} - \mathbf{r_2}\right)\delta\left(\mathbf{r_2} - \mathbf{r_3}\right)$. If the assumption is made that the neutron-star medium is spin-saturated [10], the three-body term becomes a density dependent two-body interaction [9] that we generalize to a modified Skyrme interaction that replaces the linear dependence on the density with a power-law index σ. [11]

$$v_{12}^{(3)'} = \frac{1}{6}t_3\left(1 + \hat{P}_s\right)\delta\left(\mathbf{r_1} - \mathbf{r_2}\right)\rho^{\sigma}\left(\frac{\mathbf{r_1} + \mathbf{r_2}}{2}\right). \quad (3)$$

A value of $\sigma = 1/3$ is a common choice [12, 13]. However, in the present approach we treat σ as a free parameter constrained by the skewness coefficient and observed neutron star properties [14].

All quantities and coefficients for symmetric nuclear matter are obtained from the usual relations. The pressure, is $P = n^2\frac{\partial}{\partial n}\left(\frac{E}{A}\right)$. The volume compressibility of symmetric nuclear matter is calculated from the derivative of the pressure: $K = 9\left(\frac{\partial P}{\partial n}\right) = 18\frac{P}{n} + 9n^2\frac{\partial^2}{\partial n^2}\left(\frac{E}{A}\right)$. The skewness coefficient, is from the third derivative of the free energy per nucleon $Q_0 =$

$27n^3 \frac{\partial^3}{\partial n^3}\left(\frac{E}{A}\right)$. Applying the saturation condition $P(n = n_0) = n^2 \frac{\partial}{\partial n}\left(\frac{E}{A}\right)|_{n=n_0} = 0$, one obtains [6] a system of four equations in terms of t_0, $(3t_1 + 5t_2)$, t_3, and σ. Solving this system for σ yields

$$\sigma = \frac{\frac{9}{5}T_{F_0} - 2K_0 + Q_0 - 45E_0}{3K_0 + 45E_0 - \frac{27}{5}T_{F_0}} , \tag{4}$$

where the subscript zero denotes values at the saturation density.

For our purposes, we adopt inferred values of n_0, E_0, K_0, Q_0 from the literature and use these to determine the Skyrme model parameters. We also demand that these parameters allow neutron star masses $\geq 1.97 \pm 0.04\ M_\odot$ [14]. The saturation density $n_0 \approx 0.16$ fm^{-3} and the binding energy per nucleon $E_0 = -16$ MeV are reasonably well established [15]. The determination of the compressibility parameter from experimental data on the giant monopole resonance on finite nuclei has been a long standing conundrum. For our purposes we adopt the median value and uncertainty from Ref. [16], i.e. $K_0 = 240 \pm 10$ MeV as this is appropriate for the Skyrme force approach employed here. Solving the saturation conditions self consistently, we therefore determine the best range for the nuclear compressibility consistent with the results of [16].

There is even more uncertainty in the skewness parameter Q_0. Breathing mode data [17] implies $Q_0 = -700 \pm 500$ MeV. Using the range for K_0 given by [16] and solving the saturation conditions, we find [6] a skewness coefficient of $Q_0 = -390 \pm 90$ MeV consistent within the range given in Ref. [16]. The fiducial NDL EoS is then constructed [6] using the The Skyrme coefficients $t_0 = -1718$ MeVfm^3, $(3t_1 + 5t_2) = -102$ MeVfm^5, $t_3 = 13226$ MeVfm$^{3\sigma+3}$, and $\sigma = 0.369$.

The density dependence of the symmetry energy beyond saturation is highly uncertain. For many Skyrme models the symmetry energy either saturates at high densities, or in the worst case becomes negative. This results in a negative pressure deep inside the neutron star core. We implemented [6] a linearly increasing function of density. The symmetry energy at saturation was determined by the difference between the energy per particle for pure neutron matter and that of symmetric matter at $T = 0$ MeV. For all relevant parameter sets the NDL EoS symmetry energy at saturation is $S_0 = 30.5$ MeV [6].

3. QCD and the EoS

For sufficiently high densities and/or temperature a transition from hadronic matter to quark-gluon plasma (QGP) can occur [18]. Progress [19] in lattice gauge theory (LGT) has shown that at high temperature and low density a deconfinement and chiral symmetry restoration occur simultaneously. In particular, it has been found [19] that the order parameters for deconfinement and chiral symmetry restoration changes abruptly for temperatures of $T = 145-170$ MeV [20, 21] as a smooth crossover.

At low density the hadron phase can be approximated as a pion-nucleon gas, while the QGP phase can be approximated in a bag model as a non-interacting relativistic gas of quarks and gluons [22]. The LGT results then imply a range for the QCD vacuum energy of $165 \leq B^{1/4} \leq 225$ MeV. Also requiring that the maximum mass of a neutron star exceed $1.97 \pm 0.04\ M_\odot$ [14] implies a value for $B^{1/4}$ near the top end of that range.

For the description of quark matter we utilize a bag model with 2-loop corrections, and construct the EoS from a phase-space integral representation over scattering amplitudes. We allow for the possibility of a coexistence mixed phase in a 1st order transition or a simple cross over transition. It is convenient to compute the QGP in terms of the grand potential, $\Omega(T, V, \mu)$, where the grand potential for the quark-gluon plasma takes the form:

$$\Omega = \sum_i (\Omega^i_{q0} + \Omega^i_{q2}) + \Omega_{g0} + \Omega_{g2} + BV. \tag{5}$$

Where q_0 and g_0 denote the 0th-order bag model thermodynamic potentials for quarks and gluons, respectively, and q_2 and g_2 denote the 2-loop corrections. In most calculations sufficient accuracy is obtained by using fixed current algebra masses (e.g. $m_u \sim m_d \sim 0$ GeV, $m_s \sim 0.1 - 0.3$ GeV). For this work we chose the strange quark mass to be $m_s = 150$ MeV and a bag constant $B^{1/4} = 165 - 220$ MeV. The quark contribution to the thermodynamic potential is given [18] in terms of a sum of the ideal gas contribution plus a two loop correction from phase-space integrals over Feynman amplitudes [23].

Fig. 1 compares [6] the neutron star mass radius relation for the NDL EoS for: 1) a hadronic EoS with 3-body forces (solid line); 2) a first order QCD transition with $B^{1/4} = 220$ MeV; and 3) a simple QCD cross over transition. Also, shown for comparison are results from the LS180 EoS, Shen EoS and the original Bowers & Wilson EoS. Note, that all three versions of the NDL EoS easily accommodate a maximum neutron star mass $\geq 1.97 \pm 0.04\ M_{\odot}$, however, the hadronic version must have the 3-body forces at high baryon density. A first order phase transition to a QGP is consistent with the high maximum neutron star mass constraint [14] for a bag constant $B^{1/4} > 220$ MeV. This imposes a low baryon density transition temperature of $T_c = 158$ MeV [22] which is consistent with the current range of crossover temperatures determined from LGT [19].

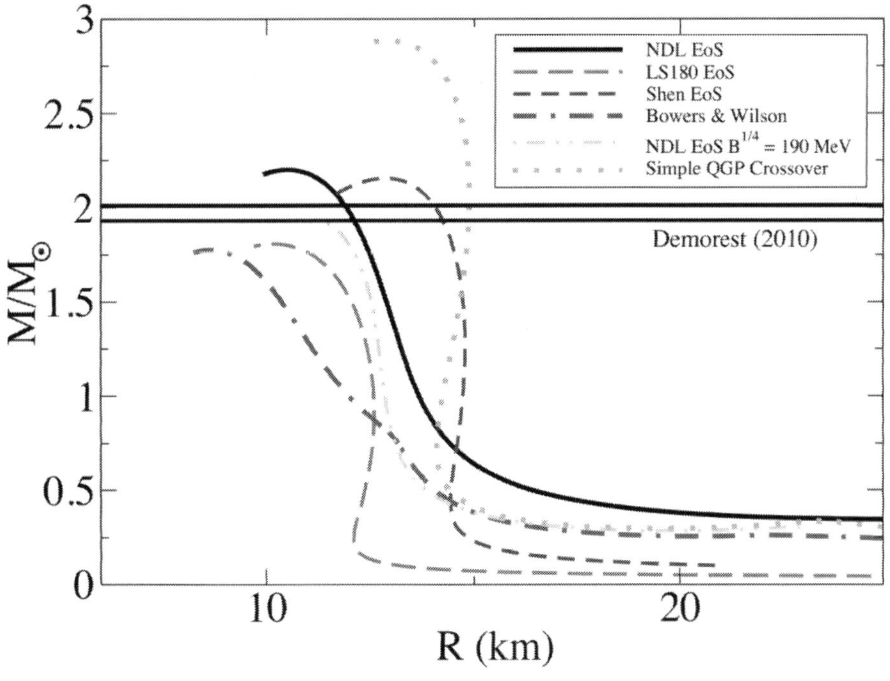

Figure 1. Mass-radius relations for the EoS of Shen [3], Lattimer & Swesty (LS180) [2], Bowers & Wilson [7], compared with the new NDL EoS [6] with and without a mixed phase transition or a simple crossover transition to a QGP. Note that the soft LS180 hadronic EoS and the previous Bowers & Wilson EoS without 3-body forces cannot satisfy the astrophysical constraint of a maximum neutron star mass $\geq 1.97 \pm 0.04\ M_{\odot}$ [14] as shown.

4. Magnetic Fields and The Eos

We have also considered the neutron star equation of state and neutrino transport in the presence of strong magnetic fields [25]-[30]. Indeed, magnetic fields are everywhere in Nature and frequently play a role in astrophysical phenomena. In particular the existence of magnetars and magnetar flares [31, 32, 33], along with the observed asymmetry in supernova explosions and the observed pulsar kick velocities all suggest the strong magnetic fields play an important role in supernova explosions and the formation of proto-neutron stars [34, 35, 36]. In view of this we have undertaken studies of a variety of phenomena.

In [25] we considered the role that strong interior magnetic fields ($B \sim 10^{17}$ G) would have on neutron star structure and stability. We considered the nuclear equation of state for an ideal *npe* gas in a strong magnetic field. In particular, we calculated the proton concentration, the threshold densities for neutron, muon, and pion production and pion condensation in a strong magnetic field both without and with the effect of the nucleon anomalous magnetic moments. It was shown [25] that the higher Landau levels are significant at high density in spite of the existence of a very strong magnetic field. In particular, at high density, the proton concentration approaches the nonmagnetic limit. In particular, we have obtained the neutron appearance threshold density in a magnetic field when the nucleon anomalous magnetic moment is included. We also have shown [25] that the muon and pion threshold densities are not affected by magnetic fields for $B < 10^{17}$ G. We also obtained an equation of state for a pion condensate in strong magnetic fields. We found [25] that magnetic fields reduce the amount of pion condensation. However, we still could find distinguishable effects from a pion condensate in strongly magnetized neutron stars. In addition, we demonstrated an oscillatory behavior of the adiabatic index in both strongly magnetized n, p, e and $n, p.e, \mu, \pi$ gases at high density. Here we speculated that this behavior might lead to an interior pulsational instability.

In [26] we investigated the possibility that soft gamma-ray repeaters (SGRs) and anomalous X-ray pulsars (AXPs) might be observational evidence for a magnetic phase separation in magnetars. We studied such magnetic domain formation as a new mechanism for SGRs and AXPs in which magnetar matter separates into phases containing different flux densities. We identified the parameter space in matter density and magnetic field strength at which there is an instability for magnetic domain formation. We showed that such instabilities will likely occur in the deep outer crust for the magnetic Baym, Pethick, and Sutherland (BPS) model and in the inner crust and core for magnetars described in relativistic Hartree theory. Moreover, we estimated that the energy released by the onset of this instability is comparable with the energy emitted by SGRs.

In [27] this has recently been extend to a study a new quantum hadrodynamic (QHD) equation of state for neutron stars (with and without hyperons) in the presence of strong magnetic fields. The parameters were constrained by the condition that deduced neutron star masses and radii that must be consistent with the recent observations [14] of a high mass neutron star. The calculated adiabatic index for these magnetized neutron stars exhibited the same rapid changes with density. This was hypothesized to provide possible insight into the mechanism of star-quakes and flares in magnetars. We also investigated the strong magnetic field effects on the moments of inertia of neutron stars. The change of the moments of inertia associated with emitted magnetic flares was shown to match well with observed glitches in some magnetars.

In [29, 30] we explored a perturbative calculation of neutrino scattering and absorption in hot and dense hyperonic neutron-star matter in the presence of a strong magnetic field. We found that the absorption cross-sections show a remarkable angular dependence in that the neutrino absorption strength is reduced in a direction parallel to the magnetic field and enhanced in the opposite direction. This asymmetry in the neutrino absorption can be as much as 2 % of the entire neutrino momentum for an large interior magnetic field. We estimate the associated pulsar kick velocities associated with this asymmetry in a fully relativistic mean-field theory formulation and show that the kick velocities are comparable to observed pulsar velocities. In [30] we have extended this calculation to include a toroidal magnetic field configuration. In this case, there can be an asymmetric emission of neutrino momentum along the magnetic field lines that are in the direction of the neutron star spin. This can substantially accelerate the spin down of a neutron star in the early cooling phase, $\sim$ 10 sec after core bounce. This is to be compared with [28] in which we considered the spin down of a neutron star purely from the outflow of neutrinos without a magnetic field.

Acknowledgments

Work at the University of Notre Dame is supported by the U.S. Department of Energy under Nuclear Theory Grant DE-FG02-95-ER40934. One of the authors (N.Q.L.) was supported in part by the National Science Foundation through the Joint Institute for Nuclear Theory (JINA)

References

[1] Lattimer J M 2012 *Ann. Rev. Nucl. and Part. Sci.* **62** 485
[2] Lattimer J M and Swesty F D 1991 *Nuclear Physics* A **535** 331
[3] Shen H, Toki H, Oyamatsu K and Sumiyoshi K 1998 *Nuclear Physics* A **637** 435
[4] Shen H, Toki H, Oyamatsu K and Sumiyoshi K 1998 *Progress of Theoretical Physics* **100** 1013
[5] Hempel M and Schaner-Bielich J 2010 *Nuclear Physics* A **837** 210
[6] Meixner M, Olson J P, Mathews G J, Lan N Q and Dalhed H E 2013 Submitted to *Phys. Rev.* C
[7] Bowers R L and Wilson J R 1982 *Phys. Rev.* C **50** 115
[8] Wilson J R and Mathews G J 2003 *Relativistic Numerical Hydrodynamics* (Cambridge University Press)
[9] Vautherin D and Brink D M 1972 *Phys. Rev.* C **5** 626
[10] Ring P and Schuck P 2000 *The nuclear many-body problem* (Springer)
[11] Mansour H M M 1990 *Acta Physica Polonica* B **21** 741
[12] Kohler H S 1965 *Phys. Rev.* **138** 831
[13] Krivine H, Treiner J and Bohigas O 1980 *Nuclear Physics* A **336** 155
[14] Demorest P B, Pennucci T, Ransom S M, Roberts M S E and Hessels J W T 2010 *Nature* **467** 1081
[15] Li BA and Ko C M 1997 *Nuclear Physics* A **618** 498
[16] Colo G, van Giai N, Meyer J, Bennaceur K and Bonche P 2004 *Phys. Rev.* C **70** 024307
[17] Farine M, Pearson J M and Tondeur F 1997 *Nuclear Physics* A **615** 135
[18] McLerran L 1986 *Reviews of Modern Physics* **58** 1021
[19] Kronfeld A S 2012 *Ann. Rev. Nucl. and Particle Sci.* **62** 265
[20] Borsanyi S, et al 2012 *Journal of High Energy Physics* **8** 126
[21] Bazavov A, et al 2012 *Phys. Rev.* D **86** 094503
[22] Fuller G M, Mathews G J and Alcock C R 1988 *Phys. Rev.* D **37** 1380
[23] Kapusta J I 1978 *High temperature matter and heavy ion collisions* Ph.D. thesis, Univ. California , Berkeley
[24] Braithwaite J and Spruit H C 2004 *Nature* **431** 891
[25] Suh I-S and Mathews G J 2001 *Astrophys. J.* **546** 1126
[26] Suh I-S and Mathews G J 2010 *Astrophys. J.* **717** 843
[27] Ryu C-Y, Cheoun M-K, Maruyama T and Mathews G J 2012 *Astroparticle Physics* **38** 25-30
[28] Ryu C-Y, Maruyama T, Kajino T, Mathews G J and Cheoun M K 2012 *Phys. Rev.* C **85** 045803
[29] Maruyama T, Yasutake N, Cheoun M-K, Hidaka J. Kajino T and Mathews G J 2012 *Phys. Rev.* D **86** 123003
[30] Maruyama T, Hidaka J. Kajino T, Yasutake N, Cheoun M-K, Ryu C Y and Mathews G J 2013 Submitted to *Phys. Rev. Lett.* (*Preprint* arXiv:1301.7495 [astro-ph])
[31] Duncan R C and Thompson C 1992 *Astrophys. J. Lett.* **392** L19
[32] Kouveliotou C, et al 1998 *Nature* **393** 235
[33] Hurley K, et al 1999 *Astrophys. J.* **510** L111
[34] Lyne A G and Lorimer D R 1994 *Nature* **369** 127
[35] Pączyński B 1992 *Acta. Astron.* **41** 145
[36] Chanmugam G 1992 *Ann. Rev. Astron. Astrophys.* **30** 143

Nuclear weak interactions, supernova nucleosynthesis and neutrino oscillation

Toshitaka Kajino

National Astronomical Observatory, 2-21-1 Osawa, Mitaka, Tokyo 181-8588, Japan,
Department of Astronomy, Graduate School of Science, University of Tokyo, Hongo 7-3-1, Bunkyo-ku, Tokyo 113-0033, Japan

E-mail: kajino@nao.ac.jp

Abstract. We study the nuclear weak response in light-to-heavy mass nuclei and calculate neutrino-nucleus cross sections. We apply these cross sections to the explosive nucleosynthesis in core-collapse supernovae and find that several isotopes of rare elements ^{7}Li, ^{11}B, ^{138}La, ^{180}Ta and several others are predominantly produced by the neutrino-process nucleosynthesis. We discuss how to determine the suitable neutrino spectra of three different flavors and their anti-particles in order to explain the observed solar system abundances of these isotopes, combined with Galactic chemical evolution of the light nuclei and the heavy r-process elements. Light-mass nuclei like ^{7}Li and ^{11}B, which are produced in outer He-layer, are strongly affected by the neutrino flavor oscillation due to the MSW (Mikheyev-Smirnov-Wolfenstein) effect, while heavy-mass nuclei like ^{138}La, ^{180}Ta and r-process elements, which are produced in the inner O-Ne-Mg layer or the atmosphere of proto-neutron star, are likely to be free from the MSW effect. Using such a different nature of the neutrino-process nucleosynthesis, we study the neutrino oscillation effects on their abundances, and propose a new novel method to determine the unknown neutrino oscillation parameters, θ_{13} and mass hierarchy, simultaneously. There is recent evidence that some SiC X grains from the Murchison meteorite may contain supernova-produced neutrino-process ^{11}B and ^{7}Li encapsulated in the grains. Combining the recent experimental constraints on θ_{13}, we show that although the uncertainties are still large, our method hints at a marginal preference for an inverted neutrino mass hierarchy for the first time.

1. Introduction

Neutrino interactions with atomic nuclei play the critical roles in various scales of astronomical and cosmological phenomena. Still unknown mass and oscillation properties of neutrinos take the important keys to resolve many fundamental questions in particle physics and astrophysics such as why baryon- and lepton-symmetries are broken in the Universe, why we need unified theory beyond the standard model of elementary particles, why the core-collapse supernovae explode, etc. We discuss in this article how to determine the unknown neutrino oscillation parameters from the studies of element synthesis in supernovae [1, 2].

2. Core-Collapse Supernovae and Neutrino Spectra

Core-collapse supernovae emit intensive flux of three flavor neutrinos with total energy $\sim 3.0 \times 10^{53}$ ergs [3]. This total energy is almost equal to the gravitational binding energy of a 1.4 $M_{\odot}$ neutron star.

2.1. Supernova Model
In our supernova model [4] the neutrino energy spectra are assumed to obey Fermi-Dirac distributions with zero-chemical potentials [5], and the explosion energy is set to be $\sim 0.3\%$ of total neutrino energy as assumed in the literatures. Nucleosynthesis yields generally depend on supernova models with different zero-age main-sequence masses, metallicities, and explosion energies. Since the neutrino-induced nucleosynthesis is a very weak process, one can adopt a post-processing method in that we calculate only the explosive nucleosynthesis in well studied supernova model for SN1987A, and then the contribution from the other supernova models scale to those calculated by Woosley and his collaborators [6, 7].

Neutrino temperatures depend on the dynamics of core-collapse and explosion. We here confine ourselves to the core-collapse supernovae which form the neutron star as a remnant. Although the black-hole associated explosion [8, 9] is another source of neutrinos, it is known to have a minor contribution due to the Salpeter initial mass function of stellar formation.

The neutrino spectra for $\nu_{\rm e}$ and $\bar{\nu}_{\rm e}$ are determined by the microscopic physics of high-density nuclear matter which has saturation properties at neutron-rich environment, leading to a neutrino temperature more or less independent of the initial supernova conditions. Although ν_e and $\bar{\nu}_{\rm e}$ interact with supernova matter with different proton and neutron number abundances through both charged- and neutral-current interactions, $\nu_{\mu,\tau}$ and $\bar{\nu}_{\mu,\tau}$ interact only through the neutral-current interaction. This results in a hierarchy of different neutrino temperatures $T_{\nu_{\rm e}}<T_{\bar{\nu}_{\rm e}}<T_{\rm x}$ for different mean-free paths and diffusion coefficients, where x stands for μ or τ.

2.2. Galactic Chemical Evolution and Neutrino Temperature for $\nu_{\mu,\tau}$ and $\bar{\nu}_{\mu,\tau}$
The supernova ν-process is known to produce ^{7}Li and ^{11}B in amounts comparable to those produced in the Galactic cosmic-ray spallation process and the stellar nucleosynthesis in small-to-intermediate mass stars [11]. The neutrino temperature for $\nu_{\mu,\tau}$ and $\bar{\nu}_{\mu,\tau}$ brings about the production of ^{11}B from supernovae appropriate for the observed Galactic chemical evolution of the light elements. The temperature of $\nu_{\mu,\tau}$ and $\bar{\nu}_{\mu,\tau}$ is thus estimated to be $T_{\nu_{\mu,\tau}} = 6.0 \pm 1.0$ MeV [5]. This temperature is also severely constrained from the measured isotopic ratio $(^{11}{\rm B}/^{10}{\rm B})_{\rm sun} = 4.5 \pm 0.1$ in the solar system abundance. This ratio clearly indicates the fact that more than 50% of ^{11}B arises from the supernova ν-process because ^{10}B is almost purely created in Galactic cosmic-ray spallation which results in $(^{11}{\rm B}/^{10}{\rm B})_{\rm GCR} \approx 2.0$.

2.3. Nucleosynthesis and Neutrino Temperature for $\nu_{\rm e}$ and $\bar{\nu}_{\rm e}$
We use nucleosynthesis of ^{180}Ta in O-Ne-Mg layer to estimate the temperatures $\nu_{\rm e}$ and $\bar{\nu}_{\rm e}$. It is currently believed that two isotopes ^{138}La and ^{180}Ta among the heavy elements may be predominantly synthesized by the ν-process [12]. Although the calculated result can reproduce the solar abundance of ^{138}La with charged current reactions for $\nu_{\rm e}$ temperature of ≈ 4 MeV, it overproduces the abundance of ^{180}Ta. Here we investigate the possibility that this overestimate originates from the unique feature that the naturally occurring abundance of ^{180}Ta is actually a meta-stable 9^- isomer with half-life of $\geq 10^{15}$ yr, while the true ground state is a 1^+ unstable state which β-decays with a half-life of only 8.15 hr (see Figure 1).

In the ν process, low-spin excited states in ^{180}Ta are strongly populated from ^{180}Hf by Gamow-Teller transitions and subsequently decay preferentially to the 1^+ ground state. However, in a high temperature photon bath, the meta-stable isomer is excited from the ground state by (γ,γ') reactions through highly excited states. Therefore, the final isomeric branching ratio should be evaluated by a time-dependent calculation. Our supernova nucleosynthesis calculation by including these complicated photon processes based on the measured nine linking transitions [13] shows that the isomeric residual population ratio turns out to be ≈ 0.39 [14]. We stress that the isomer ratio is almost independent of the astrophysical parameters of supernovae such as the peak temperature, the temperature time constant, the supernova neutrino energy spectrum, and

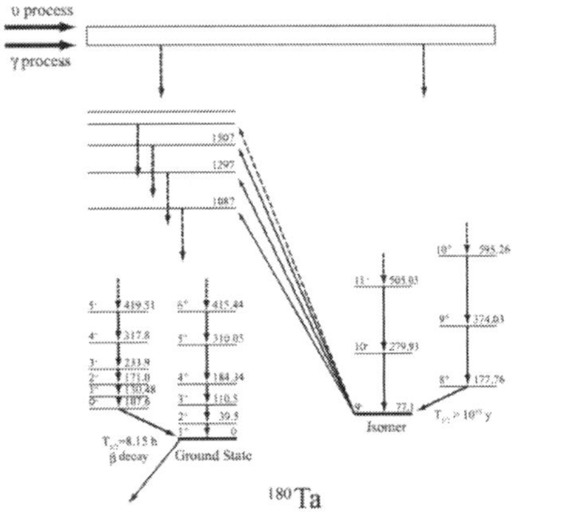

Figure 1. Partial nuclear level scheme of ^{180}Ta [14]. The ground state decays via β-decay with a half-life of 8.15 h, while the isomer is a meta-stable state. The measured [13] excitation energies of states are indicated.

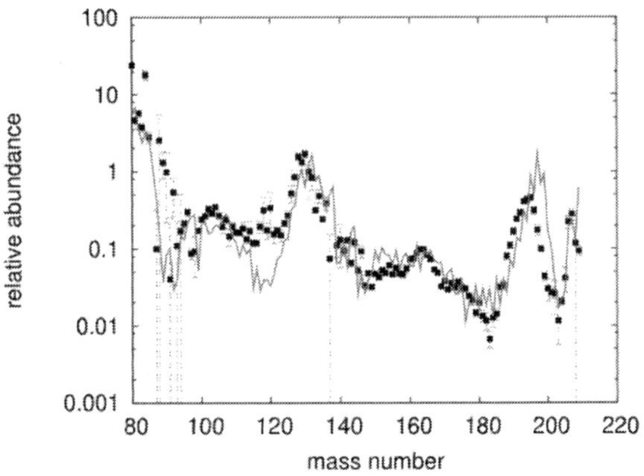

Figure 2. Abundance pattern (solid line) of the r-process nucleosynthesis [16] compared with the solar system r-process abundance (points).

the explosion energy. We finally obtain the concordant result between the observed solar system abundances of ^{138}La and ^{180}Ta and the theoretical predictions for $T_{\nu_e} \approx T_{\bar{\nu}_e} = 4.0$ MeV [14].

2.4. *R-Process Nucleosynthesis for Breaking Degenberacy* $T_{\nu_e} < T_{\bar{\nu}_e}$

We here finally try to resolve a weak degeneracy between T_{ν_e} and $T_{\bar{\nu}_e}$ by imposing a neutron-rich condition $0.2 < Y_e < 0.5$ for successful r-process heavy element synthesis. Since the r-process is presumed to occur deep inside the iron-core near the atmosphere of proto-neutron star, the environment is still very neutron-rich. The mean-free path or equivalently the diffusion coefficient for ν_e is shorter than that of $\bar{\nu}_e$ due to the different frequencies of charged-current interactions between $\nu_e + n \to p + e^-$ and $\bar{\nu}_e + p \to n + e^+$. Therefore, the neutrino-sphere for $\bar{\nu}_e$ is deeper than that for ν_e, hence resulting in $T_{\nu_e} < T_{\bar{\nu}_e}$. We find that only when we set $T_{\nu_e} = 3.2$ MeV and $T_{\bar{\nu}_e} = 4.0$ MeV, the calculated abundance pattern of r-process nucleosynthesis can well explain the observed solar-system r-process abundances [15, 16], as displayed in Figure 2.

3. Supernova Nucleosynthesis and Neutrino Oscillation

Neutrinos are emitted during $\sim$10 s in the supernova explosion. Both neutral and charged current interactions on abundant nuclei, ^{4}He and ^{12}C, spall them into free nucleons and light nuclei. Subsequently the explosive nucleosynthesis similar to the hot Big-Bang nucleosynthesis occurs at high-temperatures and densities when the shock arrives. These neutrino-induced reactions, called the ν-processes, affect the abundances of the light elements ^{6,7}Li, ^{9}Be and ^{10,11}B [10].

Figure 3 shows the major nucleosynthesis paths to produce ^{7}Li and ^{11}B during supernova explosions. The reactions, ^{4}He$(\nu,\nu'p)^3$H and ^{4}He$(\nu,\nu'n)^3$He are important for the production of ^{7}Li through ^{3}H$(\alpha,\gamma)^7$Li and ^{3}He$(\alpha,\gamma)^7$Be$(e^-,\nu_e)^7$Li processes. If the ν-^{4}He reaction cross sections are enhanced due to the neutrino oscillations as to be discussed later, not only the abundance of ^{7}Li but also that of ^{11}B increase through the radiative alpha-capture reactions ^{7}Li$(\alpha,\gamma)^{11}$B and ^{7}Be$(\alpha,\gamma)^{11}$C$(e^+\nu_e)^{11}$B. The neutral current reactions, ^{12}C$(\nu,\nu'p)^{11}$B and ^{12}C$(\nu,\nu'n)^{11}$C, also are important for the production of ^{11}B. Details are discussed in [1, 2, 5].

3.1. Weak Interaction Cross Sections

As for the ν-induced reaction cross sections, there are only LSND experimental data for ν-^{12}C [18]. We should therefore totally rely on the theoretical estimate of neutrino-nucleus cross sections. We calculate neutrino-nucleus cross sections for ^{4}He, ^{12}C, ^{138}Ba, ^{180}Hf and many other nuclei in order to study the ν -induced nucleosynthesis of LiBeB isotopes, ^{138}La, ^{180}Ta and r-process elements. For this purpose, we use nuclear shell model [19] and quasi-particle random phase approximation (QRPA) [20]. Nuclear shell model is one of the most reliable models to describe the Gamow-Teller (GT) and spin-dipole transitions at relatively low excitation energies, while QRPA is suitable at relatively high excitation energies. Therefore, we need both models to calculate not only the GT but also the spin-dipole and higher-multipole transition probabilities for the supernova neutrinos energy up to 100 MeV. See refs. [19, 20] for details of calculations and applications of the cross sections to several astrophysical weak nuclear processes.

3.2. Neutrino Mixing Parameters

Neutrino-flavor matter oscillations affect the nuclear reactions induced by charged current interactions, ν_e-A and $\bar{\nu}_e$-A, while the neutral current interactions do not change. Flavor oscillations are described by three mixing angles θ_{12}, θ_{23}, and θ_{13} plus a CP-violating phase δ_{CP}. Solar, atmospheric, and reactor neutrino oscillation measurements [21, 22, 23, 24, 25, 26] have provided information on the neutrino mass differences, i.e $\Delta m^2_{12} \equiv |m_1^2 - m_2^2| = 7.9 \times 10^{-5}$ eV [25] and $\Delta m^2_{13} \approx |\Delta m^2_{23}| \approx 2.4 \times 10^{-3}$ eV^2 [26, 22]. However, these cannot determine the mass hierarchy, i.e. whether $\Delta m^2_{23} > 0$ (normal) or $\Delta m^2_{23} < 0$ (inverted) is the correct order.

Mixing angles θ_{12} and θ_{23} were also determined precisely. θ_{13} has become available only recently. The three best current measurements are $\sin^2 2\theta_{13} = 0.092 \pm 0.016(\text{stat}) \pm 0.005(\text{syst})$ [27], $\sin^2 2\theta_{13} = 0.113 \pm 0.013(\text{stat.}) \pm 0.019(\text{syst.})$ [28], and $\sin^2 2\theta_{13} = 0.086 \pm 0.041(\text{stat.}) \pm 0.030(\text{syst.})$ [29]. These results are consistent with the previously reported upper limit $\sin^2 2\theta_{13} < 0.12(0.20)$ from the MINOS collaboration [26] for the normal (inverted) hierarchy, and with $0.03(0.04) < \sin^2 2\theta_{13} < 0.28(0.34)$ at the 90% C.L. from the T2K collaboration [30]. However, the data do not yet determine the mass hierarchy.

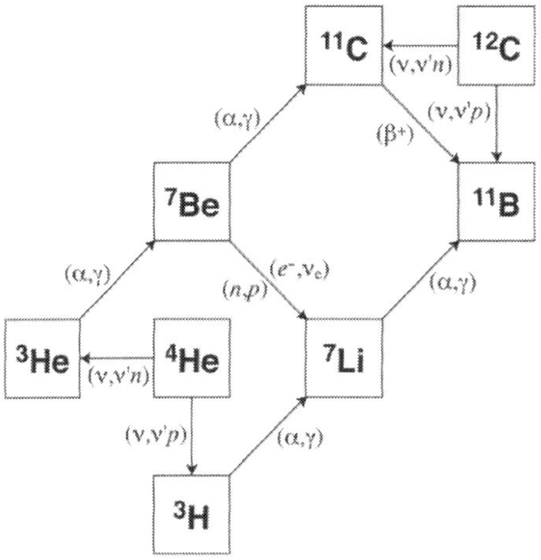

Figure 3. Nucleosynthesis path of light elements ^{7}Li and ^{11}B during supernova explosions [5].

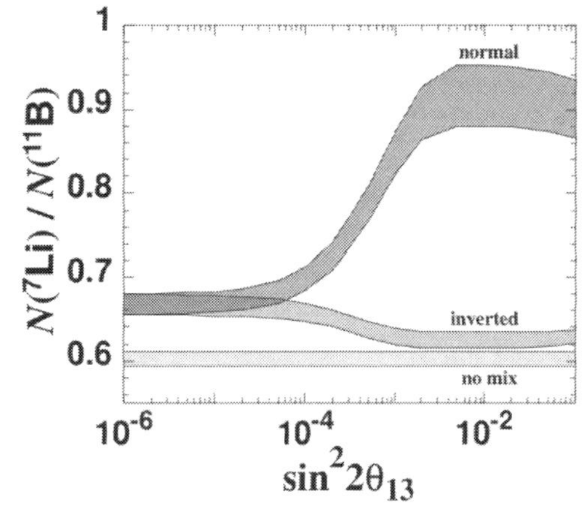

Figure 4. The shaded ranges include the uncertainties of neutrino energy spectra deduced from the calculations using three sets of neutrino temperatures and total neutrino energies [1].

3.3. New Constraint on Neutrino Mass-Hierarchy

Since supernova neutrino temperatures for $\nu_{\mu,\tau}$ and $\bar{\nu}_{\mu,\tau}$ are the same, energy spectra of these neutrinos are degenerate to one another. In this case any observables which arise from the neutrino oscillation do not depend on the CP-violating phase δ_{CP} [31, 32]. Therefore, the supernova nucleosynthesis is only sensitive to the mixing angle θ_{13} and the mass hierarchy.

Figure 4 shows the predicted number-abundance ratio of ^{7}Li/^{11}B as a function of mixing angle θ_{13} for both mass hierarchies [1, 2, 5] . The uncertainty due to neutrino spectra is included as shaded regions. We should note that although uncertainties in the ν-process cross sections still remain, we find that they are largely canceled out when we take the ^{7}Li/^{11}B ratio. This is because ^{7}Li and ^{11}B are mainly produced through the ν-process from ^{4}He and the dependence of their yields on the ν-process reaction rates is similar to each other. Therefore, the ratio is almost independent of the Hamiltonians used in the calculations of the ν-process cross sections [1, 2].

The ^{7}Li/^{11}B ratio in the case of adiabatic 13-mixing resonance and normal hierarchy is remarkably larger than that without neutrino oscillations, even with the spectral uncertainties included. Thus, the enhancement of observed ^{7}Li/^{11}B ratio may constrain the lowest value of θ_{13} and eliminate the possibility of inverted hierarchy. Several long-baseline experiments in particle physics have determined $\sin^2 2\theta_{13} \approx 10^{-1}$. Our proposed nucleosynthetic method as displayed in Figure 4 would provide an independent confirmation [1, 2, 5].

Fujiya et al. [33] have recently discovered SiC X grains from the Murchison meteorite which contains most likely supernova-produced ν-process ^{11}B and ^{7}Li. We apply their meteoritic constraint by combining with the recent experimentally determined θ_{13} value to our predicted ^{7}Li/^{11}B ratio, as illustrated in Figure 4. We examined the possible implications of these new results based upon a Bayesian analysis of the uncertainties in the measured meteoritic material and the associated supernova nucleosynthesis models [34], and obtained that all these data are marginally more consistent with the inverted neutrino mass hierarchy [35].

4. Summary and Discussions

We studied the ν-process nucleosynthesis in core-collapse supernovae. We first found that average neutrino temperatures suitable for explaining the observed solar system abundances, meteoritic isotope ratios, and Galactic chemical evolution of the light-to-heavy mass elements, turn out to be $T_{\nu_e} = 3.2$ MeV, $T_{\bar{\nu}_e} = 4.0$ MeV, and $T_{\nu_{\mu,\tau}} = T_{\bar{\nu}_{\mu,\tau}} = 6.0$ MeV, respectively. Some of these isotopes like ^{7}Li, and ^{11}B are predominantly produced by the charged current interactions in the He-layer, and are affected strongly by the neutrino flavor oscillation due to the matter effects (MSW-effects). With the use of this specific feature, we proposed a new method to determine the unknown neutrino mass hierarchy along with the mixing angle θ_{13}. Recent experimental constraints on θ_{13} and the first detection of ^{11}B and ^{7}Li encapsulated in the SiC X grains from the Murchison meteorite lead to marginal preference for an inverted neutrino mass hierarchy in our proposed method.

A recent high-resolution spectroscopic observation of supernova remnant has succeeded in the determination of lithium isotopic abundances [36]. Observational efforts [37, 38, 39] to obtain Li and B abundances in stars, which have been formed in regions directly affected by prior generations of massive stars and subsequent supernovae, also are underway. These would detect the signature of the supernova ν-process.

Supernovae are the unique source in nature that provides three flavors of energetic neutrinos. The neutrino interactions with abundant nuclei produce rare isotopes such as ^{7}Li, ^{11}B, ^{23}Na, ^{40}Ar, ^{50}V, rare Mn-Fe-Co-Ni isotopes, ^{92}Nb, ^{98}Tc, ^{138}La, ^{180}Ta, and many others. For the studies of supernova ν-process, however, one needs to know the weak interactions between energetic neutrinos and these nuclei. At the moment, we still do not have efficient neutrino-beam whose intensity is high enough to measure the neutrino-nucleus reaction cross sections in the laboratory experiments. It is therefore highly desirable to study the photo-induced [40] and charge-exchange reactions [41] in order to estimate the nuclear weak matrix elements. High precision theoretical models [19, 20] to calculate these weak matrix elements are also desirable.

The combination of supernova nucleosynthesis and nuclear physics studies may ultimately provide precise constraints on still poorly known neutrino mass hierarchy.

I would like to dedicate this paper to Prof. Otsuka who is my senior from Prof. A. Arima's school of the University of Tokyo, and also is currently one of my strong collaborators in the studies of nuclear structure and the applications to several nuclear astrophysical problems. This work has been supported in part by Grants-in-Aid for Scientific Research (20244035, 20105004, 24340060) of the Ministry of Education, Culture, Sports, Science and Technology of Japan.

References

[1] Yoshida T *et al.* 2008 *Ap. J.* **686** 448–66; 2009 *Phys. Rev.* D **80** 125032; and references therein
[2] Kajino T *et al.* 2008 *Mod. Phys. Lett.* A **23** 1409–18
[3] Hirata K *et al.* (Kamiokande Collaboration) 1987 *Phys. Rev. Lett.* **58** 1490–3
[4] Shigeyama T and Nomoto K 1990 *Ap. J.* **360** 242–56
[5] Yoshida T, Kajino T and Hartmann D H 2005 *Phys. Rev. Lett.* **94** 231101; Yoshida T *et al.* 2006 *Phys. Rev. Lett.* **96** 091101; 2006 *Ap. J.* **649** 319–31
[6] Hofman R D, Woosley S E and Weaver A A 2001 *Ap. J.* **549** 1085–92
[7] Rauscher T, Heger A, Hofman R D and Woosley S E 2002 *Ap. J.* **576** 323–48
[8] Sumiyoshi K, Yamada S and Suzuki H 2008 *Ap. J.* **688** 1176–85
[9] Nakamura K, Yoshida T, Shigeyama T and Kajino T 2010 *Ap. J.* **718** L137–40
[10] Woosley S E, Hartmann D H, Hofman R D and Haxton W C 2000 *Ap. J.* **356** 272–301
[11] Ryan S G *et al.* 2001 *Ap. J.* **549** 55–71
[12] Heger A *et al.* 2005 *Phys. Lett.* B **606** 258–64
[13] Belic D *et al.* 2002 *Phys. Rev.* C **65** 035801
[14] Hayakawa T *et. al.* 2010 *Phys. Rev.* C **81** 052801(R); 2010 C **82** 158801
[15] Yoshida T, Terasawa M, Kajino T and Sumiyoshi K 2004 *Ap. J.* **600** 204–13
[16] Otsuki K, Tagoshi H, Kajino T and Wanajo S 2000 *Ap. J.* **533** 424–39
[17] Nakamura K, Sato S, Harikae S, Kajino T and Mathews G J 2012 *Ap. J.* submitted
[18] Athanassopoulos C *et al.* (LSND Collaboration) 1997 *Phys. Rev.* C **55** 2078–91
[19] Suzuki T *et al.* 2006 *Phys. Rev.* C **74** 034307; 2009 *Phys. Rev.* C **79** 061603(R); 2011 *Phys. Rev.* C **83** 044619; 2012 *Phys. Rev.* C **85** 015802; 2012 C **85** 048801; 2012 *Phys. Rev.* C **86** 015502
[20] Cheoun M K *et al.* 2010 *Phys. Rev.* C **81** 028501; 2010 *Phys. Rev.* C **82** 035504; 2010 *J. Phys.* G **37** 055101; 2011 *Phys. Rev.* C **83** 028801; 2012 *Phys. Rev.* C **85** 065807
[21] Ashie Y *et al.* (Super-Kamiokande Collaboration) 2005 *Phys. Rev.* D **71** 112005
[22] Hosaka J *et al.* (Super-Kamiokande Collaboration) 2006 *Phys. Rev.* D **73** 112001
[23] Aharmim B *et al.* (SNO Collaboration) 2005 *Phys. Rev.* C **72** 055502
[24] Ahn M H *et al.* (K2K Collaboration) 2006 *Phys. Rev.* D **74** 072003
[25] Abe S *et al.* (KamLAND Collaboration) 2008 *Phys. Rev. Lett.* **100** 221803
[26] Adamson P *et al.* (MINOS Collaboration) 2011 *Phys. Rev. Lett.* **107** 181802
[27] An F P *et al.* (Daya Bay Collaboration) 2012 *Phys. Rev. Lett.* **108** 171803
[28] Ahn J K *et al.* (RENO Collaboration) 2012 *Phys. Rev. Lett.* **108** 191802
[29] Abe Y *et al.* (Double Chooz Collaboration) 2012 *Phys. Rev. Lett.* **108** 131801
[30] Abe K *et al.* (T2K Collaboration) 2011 *Phys. Rev. Lett.* **107** 041801
[31] Kuo T and Pantaleone J 1987 *Phys. Lett.* B **198** 406–10
[32] Kimura K, Takamura A and Yokomakura H 2002 *Phys. Lett.* B **537** 86–94; 2002 *Phys. Rev.* D **66** 073005; Yokomakura H, Kimura K and Takamura A 2002 *Phys. Lett.* B **544** 286–94
[33] Fujiya W, Hoppe P and Ott U 2011 *Ap. J.* **730** L7
[34] Austin S M, Heger A and Tur C 2011 *Phys. Rev. Lett.* **106** 152501
[35] Mathews G J, Kajino T, Aoki W and Fujiya W 2012 *Phys. Rev.* D **85** 105023
[36] Taylor C J *et al.* 2012 *Ap. J.* **750** L15
[37] Primas F, Dancan D and Thorburn J 1998 *Ap. J.* **506** L51–5
[38] Rebull L *et al.* 2000 *The First Stars* (Springer) p 176
[39] Aoki A *et al.* with Kajino T (SUBARU HDS Collaboration) 2013 private communication
[40] Shima T *et al.* 2005 *Phys. Rev.* C **72** 044004
[41] Sasano M *et al.* 2011 *Phys. Rev. Lett.* **107** 202501

International Symposium on Exotic Nuclear Structure From Nucleons (ENSFN 2012) IOP Publishing
Journal of Physics: Conference Series **445** (2013) 012025 doi:10.1088/1742-6596/445/1/012025

Implementing the r-process in metal-poor stars via black hole collapse and relevance to the light element enhancement

Michael Famiano[1],Richard N Boyd[2],Toshitaka Kajino[3], Bradley Meyer[4],Yuko Motizuki[5] and Ian Roederer[6]

[1]Western Michigan University, Kalamazoo, MI, USA
[2]Sonoma County Center for Astrophysics, Windsor, CA, USA
[3]National Astronomical Observatory of Japan, Mitaka, Tokyo, Japan
[4]Clemson University, Clemson, SC, USA
[5]The Institute of Physical and Chemical Research, Wako, Saitama, Japan
[6]Carnegie Observatories, Pasadena, CA, USA

E-mail: `michael.famiano@wmich.edu`

Abstract. Recent data compilations of abundances of Strontium and Barium in Extremely Metal Poor Stars show that large fluctuations exist in the ratio of the abundances of those elements. Two models exist for explaining those fluctuations, as well as the apparent truncation of data for those elements for stars having metallicity of [Fe/H] < -3.5. We study the factors that place upper limits on the logarithmic ratio [Sr/Ba] in the two models. A model has been developed in which the collapse of type II supernovae may produce a pronounced [Sr/Ba] enhancement in single-site r-process enriched stars. This model is consistent with galactic chemical constraints of light-element enrichment in metal-poor stars.

1. Introduction

Recent studies have indicated that a population of metal-poor stars exists with abundance patterns that do not fit the universal r-process template.[1, 2, 3] A few explanations have been proposed for this. Multiple r-process sources have been proposed with contributions from each source weighted parametrically.[4] In prior work, we have addressed the issue of light-element enrichment via the so called "tr-process"[5] in which ejection of r-process material is "truncated" via fallback below the mass cut of an accretion-induced black hole. We now attempt to address the salient features of the observed [Sr/Ba] distributions in metal poor stars[1] via simple hydrodynamics considerations. Specifically, we examine the Sr and Ba yields in a Population II star produced from a r-process source.

Mixing has been studied in two models. In the first model, turbulent mixing above and below the mass cut of an accretion-induced black hole is addressed and considered in a simple network calculation. In the second model, turbulence is responsible for density fluctuations in a single trajectory resulting in inhomogeneous ejection of material. In both cases, spherically symmetric hydrodynamics results[6] have been used to emulate mixing and turbulent inhomogeneities in the ejecta. While mixing between adjacent trajectories can be accomplished in a spherical model, inhomogeneous zones within a trajectory are simulated by subdivision on individual trajectories.

In addition to sharp dynamic mass cuts which limit the r-process ejecta, neutrino effects have been taken into account in calculating the effect of black hole formation on the r-process. In this regime, the neutrino emission spectra are altered as the neutrinosphere falls below the event horizon. Since the electron and anti-electron neutrinos do not have the same mean-free paths, and hence different energy spectra corresponding to different radii in the nascent neutron star, a common event horizon will result in energy spectra that differ from the usual scenario of an r-process in the environment surrounding a nascent neutron star.

In this study, we present the result of dynamic mass cuts in a collapse model. The ejection of Sr and Ba is considered, and a model which can explain the enrichment of light elements at low metallicity is presented.

2. Dynamic Collapse of r-Process Zones

To study the truncated r-process in the neutrino- driven wind, we applied the basic idea behind the study of Woosley et al.,[6] which assumed that the r-process occurred in the neutrino wind from a core collapse supernova. In that study, a succession of 40 "trajectories" (that is, thin shell wind elements), all assumed to have originated deep within the (assumed spherically symmetric) star, but having initially different conditions of density, temperature, entropy, and electron fraction, were emitted into the interstellar medium, thus contributing to the total r-process nucleosynthesis. The bubble was evolving in time, so that the conditions under which the individual trajectories were processed changed with the identity of the trajectory. We assumed that the different trajectories were emitted from the star successively, but ceased to be emitted when the collapse to the black hole occurred. Trajectories falling inside the collapse do not escape.

Truncating the r-process at higher trajectories has the effect of terminating the r-process as expected. Comparison to the low-metallicity star HD122563[2] indicates that a physical truncation of the r- process below the mass cut of an accretion- induced black hole may produce a light- element enriched r-process.

Our calculations used a network code based on *libnucnet*, a library of C codes for storing and managing nuclear reaction networks.[7] The nuclear and reaction data for the calculations were taken from the JINA reaclib database.[8] We performed calculations for trajectories 24 - 40 in the [6] hydrodynamics model. For each trajectory, reaction network calculations were performed for $T_9 < 2.5$ using initial abundances from the [6] results. Our calculations were simplified by assuming an initial abundance of massive nuclei from a single nuclear seed of mass equal to the average mass at $T_9 = 2.5$ and an atomic number derived from the average mass number, the Y_e, and the neutron and alpha mass fractions at $T_9 = 2.5$ in reference [6]. This is justified by the sharply peaked heavy-nuclide distribution near $T_9 = 2.5$. An adiabatic expansion was assumed for each trajectory, with entropy constant within a trajectory but varying between trajectories, again consistent with reference [6], for times at which $T_9 < 2.5$. The material expanded at constant velocity on a time scale consistent with that from prior calculations. Each calculation was continued until the abundance distribution versus mass number had frozen out.

Previous results have indicated that a truncation in the r-process in type II supernovae due to a collapse of the protoneutron star into a black hole may be responsible for an enrichment of light elements in some metal-poor halo stars produced by a single-site r-process.[5] This "tr-process" could be caused by either a dynamic collapse of material below the black hole mass cut or by a change in the neutrino luminosity resulting in an increased Y_e in the early nucleosynthesis.

We adopt a model based on prior calculations in which the Sr and Ba production in an r-process is directly related to the post-bounce time of black hole formation as shown in Figure 1. As with the initial tr-process model, we assume that the r-process is universal in post-bounce ejection time of r-processing zones in a type II supernova independent of the mass of the progenitor star. This simplification will be addressed in future work. As the goal of this work

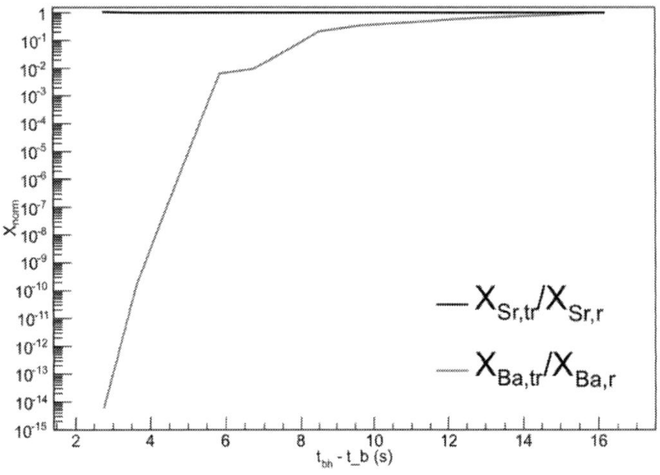

Figure 1. Sr and Ba mass fraction in an r-process site as a function of the post-bounce black hole collapse time in an r-process site.

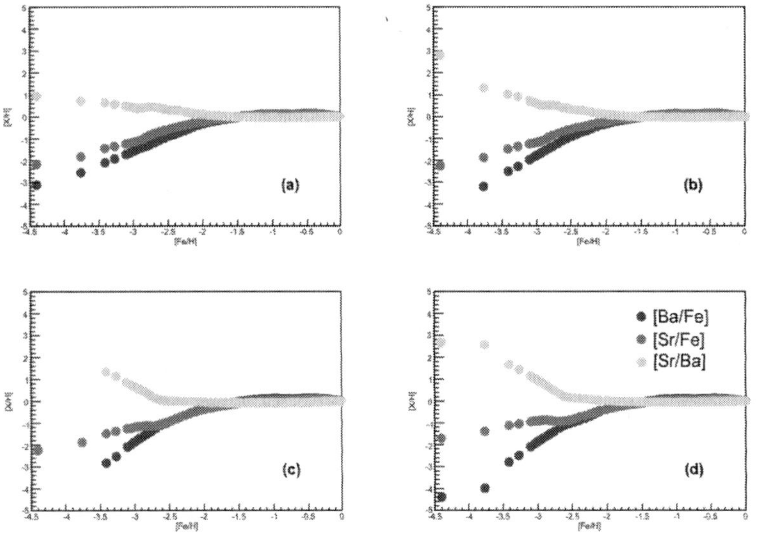

Figure 2. Results of the GCE calculation showing [Sr/Ba] as a function of metallicity single-site halo stars produced in a tr-process for various assumptions of the nuclear EOS. These are (a)LS375, (b)Shenn, (c)LS220, and (d) the *ad hoc* parametrization. The assumption of an LS375 EOS closely resembles the case of no tr-process.

is to examine the sensitivity of the EOS dependence of the ejection prior to the post-bounce collapse, we maintain this assumption of universality in this simplified model. Using the model of Boyd et al.,[5] the Sr and Ba mass fractions normalized to those of a complete r-process have been determined. A value of one indicates that the resultant mass fraction in the r-process site is equal to that of a complete r-process (which produces the A$\sim$195 peak). As Sr is produced very early on in the r-process evolution, only extremely early collapse times result in a significant reduction of the Sr production, while the Ba abundance is built up more slowly in time; delayed black hole collapse can cut off the later ejected mass shells, which produce a more robust Ba abundance. We use this model to parametrize the Sr and Ba production as a function of collapse time in all collapse scenarios.

In order to relate the collapse time to the progenitor, results from the spherical collapse code GR1D[9] were used to study the relationship between the progenitor star's mass and metallicity and the Sr and Ba yields in massive stars. The relationship between the stellar progenitor and the compactness has been shown in previous works,[9], and the collapse time depends on a compactness in a power law relationship. This relationship has been determined for various progenitor masses and metallicities, and varies only slightly with the nuclear EOS.

Given the results from the network calculation and the collapse model, the Sr and Ba yields (normalized to a full r-process) in a tr-process can be determined as a function of progenitor mass. It is noted that these yields are minimum yields. That is, in the non-rotating spherical models employed here, the collapse times are minimum collapse times. Longer collapse times may result from asymmetric explosions, rotation, and neutrino heating.

The Sr and Ba yields as a function of progenitor mass and metallicity are placed in a galactic chemical evolution model to determine the metallicity relationship of [Sr/Ba]=$\log(Sr/Ba) - \log(Sr/Ba)_{\odot}$.[11] Results for single-site halo events have been determined. In this model, Sr and Ba yields in massive stars for a "universal" are taken from the results of references [12] and [13]. In the assumption of a single-site tr-process star, the Sr and Ba yields are scaled from the yields of [12] and [13] according to the mass and metallicity of the progenitor for each EOS.

3. Results

Assuming a tr-process model resulting in a Sr and Ba production yield, the single-site [Sr/Ba] values as a function of metallicity [Fe/H] are shown in Figure 2. It is seen that a softer EOS results in a progressively larger [Sr/Ba] at lower metallicity (corresponding to earlier times in the galactic evolution produced by more massive stars).

If this production is a result of the tr-process, then the EOS dependence of the collapse time has a direct affect on the Sr and Ba production. Since the plotted values of [Sr/Ba] are maximum values, then it would appear that in this model that a very stiff EOS cannot produce the upward scatter in [Sr/Ba]. A softer EOS is necessary. It's also noted that the upper limit of observed [Sr/Ba] values provides an observed lower limit on the softness of the EOS.

While further work is in progress to examine some of the finer details of this effect, it is fascinating to note that the effect of a truncated r-process on GCE may be directly related to the nuclear equation of state, and astronomical observations of [Sr/Ba] ratios in metal-poor stars may provide an upper limit on the stiffness of the nuclear EOS.

References

[1] Aoki W et al 2000 *ApJ* **536** L97-L100
[2] Honda S et al 2006 *ApJ* **643** 1180 - 89
[3] Roederer I U et al 2010 *ApJ* **724** 975 - 93
[4] Qian Y Z and Wasserberg G J 2008 *ApJ* **687** 272 - 86
[5] Boyd R N 2012 *ApJ* **744** L14 - L17
[6] Woosley S E et al 1994 *ApJ* **433** 229 - 46
[7] Meyer B A and Adams D C 2007 *Met. Planet. Sci.* **42** 5215
[8] Cyburt R H et al 2010 *ApJS* **189** 240 - 52
[9] O'Connor E and Ott C D 2010 *Class. Quant. Grav.* **27** 114103
[10] Woosley S E et al 2002 *Rev. Mod. Phys.* **74** 1015 - 71
[11] Timmes F X et al 1995 *ApJS* **98** 617 - 58
[12] Ishimaru Y et al 2004 *ApJ* **600** L47 - L50
[13] Cescutti G et al 2006 *A&A* **448** 557 - 69

Deformation and shape coexistence in neutron-rich nuclei at N~40

S M Lenzi
Dipartimento di Fisica e Astronomia, Università di Padova and INFN, Padova, Italy

E-mail: lenzi@pd.infn.it

Abstract. The A ~ 60-70 mass region of neutron-rich nuclei with $Z \leq 28$ presents several interesting properties when approaching N=40. At this subshell closure a new region of deformation develops. From the experimental side, these nuclei can be populated at relatively high spin by means of multi-nucleon and deep-inelastic collisions. Selected results on the gamma-ray spectroscopy of the neutron-rich nuclei in this mass region are discussed. In particular, the shape coexistence in ^{67}Co at low excitation energy, populated by bombarding a ^{238}U target with a 460 MeV ^{70}Zn beam at the Legnaro National Laboratory is interpreted by means of shell model calculations in a large valence space. These calculations are extended to the description of excited states in Zn isotopes.

1. Introduction

Far from the valley of beta stability, the nuclear shell structure undergoes important and substantial modifications. For many decades, theoretical models in nuclear structure were constructed relying on the experimental information obtained for nuclear systems near the stability line. The continuous experimental developments allow nowadays the study of exotic nuclei far from stability. In particular, neutron-rich nuclei are of current interest because of clear experimental indications that the extrapolation of the traditional shell model to this region of the isotope table is not straightforward. Unexpected modifications to the shell structure have been already encountered and there is evidence that some magic numbers change when increasing the neutron number, in particular in light and medium-light nuclei [1-4]. The observed changes help to shed light on specific terms of the effective nucleon-nucleon interaction and to improve our knowledge of the nuclear structure evolution towards the drip lines. In particular, it has been shown that the monopole part of the tensor force of the proton-neutron interaction is the main responsible of the shell evolution [2].

In the last few years, detailed nuclear structure information has become available both with stable and radioactive beams in Cr, Mn, Fe and Co isotopes in the mass region around N=40, where rapid changes of the nuclear shape occur along isotopic and isotonic chains. The knowledge of excited states in these neutron-rich nuclei is rather limited due to the difficulties in the production and identification of isotopes by conventional means. The first results have been obtained using beta-decay reactions. These investigations are, however, limited to a small range of spins and excitation energy. Other methods with radioactive beams consist on the use of knockout reactions and (p,p') inelastic scattering that allow to populate low-lying states in nuclei far from stability. On the other hand, a very successful method to populate relatively high-spin states in moderately neutron-rich nuclei consists on the use of binary reactions between stable neutron-rich nuclei, such as multi-nucleon transfer and deep-inelastic collisions, combined with modern gamma-ray arrays and magnetic spectrometers.

In this contribution, some recent experimental findings in this mass region are discussed in terms of the interacting shell model in a wide valence space.

2. Experimental details

A research program on nuclear structure of neutron-rich nuclei far from stability has been carried out at LNL, in Italy, with the gamma-ray array CLARA [5] and the tracking array AGATA [6], coupled to the magnetic spectrometer PRISMA [7] to study the gamma decay of excited nuclei populated by multi-nucleon transfer and deep-inelastic collisions. With these setups both gamma-ray spectroscopy and lifetime measurements have been carried out in several mass regions of the table of nuclides.
In particular, the mass region A~60-70 at N~40 has been object of several experiments. Beams of ^{64}Ni and ^{70}Zn were used to bombard targets of ^{238}U and for the first time, excited states could be identified in several neutron-rich nuclei of mass A~60 south of the "doubly magic" ^{68}Ni. Very rapid changes in shape and in collectivity have been observed in V, Ti, Cr, Mn, Fe and Co isotopes.

These experimental setups allow to measure the mass A and atomic number Z of the recoiling projectile-like nuclei. With the use of a thin U target short-living states can be observed. However, the limited solid angle of the magnetic spectrometer allows to collect only a small fraction of the cross section and therefore the statistics of gamma rays results insufficient to perform coincidence analysis. With the aim of measuring γ-γ coincidences we have performed an experiment with the GASP gamma-array [8] alone, using a thick U target. Such type of experiment produces a wide range of target-like and projectile-like fragments that are stopped at the target and cannot be identified. The resulting gamma ray spectra are therefore very complex. By using triple gamma coincidence data, useful information can be obtained for the strongest channels when at least one gamma transition is known.

3. Selected results

In multi-nucleon transfer and deep-inelastic experiments, several isotopes of an element can be produced. This allows to follow the evolution of the nuclear structure along isotopic and isotonic chains. In the mass A~60 region, neutron-rich Fe and Cr isotopes pass from spherical shapes at the new magic number N = 32 to very deformed shapes approaching N = 40 [9-11]. With two protons less than ^{68}Ni, the nucleus ^{66}Fe shows a well deformed prolate structure. This has been further confirmed by the measurement of the lifetime of the 2^+ state [12]. More recently, the energy of the 2^+ and 4^+ states have been measured in ^{64}Cr [11], the nucleus that seems to be the most deformed in the region, situated in the centre of the island of inversion at N~40.

To interpret the structure of these nuclei and its evolution, the best model to apply is the interacting shell model that allows to describe with good accuracy the spectroscopic properties. The problem here is the choice of the model space. Several studies have shown the inadequacy of the fp shell, and even the fpg model space to account for the development of collectivity in this mass region [9,10,13], In a recent publication [14] a new interaction has been developed that is able to reproduce very well the variety of shapes and collectivity in Cr, Fe and Ni isotopes around N=40. This is possible due to the large, appropriate model space chosen. In fact, as stated in ref. [15], the deformation can be accounted for by considering the pseudo-SU(3) plus quasi-SU(3) model spaces in different regions of the table of nuclides: The deformation can be generated by the interplay between the quadrupole force and the central field in the subspace consisting on the lowest $\Delta j = 2$ orbitals of a major shell. The new effective interaction, LNPS, is based on renormalized realistic interactions and monopole corrections in the model space consisting on the full pf shell for the protons and the $p_{3/2}$, $p_{1/2}$, $f_{5/2}$, $g_{9/2}$ and $d_{5/2}$ orbits for the neutrons [14]. It has been shown that only with the inclusion of both the neutron $g_{9/2}$ and $d_{5/2}$ orbits it is possible to obtain the correlations necessary for the development of the island of inversion around ^{64}Cr. In a forthcoming work, these results are compared with IBM-2 calculations and a very good agreement is found [16].

In Cr isotopes, the deformation develops already at N=38, while the Fe isotopes become well deformed at N=40. In both cases the deformation stabilizes al least up to N=42. While the excitation energy of the first 2^+ state in even-even nuclei can give a direct insight on the shape changes and the development of collectivity, this is not straightforward in odd and odd-odd mass isotopes. These constitute also a good test of the robustness of the interaction. In our experiments we have populated several Co isotopes, situated between the semi-magic Ni isotopes and the well deformed Fe nuclei. These isotopes follow a very regular behaviour at low excitation energy that can be interpreted in terms of a week coupling of the proton hole to the Ni core. Previously to the present investigation, a sudden development of deformation in ^{67}Co has been suggested in Ref. [16]. This was based on a beta-decay study where a $1/2^-$ state was observed at very low excitation energy. It was argued that such a state would be formed by exciting one proton from the $f_{7/2}$ orbital.

It is thus interesting to interpret the results in terms of the shell model. The experimental and theoretical energy levels of ^{67}Co are reported in Fig. 1. The structure based on the ground state is well described and can be interpreted in terms of the weak coupling with the states in ^{68}Ni. The excited structure on the left of the g.s. correspond to a well deformed band with an intrinsic configuration that confirms the two proton-hole character. The calculations are therefore able to describe at the same time both the spherical and well deformed configurations, a very challenging task for any theoretical model.

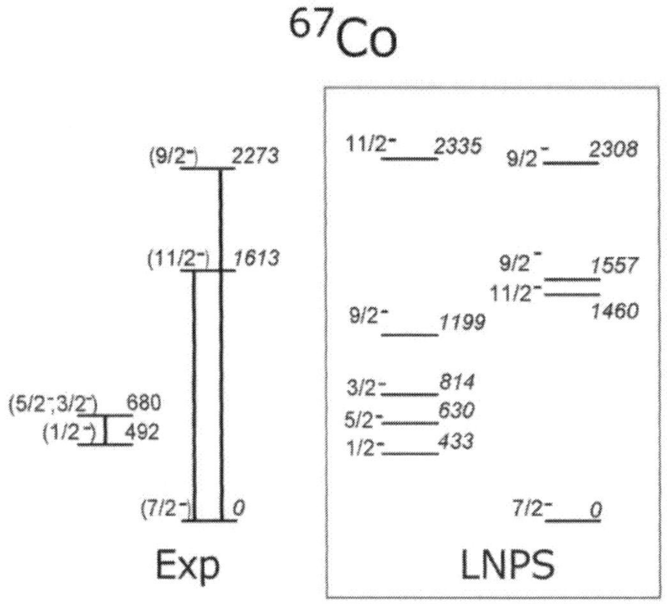

Figure 1. Experimental and theoretical levels in ^{67}Co from ref. [18].

Journal of Physics: Conference Series **445** (2013) 012026 doi:10.1088/1742-6596/445/1/012026

More recently, several experiments with the tracking array AGATA have been performed at LNL. Many interesting results were obtained, in particular with lifetime measurements using the differential plunger method. Although these results are not yet published, it can be anticipated that shell model calculations using the LNPS interaction reproduce with good accuracy the transition probabilities in neutron-rich Co isotopes [19].

Another interesting region studied with AGATA is that of the neutron-rich Zn isotopes [20]. Here the protons occupy the $p_{3/2}$ shell while neutrons are filling the $g_{9/2}$ orbital. Also in this case the LNPS interaction can give a good description of the spectroscopy. In Fig. 2, the calculated excitation energy of the first 2^+ and 4^+ states are reported in comparison with the experimental data taken from the literature. The agreement is very good all along the isotopic chain.

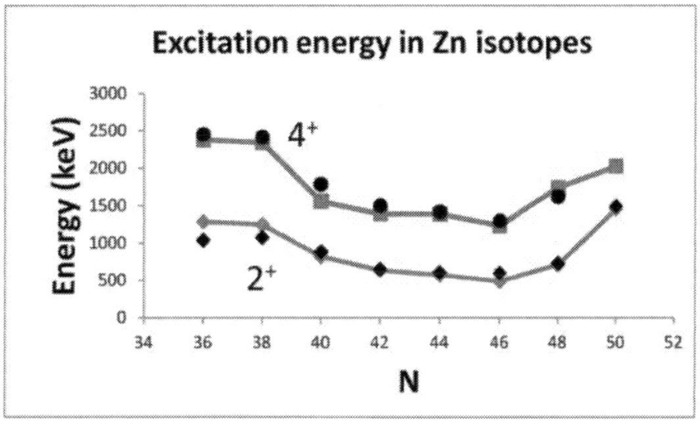

Figure 2. Excited states in Zn isotopes. Experimental values are shown with full circles and diamonds. Theoretical results are connected by coloured lines.

4. Conclusions

The development of deformation in the neutron-rich nuclei around N=40, Z<28, is a subject of current interest from both the experimental and the theoretical points of view. Data are being obtained by means of different experimental techniques. In this contribution some results recently obtained with the CLARA and AGATA gamma arrays coupled to the PRISMA magnetic spectrometer at LNL have been briefly discussed. In particular, it has been shown that ^{67}Co presents shape coexistence phenomena near the ground state. These different structures are very well reproduced by recent shell model calculations using the effective interaction LNPS [14]. A good description is also given for heavier nuclei such as the neutron-rich Zn isotopes.

Of course there are several experimental groups investigating this mass region, in particular in USA and in Japan, where the radioactive beams allow to explore regions farther than the valley of stability. This opens very exciting perspectives to our knowledge of the nuclear interaction.

Acknowledgments
I want to thank prof. T. Otsuka for the fruitful and friendly collaboration we have since many years and for the fundamental contributions he has given and continues giving to the field of nuclear structure research.

This contribution reports experimental results obtained in LNL and theoretical calculations done in collaboration with several colleagues. In particular I want to thank for the experimental part F. Recchia, S. Lunardi, V. Modamio and J.J. Valiente-Dobón. For the shell model developments I am very grateful to F. Nowacki, A. Poves and K. Sieja.

References

[1] Otsuka T, Fujimoto R, Utsuno Y, Brown B A, Honma M and Mizusakil T 2001 *Phys. Rev. Lett.* **87** 082502
[2] Otsuka T, Suzuki T, Fujimoto R, Grawe H and Akaishi Y 2005 *Phys. Rev. Lett.* **95** 232502
[3] Caurier E, Nowacki F and Poves A 2002 *Eur. Phys. J.* A **15** 145
[4] Sorlin O and Porquet M G 2008 *Prog. Part. Nucl. Phys.* **61** 602
[5] Gadea A *et al.* 2004 *Eur. Phys. J.* A **20** 193
[6] Akkoyun S *et al.* 2012 *Nuclear Instruments and Methods in Physics Research* A **668** 26
[7] Stefanini A M *et al.* 2002 *Nucl. Phys.* A **701** 217c
[8] Rossi Alvarez C 1993 *Nuclear Physics News* **2** 10
[9] Marginean N *et al.* 2006 *Phys. Lett.* B **633** 696
[10] Lunardi S *et al.* 2007 *Phys. Rev.* C **76** 034303
[11] Gade A *et al.* 2010 *Phys. Rev.* C **81** 051304
[12] Rother W *et al.* 2011 *Phys. Rev. Lett.* **106** 022502
[13] Aoi N *et al.* 2009 *Phys. Rev. Lett.* **102** 012502
[14] Lenzi S M, Nowacki F, Poves A and Sieja K 2010 *Phys. Rev.* C **82** 054301
[15] Zuker A P, Retamosa J, Poves A and Caurier E 1995, Phys. Rev. C **52** R1741
[16] Kotila J and Lenzi S M *to be published*
[17] Pauwels D *et al.* 2009 *Phys. Rev.* C **79** 044309
[18] Recchia F *et al.* 2012 *Phys. Rev.* C **85** 064305
[19] Modamio V *et al.* 2013 *to be submitted*
[20] Louchart C *et al.* 2013 *to be published*

Journal of Physics: Conference Series **445** (2013) 012027 doi:10.1088/1742-6596/445/1/012027

Skyrme energy-density functional approach to collective modes of excitation in exotic nuclei

K Yoshida

Department of Physics, Graduate School of Science and Technology, Niigata University, Niigata 950-2181, Japan

E-mail: kyoshida@nt.sc.niigata-u.ac.jp

Abstract. Low-frequency collective modes of excitation in neutron-rich nuclei are investigated in the framework of the nuclear energy-density functional method. It is shown that the collective Hamiltonian approach gives the quantitative description of the low-lying states in transitional nuclei with the Skyrme and pairing energy-density functionals as a microscopic input. The inertial functions for large-amplitude vibration and rotation are evaluated by the local normal modes along the axial quadrupole collective coordinate. The time-odd components of the mean fields are fully included in the derived masses.

1. Introduction

Density functional theory has been widely used to describe a variety of quantum many-body systems, including nuclear many-body systems [1]. Ground-state properties including nuclear deformation are described well by the modern nuclear energy-density functional (EDF) method. Experimental evidences of the nuclear shape change are related to the low-lying quadrupole collectivity, such as the ratio of the excitation energy of 2_1^+ and 4_1^+ states, the reduced transition probability $B(E2; 2_1^+ \to 0_1^+)$, etc.

Magic number or shell closure is an essential concept in understanding the stability against deformation. With the advent of RI beam technology, physics of neutron-rich nuclei has been one of the main current subjects in nuclear physics. It has been revealed that the conventional magic numbers disappear and new magic numbers appear instead in neutron-rich nuclei. Quite recently, the collectivity around $N = 40$ has been extensively studied both experimentally and theoretically. The observed small excitation energy of the 2_1^+ state in neutron-rich Cr isotopes indicates that the deformation develops toward $N = 40$ [2, 3, 4, 5, 6].

In this paper, we demonstrate that the Skyrme-EDF method describes well the low-lying collective states in nuclei undergoing the shape-phase transition.

2. Skyrme-EDF approach for collective excitations using parallel computers

To describe the nuclear deformation and the pairing correlations simultaneously in good account of the continuum, we solve the Hartree-Fock-Bogoliubov (HFB) equations [7] in the coordinate space using cylindrical coordinates. The Skyrme and the density-dependent contact interactions are employed for the particle-hole and particle-particle channels, respectively. We assume axial and reflection symmetries in the ground state. Since the parity (π) and the magnetic quantum number (Ω) are good quantum numbers, the HFB Hamiltonian takes a block diagonal form with respect to each (Ω^π, q) sector, where q stands for neutron or proton. The HFB equations for each sector are solved independently using parallel computers. Then, the densities and HFB Hamiltonian are updated, which requires the communication among processors. The modified Broyden's method [8] is utilized to calculate new densities. The qp states are truncated according to the qp energy cutoff at $E_\alpha \leq 60$ MeV.

We introduce the additional truncation for the quasiparticle-random-phase approximation (QRPA) calculation in terms of the two-quasiparticle (2qp) energy as $E_\alpha + E_\beta \leq 60$ MeV to reduce the number of 2qp states. The calculation of the QRPA matrix elements in the qp basis is performed in parallel computers. In the present calculation, all the matrix elements are real. For diagonalization of the matrix, we use the ScaLAPACK PDSYEV subroutine. Details of the setup for the HFB+QRPA calculation are given in Ref. [9].

To describe the evolution of the transitional quadrupole collectivity, we need to go beyond the QRPA because the dynamical change of the mean-field potential associated with the large-amplitude collective motion should be taken into account. We thus take the quadrupole collective Hamiltonian approach with the local QRPA (LQRPA) method [10] in the present paper. For evaluation of the collective masses, the LQRPA equations are solved on the constrained HFB states along the collective coordinate. As a collective coordinate, we take the axially symmetric quadrupole deformation β. The collective coordinate is discretized into 20-30 mesh points in the present calculation. The LQRPA equations are solved independently on the discretized points of the collective coordinate. We confirm that the physical quantities calculated by using 20 discretized points get already converged. Details of the model employed and the calculation scheme are given in Ref. [11].

The similar attempts of the EDF-based collective Hamiltonian starting from the Gogny interaction [12, 13], the relativistic Lagrangian [14, 15] and the Skyrme interaction [16] have been made for description of the large-amplitude collective motion. The cranking approximation, however, has been applied to calculate the inertial functions [12, 13, 14, 15, 16], and the time-odd components of the mean field remain largely unexplored except for a few attempts in the adiabatic time-dependent Hartree-Fock-Bogoliubov theories [17, 18, 19]. Our method includes the time-odd mean fields in the inertial functions.

3. Low-lying states in Cr isotopes: Large-amplitude dynamics

We investigated the deformation properties in neutron-rich Cr isotopes with $N = 34 - 44$ employing the constrained HFB method. In ^{58}Cr and ^{60}Cr, the collective potentials are soft against the β deformation even we get the HFB minima at $\beta = 0$. Beyond $N = 38$, the prolate minimum gradually develops up to $N = 44$. We see a shoulder or a shallow minimum at the oblately deformed region as well.

To investigate the effects of the softness of the collective potential on properties of low-lying states in Cr isotopes, the collective Hamiltonian approach with the LQRPA method was employed [11]. When the neutron number increases from ^{58}Cr, the excitation energy of the 2_1^+ state drops toward ^{62}Cr. The ratio of the E_{4^+} to E_{2^+} $(R_{4/2})$ and the transition probability $B(E2; 2_1^+ \to 0_1^+)$ grow up at the same time. These features are consistent with the experimental results [4, 5, 6], and it clearly shows the onset of deformation at $N \sim 38$.

We show in Fig. 1(a) the vibrational wave functions squared and in Fig. 1(b) the probability densities that one finds a shape with a specific value of β in the ground state of $^{58-68}$Cr. The spherical components seen in Fig. 1(a) disappear in Fig. 1(b) due to the moment of inertia $\mathcal{J}(\beta)$ in the volume element.

We discuss the shape evolution of the ground state in detail. In ^{58}Cr, the density is distributed around $\beta = 0$. But it is not sharply localized at the spherical minimum of the potential, which reflects the softness of the collective potential against the deformation. When two neutrons are added to ^{58}Cr, broadening of the density distribution can be seen. When two more neutrons are

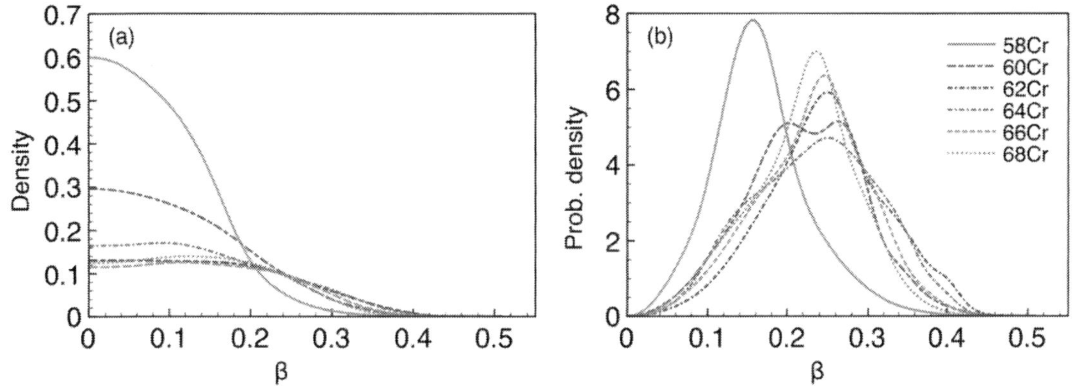

Figure 1. (a) Density and (b) probability-density distributions of the 0_1^+ state in $^{58-68}$Cr.

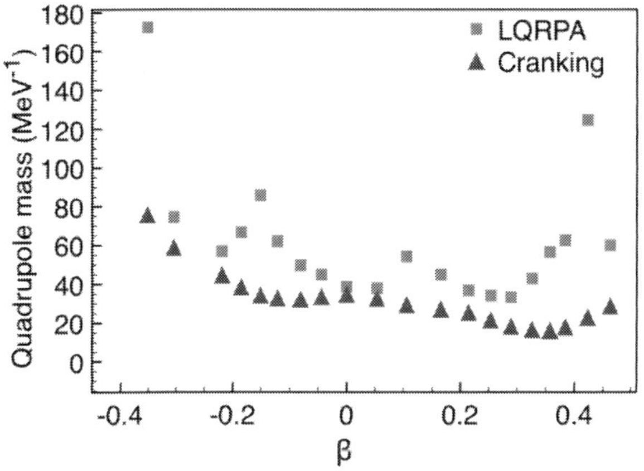

Figure 2. Quadrupole masses as functions of deformation parameter calculated by the LQRPA and the cranking formula in ^{64}Cr.

added, the distribution moves toward a prolately deformed minimum with a large spreading. We see that the distributions in $^{62-68}$Cr spread out from the spherical shape to deformed shape with $\beta \simeq 0.4$. We see that the peak in Fig. 1(b) moves toward a larger β in going from ^{58}Cr to ^{62}Cr, and that the peak becomes sharper in ^{66,68}Cr. In both figures, we see that the distribution of ^{60}Cr is in between that of ^{58}Cr and of $^{62-68}$Cr. We can thus say that ^{60}Cr is located close to the critical point of the shape-phase transition in neutron-rich Cr isotopes and the large-amplitude dynamics plays a dominant role.

In our approach, the time-odd components of the mean field are fully included for the calculation of the collective masses. Figure 2 shows the calculated vibrational masses along the collective coordinate. The collective masses calculated by use of the cranking approximation are also shown, where the time-odd components are neglected. The cranking masses show a smooth behavior as functions of the deformation β. In contrast to the cranking masses, the LQRPA masses strongly depend on the deformation, microscopic structure of the collective mode. And the LQRPA masses are larger than the cranking masses.

We see the effects of the time-odd components of the mean field in the excited states. The excited 0_β^+ state in ^{64}Cr is calculated to appear at 1.90 MeV with the LQRPA mass, while we get the 0_β^+ state at 3.38 MeV with the cranking mass. This is due to the lager LQRPA mass than the cranking mass; the collective kinetic energy is much reduced thanks to the time-odd mean field.

4. Summary

We have developed a new framework of the microscopic model for the large-amplitude collective motion based on the nuclear EDF method. The collective masses and the potential appearing in the quadrupole collective Hamiltonian are evaluated by employing the constrained HFB and local QRPA approach, where the time-odd components of the mean field are fully taken into account. We applied this new framework to the shape-phase transition in neutron-rich Cr isotopes around $N = 40$. The present calculation gives consistent results for the low-lying excited states with the observations and the other theoretical approaches. Investigating the collective wave functions, we reach a conclusion that ^{60}Cr is located close to the critical point of the shape-phase transition, and the onset of deformation takes place at $N = 38$. The large-amplitude dynamics plays a dominant role in the shape changes in neutron-rich Cr isotopes.

Systematic calculations with the HFB+QRPA and the collective Hamiltonian approach with the LQRPA method employing the massively paralleled computers for spherical-to-deformed and light-to-heavy nuclei help us not only to understand and to predict new types of collective modes of excitation, but also to shed light on the nuclear EDF of new generations.

Acknowledgments

The work presented here is based on the collaborations with N. Hinohara, T. Nakatsukasa and K. Sato. The work is supported by Grant-in-Aid for Scientific Research (No. 23740223). The numerical calculations were performed on RIKEN Integrated Cluster of Clusters (RICC), T2K-Tsukuba and SR16000 at YITP, Kyoto University.

References

[1] Bender M, Heenen P-H and Reinhard P-G 2003 *Rev. Mod. Phys.* **75** 121
[2] Sorlin O *et al.* 2003 *Eur. Phys. J.* A **16** 55
[3] Bürger A *et al.* 2005 *Phys. Lett.* B **622** 29
[4] Zhu S *et al.* 2006 *Phys. Rev.* C **74** 064315
[5] Aoi N *et al.* 2009 *Phys. Rev. Lett.* **102** 012502
[6] Gade A *et al.* 2010 *Phys. Rev.* C **81** 051304R
[7] Dobaczewski J, Flocard H and Treiner J 1984 *Nucl. Phys.* A **422** 103
[8] Baran A, Bulgac A, Forbes M M, Hagen G, Nazarewicz W, Schunck N and Stoitsov M V 2008 *Phys. Rev.* C **78** 014318
[9] Yoshida K and Nakatsukasa T 2011 *Phys. Rev.* C **83** 021304R
[10] Hinohara N, Sato K, Nakatsukasa T, Matsuo M and Matsuyanagi K 2010 *Phys. Rev.* C **82** 064313
[11] Yoshida K and Hinohara N 2011 *Phys. Rev.* C **83** 061302R
[12] Libert J, Girod M and Delaroche J-P 1999 *Phys. Rev.* C **60** 054301
[13] Gaudefroy L *et al.* 2009 *Phys. Rev.* C **80** 064313
[14] Nikšić T, Li Z P, Vretenar D, Próchniak L, Meng J, and Ring P 2009 *Phys. Rev.* C **79** 034303
[15] Li Z P, Nikšić T, Vretenar D, and Meng J 2010 *Phys. Rev.* C **81** 034316
[16] Próchniak L and Rohoziński S G 2009 *J. Phys. G: Nucl. Part. Phys.* **36** 123101
[17] Dobaczewski J and Skalski J 1981 *Nucl. Phys.* A **369** 123
[18] Giannoni M J and Quentin P 1980 *Phys. Rev.* C **21** 2060
[19] Giannoni M J and Quentin P 1980 *Phys. Rev.* C **21** 2076

Study of nuclei around $Z = 28$ by large-scale shell model calculations

Y Tsunoda[1], **T Otsuka**[1,2,3], **N Shimizu**[2], **M Honma**[4] **and Y Utsuno**[5]

[1] Department of Physics, University of Tokyo, Tokyo 113-0033, Japan
[2] Center for Nuclear Study, University of Tokyo, Tokyo 113-0033, Japan
[3] National Superconducting Cyclotron Laboratory, Michigan State University, East Lansing, Michigan, USA
[4] Center for Mathematical Sciences, University of Aizu, Ikki-machi, Aizu-Wakamatsu, Fukushima 965-8580, Japan
[5] Advanced Science Research Center, Japan Atomic Energy Agency, Tokai, Ibaraki, 319-1195, Japan

E-mail: ytsunoda@nt.phys.s.u-tokyo.ac.jp

Abstract. We study Cr and Ni isotopes by Monte Carlo shell model (MCSM) calculations in a pfg_9d_5 shell. In the MCSM, the wave function is represented as a linear combination of angular-momentum- and parity-projected deformed Slater determinants (MCSM bases), and we can calculate eigenstates in a large model space such as the pfg_9d_5 shell. We study the intrinsic shape of nuclei by using deformations of the unprojected MCSM bases. We show calculated results of Cr and Ni isotopes and a level scheme of ^{68}Ni, and discuss the strength of the magicity of $N = 40$ in Cr and Ni isotopes and shape coexistence in ^{68}Ni.

1. Introduction

An exotic nucleus is a nucleus whose neutron number is much larger or smaller than its proton number and shows various phenomena such as the shell evolution. An exotic nucleus has different shell structure from that of a non-exotic nucleus, which causes appearance or disappearance of magic numbers. In order to treat magic numbers 28, 50 and a submagic number 40, we use a pfg_9d_5 shell, which consists of the $0f1p$ shell, $0g_{9/2}$ and $1d_{5/2}$ orbits, as a model space. This space is too large to perform conventional shell-model calculations and we use the Monte Carlo shell model (MCSM) [1]. We discuss the magicity of $N = 40$ in Cr ($Z = 24$) and Ni ($Z = 28$) isotopes and shape coexistence in a double-magic nucleus ^{68}Ni.

2. Monte Carlo Shell Model

In shell-model calculations, we diagonalize the Hamiltonian matrix in the model space to obtain eigenstates. We can diagonalize the Hamiltonian matrix directly in the model space whose m-scheme dimension is about 10^{11} at most. However, the dimension reaches more than 10^{15} in the pfg_9d_5 space. Therefore, certain approximations are necessary to perform calculations in such a large model space. In the MCSM, we approximate states with linear combinations of a small number of angular-momentum-projected, parity-projected deformed Slater determinants (MCSM bases) and diagonalize the Hamiltonian matrix in a small subspace spanned by the

MCSM bases. The MCSM wave function $|\Psi_N\rangle$ is given as

$$|\Psi_N\rangle = \sum_{n=1}^{N} \sum_{K=-J}^{J} f_{n,K}^{(N)} P_{MK}^{J\pi} |\psi_n\rangle, \tag{1}$$

$$|\psi_n\rangle = \prod_k \left(\sum_l D_{lk}^{(n)} c_l^\dagger \right) |-\rangle, \tag{2}$$

where $|\psi_n\rangle$ is a Slater determinant, $P_{MK}^{J\pi}$ is the angular-momentum and parity projector, $c_l^\dagger$ is a creation operator, and $|-\rangle$ is an inert core. The coefficients $f_{n,K}^{(N)}$ are determined by the diagonalization of the Hamiltonian in the subspace spanned by MCSM bases $P_{MK}^{J\pi}|\psi_n\rangle$. The angular-momentum and parity projection is performed by using the numerical integration with about 50,000 mesh points. We calculate on each mesh point parallelly with a supercomputer. The coefficients $D_{lk}^{(n)}$ in the MCSM bases are determined to lower the energy eigenvalues by using the auxiliary-field Monte Carlo method and the conjugate gradient method.

In addition, we use the extrapolation using the energy variance [2, 3] in order to obtain more accurate energy eigenvalues. We change the number of the MCSM bases N and calculate the expectation value of the energy $\langle H\rangle$ and the energy variance $\langle \Delta H^2\rangle = \langle H^2\rangle - \langle H\rangle^2$ of the wave function $|\Psi_N\rangle$. As N increases, $\langle H\rangle$ approaches the exact energy and $\langle \Delta H^2\rangle$ approaches zero. We fit these values by a second-order polynomial and extrapolate the energy to $\langle \Delta H^2\rangle = 0$.

3. Effective Interaction

We calculated nuclei in the pfg_9d_5 shell with the effective interaction used in Ref. [4, Sec. 4]. The two-body matrix elements (TBMEs) of the interaction consist of three parts. The TBMEs of the pf shell are those of the GXPF1A interaction [5] and the TBMEs of the f_5pg_9 shell related to the $0g_{9/2}$ orbit are those of the JUN45 interaction [6]. The GXPF1A and JUN45 interactions were determined by combining microscopically derived interactions with a minor empirical fit. The other TBMEs are from the G-matrix effective interaction [7] calculated from the chiral N3LO interaction [8]. The Coulomb interaction is not included and the isospin symmetry is conserved. We made further corrections of the single-particle energies and the monopole interaction.

4. Results

We performed the MCSM calculations of 0^+ and 2^+ yrast states of Cr and Ni even-even isotopes systematically and calculated many spin-parity states of ^{68}Ni. We took 50 MCSM bases for Cr isotopes, 80 bases for Ni isotopes except ^{68}Ni 0^+ states, and 120 bases for ^{68}Ni 0^+ states. The spurious center-of-mass motion was removed by using the prescription of Gloeckner and Lawson [9]. The effective charges are taken as $(e_p, e_n) = (1.5, 0.5)e$.

Figure 1 shows the excitation energies of 2^+ yrast states of Cr and Ni isotopes. The MCSM results reproduce the experimental values well. The excitation energies of Ni isotopes are high at $N = 40$, which suggests the magicity of $N = 40$. However, the excitation energies of Cr isotopes are low around $N = 40$ and the $N = 40$ magicity disappears. Figure 2 shows the $B(E2; 0_1^+ \to 2_1^+)$ transition probabilities of Cr and Ni isotopes. The small $B(E2)$ values of Ni isotopes and large values of Cr isotopes around $N = 40$ also indicate the $N = 40$ magicity of Ni isotopes and the missing $N = 40$ magicity of Cr isotopes. This change of the strength of $N = 40$ magicity may be caused by the change of the $N = 40$ shell gap and the $Z = 28$ magicity of Ni isotopes.

In ordinary shell-model calculations, we cannot study the nuclear deformations of 0^+ states directly because the wave functions of 0^+ states are spherical. In the MCSM, we can study deformations of 0^+ states by using the MCSM bases before angular-momentum projection.

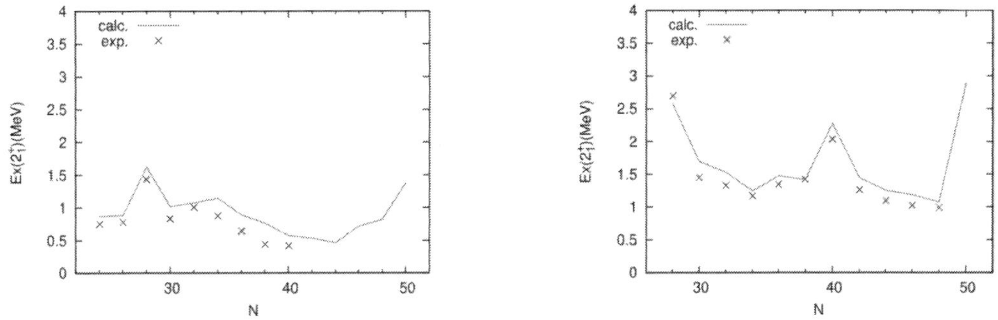

Figure 1. The excitation energies of 2_1^+ states of Cr (left) and Ni (right) isotopes. Experimental data are taken from Ref. [10].

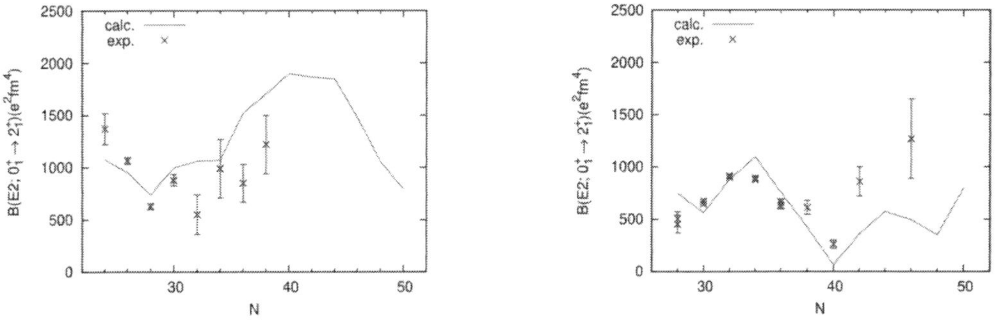

Figure 2. $B(E2; 0_1^+ \to 2_1^+)$ values of Cr (left) and Ni (right) isotopes. Experimental data are taken from Ref. [11].

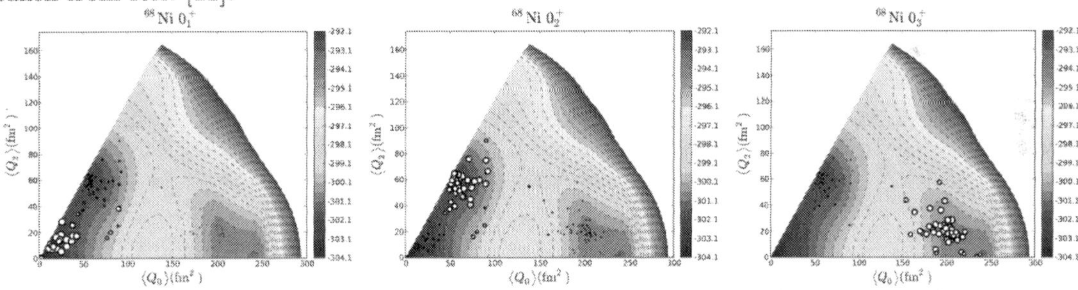

Figure 3. Total energy surfaces of the $0^+_{1,2,3}$ states of ^{68}Ni. The positions of the circles represent quadrupole deformations of the MCSM bases before projection. The areas of the circles represent the overlap probabilities of the bases and the resulting wave function.

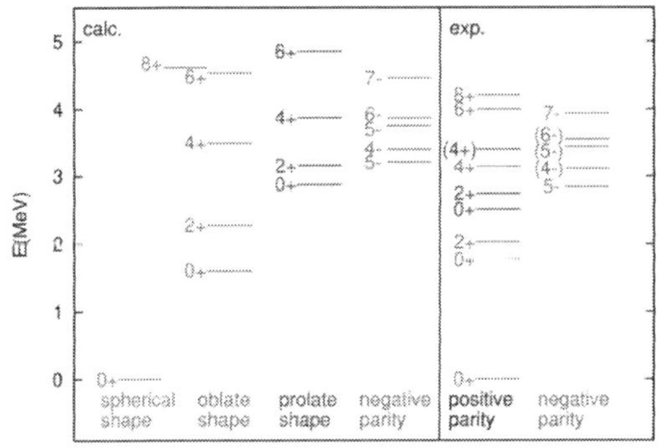

Figure 4. The level scheme of ^{68}Ni. Experimental data are taken from Ref. [12].

Figure 3 shows the total energy surface of ^{68}Ni obtained by the Q-constrained Hartree-Fock calculation [13]. There are spherical, oblate and prolate deformed three minimum points on the total energy surface. The circles on the total energy surface correspond to the MCSM bases. The quadrupole deformations of the bases are calculated before projection. The areas of the circles are proportional to the overlap probabilities of the bases and the resulting wave function. Figure 3 indicates that the 0_1^+, 0_2^+ and 0_3^+ states correspond to the spherical, oblate and prolate shapes, respectively.

Figure 4 shows the level scheme of ^{68}Ni. ^{68}Ni is one of the interesting nuclei because its proton number 28 is a magic number and its neutron number 40 is a submagic number. Our calculated results reproduce the experimental values including non-yrast or negative-parity states, although the excitation energies are slightly overestimated. Our analysis of the deformation using the unprojected bases shows that the 0_1^+ state is spherical, the 0_2^+, 2_1^+, 4_1^+ and 6_1^+ states are oblate deformed, and the 0_3^+, 2_2^+, 4_2^+ and 6_2^+ states are prolate deformed. There are oblate and prolate deformed rotational bands. The 8_1^+ state is oblate deformed but has different properties from other oblate-deformed states. The excitation energies of 6_1^+ and 8_1^+ are close and the $B(E2; 8_1^+ \rightarrow 6_1^+)$ value is small.

5. Summary

We performed MCSM calculations of Cr and Ni isotopes in the pfg_9d_5 shell and studied nuclear shapes by using deformations of the unprojected MCSM bases. We showed the change of the strength of $N = 40$ magicity in Cr and Ni isotopes and three 0^+ states in ^{68}Ni with different shapes and close excitation energies. The calculated level scheme of ^{68}Ni reproduces the experimental data including non-yrast or negative-parity states. These various properties were obtained from the same interaction and we are investigating other nuclei with this interaction.

Acknowledgments

We thank Prof. M. Hjorth-Jensen for providing the G-matrix effective interaction. This work has been supported by Global COE Program "the Physical Sciences Frontier" and HPCI Strategic Program field 5, MEXT, Japan. The numerical calculations were performed mainly on the K computer at the RIKEN AICS (Proposal number hp120284), the FX10 supercomputer at the University of Tokyo and the T2K Open Supercomputers at the University of Tokyo and the University of Tsukuba.

References

[1] Otsuka T, Honma M, Mizusaki T, Shimizu N and Utsuno Y 2001 *Prog. Part. Nucl. Phys.* **47** 319
[2] Shimizu N, Utsuno Y, Mizusaki T, Otsuka T, Abe T and Honma M 2010 *Phys. Rev.* C **82** 061305(R)
[3] Shimizu N, Utsuno Y, Mizusaki T, Honma M, Tsunoda Y and Otsuka T 2012 *Phys. Rev.* C **85** 054301
[4] Shimizu N, Abe T, Tsunoda Y, Utsuno Y, Yoshida T, Mizusaki T, Honma M and Otsuka T 2012 *Prog. Theor. Exp. Phys.* **2012** 01A205
[5] Honma M, Otsuka T, Brown B A and Mizusaki T 2005 *Eur. Phys. J.* A **25** s01 499
[6] Honma M, Otsuka T, Mizusaki T and Hjorth-Jensen M 2009 *Phys. Rev.* C **80** 064323
[7] Hjorth-Jensen M, Kuo T T S and Osnes E 1995 *Phys. Rep.* **261** 125
[8] Entem D R and Machleidt R 2003 *Phys. Rev.* C **68** 041001(R)
[9] Gloeckner D H and Lawson R D 1974 *Phys. Lett.* B **53** 313
[10] National Nuclear Data Center, information extracted from the NuDat 2 database, http://www.nndc.bnl.gov/nudat2/
[11] Pritychenko B, Choquette J, Horoi M, Karamy B and Singh B 2012 *Atomic Data and Nuclear Data Tables* **98** 798
[12] Broda R *et al.* 2012 *Phys. Rev.* C **86** 064312
[13] Ring P and Schuck P 1980 *The Nuclear Many-Body Problem* (New York: Springer-Verlag)

Journal of Physics: Conference Series **445** (2013) 012029 doi:10.1088/1742-6596/445/1/012029

Re-entrance in nuclei: competitive phenomena

H Nam and D J Dean

Oak Ridge National Laboratory, P.O. Box 2008, Oak Ridge, TN 37831, USA

E-mail: namha@ornl.gov

Abstract. Using the shell-model Monte Carlo method, we investigate how temperature and rotation affect pairing properties for nuclei in the $fp - gds$ region. The re-entrance of pairing correlations with temperature is predicted at high rotational frequencies. It manifests through an anomalous behavior of the specific heat and level density.

1. Introduction

Phase transitions, as seen in superconductivity, superfluidity or ferromagnetism, are the result of competition between the order interaction and thermal fluctuations. Ordered phases (less symmetric) usually reside at lower temperatures and transition to disordered phases (more symmetric) at higher temperatures due to thermal fluctuations. Re-entrance (or partial order) phenomena manifests itself in successive phase transitions as a function of temperature or other extensive quantity. For example, re-entrance was discovered experimentally for liquid crystals in 1975 [1], showing a higher symmetry nematic phase occurring at a lower temperature than the lower symmetry smectic phase. Interestingly, the discovery of this phenomena led to many technological applications in electronic displays. Other condensed matter systems that display re-entrant phenomena include ferromagnetic insulators [2] and orbital antiferromagnetism [3].

The properties of nuclei are governed by the nuclear interactions. The singlet-S and triplet-P channels in the nucleon-nucleon interaction generate global nuclear pairing properties. Indeed, a large even-even nucleus will exhibit specific pairing properties, such as a large pairing gap from the ground state to the first excited state [4]. The nuclear Hamiltonian also generates other phenomena such as intrinsic deformation, with an experimental signature of the energy of a band of states increasing like $E(J) \sim J(J+1)$ where J is the spin of the state. Pairing and deformation are competitive phenomena in nuclei. One can heat a nucleus thermally and induce a pairing transition. This transition occurs at around a temperature of $T = 0.7 - 1.5MeV$ where thermal fluctuations overcome the effects of pairing [5–8]. A clearly defined peak occurs in the specific heat at these temperatures. If the nucleus is well deformed, that peak structure can be smeared. Early work on thermally assisted pairing predicted by Kammuri in 1964 [9], describes a local increase of pairing correlations in a rotating nucleus with excitation energy. One can also enhance or suppress structures in a nucleus through cranking, which practically involves adding to the Hamiltonian a time-reversal breaking term ωJ_z, where ω is the cranking frequency, and J_z is the projection of spin along the axis of rotation. As the cranking frequency increases, pairing within the nucleus should decrease due to spin alignment [10].

The objective of this work is to present a short survey on our studies using the shell-model Monte Carlo method to examine how temperature and rotation affect pairing properties on the

$N = 40$ systems, ^{68}Ni, ^{70}Zn, ^{72}Ge and ^{80}Zr, which exhibit distinct phase-transitional behavior [10, 11]. Interestingly, we have found one system, ^{72}Ge in which a re-entrant phenomena occurs where at high cranking frequency there is a small window in temperature where pairing actually increases at a critical temperature [10]. This leads to a specific heat that shows a sharp decrease with temperature. We also present early results on current work investigating the phase-transitional behavior of the odd-odd $N = Z$ nuclei, ^{70}Br, ^{74}Rb and ^{78}Y.

2. The Shell-Model Monte Carlo Method

The shell-model Monte Carlo (SMMC) method [12] allows investigations of nuclei at finite temperatures with the relevant degrees of freedom included. This makes it possible to account for the thermal and quantal fluctuations which are important to describe phase transformations in finite-size systems. The SMMC approach describes nuclear observables at finite-temperature as thermal averages

$$\langle A \rangle = \frac{Tr_N(Ae^{-\beta H})}{Tr_N e^{-\beta H}} \tag{1}$$

where $\beta = 1/T$ is the inverse temperature, Tr_N is the many-body trace at fixed particle number N, and $-\beta H$ is the imaginary-time many-body propagator. For certain classes of residual nucleon-nucleon interactions [13], like the attractive pairing+quadrupole force employed in this work, the evaluation of observables is exact, subject only to statistical errors related to the Monte Carlo integration. As we are concerned here with a description of collective quadrupole and pairing correlations at relatively low energies, we have employed a pairing+quadrupole-quadrupole Hamiltonian

$$H = \sum_{jmt_z} \epsilon(j) a^{\dagger}_{jmt_z} a_{jmt_z} - \frac{G}{4} \sum_{\alpha,\alpha',t_z} P^{\dagger}_{JT=01,t_z}(\alpha) P_{JT=01,t_z}(\alpha') - \chi \sum_{\mu} (-1)^{\mu} Q_{2\mu} Q_{2-\mu}, \tag{2}$$

where $Q_{2\mu}$ is the mass quadrupole moment operator given by

$$Q_{2\mu} = \frac{1}{\sqrt{5}} \sum_{ac} \langle j_a \| \frac{dV}{dr} Y_2 \| j_c \rangle [a^{\dagger}_{j_a} \times \tilde{a}_{j_c}]^{2\mu} \tag{3}$$

with projection μ, $a^{\dagger}_{jmt_z}(a_{jmt_z})$ creates (destroys) a nucleon of isospin t_z in the orbital jm, and $\tilde{a}_{jm} = (-1)^{j+m} \hat{a}_{j-m}$, and the seniority pairing operator $P^{\dagger}$ is defined as

$$P^{\dagger}_{JT=01,t_z}(\alpha) = (-)^l [a^{\dagger}_{\alpha,t_z} \times a^{\dagger}_{\alpha,t_z}]^{JM=00,T=1} \tag{4}$$

with $\alpha = \{n, l, j\}$.

The details of our SMMC calculations follow those of Refs. [10, 11]. Namely, calculations were performed in the complete $(0f1p - 0g1d2s)$ model space for protons and neutrons. The single-particle (s.p.) energies have been determined from a Woods-Saxon potential parametrization of ^{56}Ni. Using the parameters $G = 0.106$ MeV and $\chi = 0.0104$ MeV^{-1} fm^2, we reproduce the low-energy spectrum of ^{64}Ni and ^{64}Ge. Nuclei are described by valence protons and neutrons outside the closed core of ^{40}Ca. In order to generate the angular momentum polarization, we consider the Routhian $\hat{H}^{\omega} = \hat{H} - \omega \hat{J}_z$, where the cranking frequency ω (in units of MeV) enters through the cranking term. Our SMMC calculations were performed on Jaguar, a Cray XT5 at Oak Ridge National Laboratory (ORNL) with 18,688 nodes containing dual hex-core AMD Opteron processors, totalling 224,256 cores. Simulations investigated cranking frequencies $\omega = 0.0$ to 0.5 MeV and the temperature parameter β was split into $N_{\beta}\Delta\beta$ slices with $\Delta\beta = 1/32$ MeV^{-1}. Each parameter set was run with up to 15 840 statistical samples, requiring a total of 190,080 cores for 4 hours for a full parameter study of a single nucleus.

Journal of Physics: Conference Series **445** (2013) 012029 doi:10.1088/1742-6596/445/1/012029

3. Nuclei in the *fp-gds* region

To understand the competition between nuclear pairing correlations as a function of the rotational frequency and temperature for nuclei, we first apply the SMMC method to look at known nuclei in the $fp-gds$ region where sizable pairing correlations dominate in their ground state without the effects of rotation. Studying the temperature-induced interplay between deformation and pairing in four $N = 40$ isotones, ^{68}Ni, ^{70}Zn, ^{72}Ge, and ^{80}Zr shows that shape and pairing contribute to the specific heat of the nucleus [11]. Theoretical studies to identify phase transitions in nuclei have focused on the relationship between pairing correlations and an associated peak in the specific heat $C_v = dE/d(kT)$. To assess the magnitude of pairing correlations, it is convenient to employ the $J = 0$ pair operator

$$\Delta^\dagger = \sum_{jm>0} a^\dagger_{jm} a^\dagger_{j\bar{m}}, \tag{5}$$

where we sum over orbitals with $m > 0$ and the time-reversed creation operator is $a^\dagger_{j\bar{m}} = (-)^{j-m} a^\dagger_{j-m}$. The observable $\langle \Delta^\dagger \Delta \rangle$, referred to as the $J = 0$ pairing strength, measures the number of $J = 0$, $T = 1$ pairs in the correlated nuclear state.

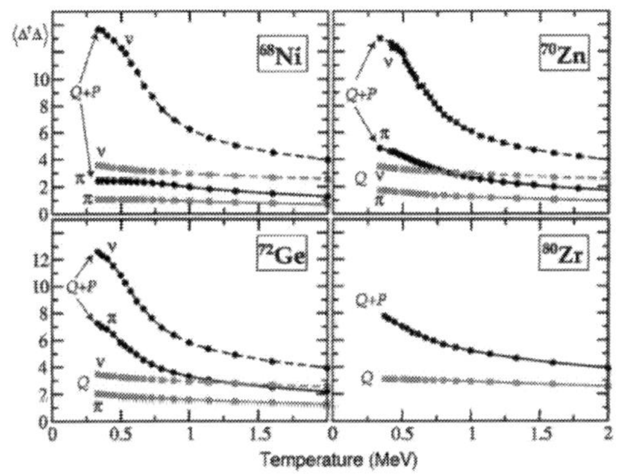

Figure 1. (color online). Neutron and proton $J = 0$ pairing strength $\langle \Delta^\dagger \Delta \rangle$ as a function of temperature for the four $N = 40$ isotones. Shown are proton and neutron pairing content with $(Q+P)$ and without (Q) the pairing term included. Taken from [11].

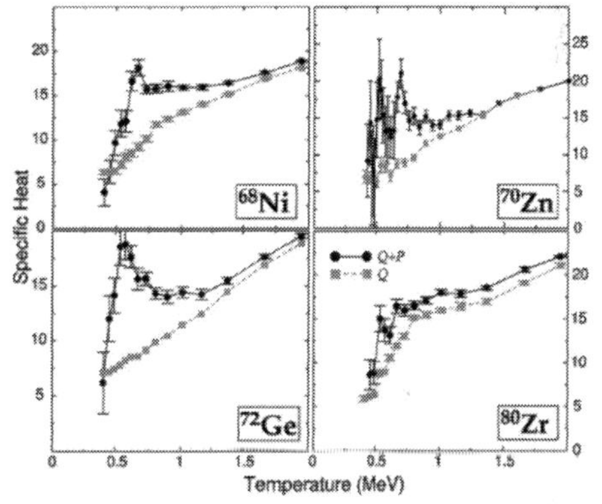

Figure 2. (color online). Specific heat calculated as a function of temperature for the four $N = 40$ isotones. The squares (dashed line) indicate calculations with only the $Q \cdot Q$ part of the interaction present, while the circles (solid line) show both $P^\dagger P$ and $Q \cdot Q$ terms included. Taken from [11].

Figure 1 shows the neutron and proton $J = 0$ pairing strength and Fig. 2 shows the corresponding specific heat calculations for the four $N = 40$ isotones. Only for the full interaction $(Q + P)$ do peaks appear in the specific heat corresponding to the increase in pairing strength at low temperatures. In the case of ^{68}Ni, proton pairing correlations are very weak and the peak in the C_v is associated with the collapse of the neutron pairing with temperature. For ^{70}Zn, the two peaks in the specific heat can be associated separately with the proton pairing and neutron pairing transitions to normal phase, which occur at different rates with temperature. The proton and neutron pairing transitions in ^{72}Ge occur in a similar range of temperatures, resulting in a pronounced single peak in C_v. In the strongly deformed ^{80}Zn, there is an absence of a sharp maxima in C_v with a pairing strength equal for proton and neutron that is significantly reduced from the lighter $N = 40$ isotones.

3.1. Re-entrance in ^{72}Ge

In the absence of rotation, the SMMC calculations [11] gives clear evidence for the breaking of isovector pairs at temperatures around $kT_c \approx 0.6$ MeV in ^{72}Ge which is reflected by a noticeable peak in the specific heat C_v. To study the effects of rotation, we employ the cranking term from $\omega = 0.0$ to 0.5 MeV. In a model of non-interacting particles, the neutrons in ^{72}Ge would completely occupy the fp shell. However, correlations induced by the residual interaction make it energetically favorable to scatter neutrons across the N=40 shell gap which is about 2.5 MeV. From Ref. [10], we see the effects of rotating the nucleus. Figure 3 shows single-neutron occupations in the wave function of ^{72}Ge as a function of rotational frequency at two temperatures: $kT = 0.47$ meV (slighly above g.s.) and 1.6 MeV (well above T_c). In the g.s. configuration, the total neutron occupation of the gds shell is about 3.5, with about 3 neutrons in the $g_{9/2}$ orbital. Upon rotating the nucleus, the s.p. cranking term $\omega\hat{j}_z$, representing the combined effect of the Coriolis and centrifugal force, generates angular momentum polarization by lifting the magnetic m-degeneracy of s.p. states.

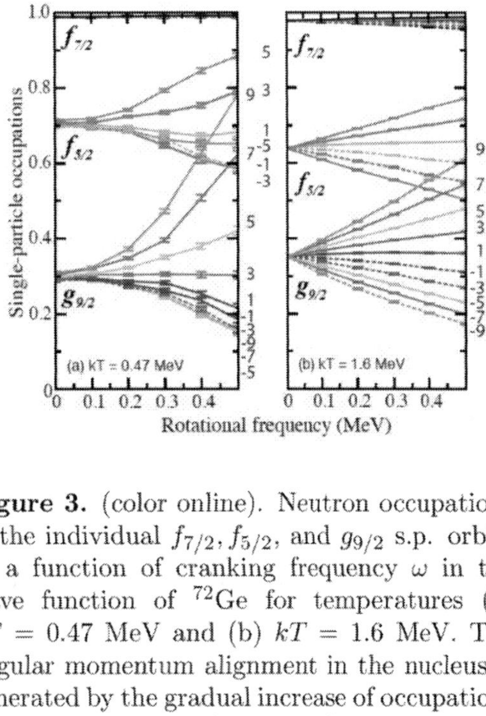

Figure 3. (color online). Neutron occupations of the individual $f_{7/2}, f_{5/2}$, and $g_{9/2}$ s.p. orbits as a function of cranking frequency ω in the wave function of ^{72}Ge for temperatures (a) $kT = 0.47$ MeV and (b) $kT = 1.6$ MeV. The angular momentum alignment in the nucleus is generated by the gradual increase of occupations of positive-m orbits (solid lines) at the expense of negative-m states (dashed lines). The doubled magnetic quantum numbers, $2m$, of individual orbits are marked. Taken from [10].

The highly symmetric pattern of s.p. occupations seen in Fig.3(a) is due to an interplay between the Coriolis force, which tries to align individual single-particle angular momenta along the axis of rotation, thus introducing spatial polarization of the system, and the symmetry-restoring pairing force. At low temperatures, the occupations of the various m substates of the $g_{9/2}$ and $f_{5/2}$ orbitals do not follow a simple thermal ordering. In fact, the occupations of the substates with largest m values are largest, as expected from energy considerations, but far from one (the extreme s.p. limit). On the other hand, the states with the lowest negative m values have very similar reduced (but nonzero) occupations, indicating that the residual population, although disfavored by the cranking term, allows the system to gain energy through pairing.

If the nucleus is heated above T_c, or if the rotation is rapid, pairing correlations are dramatically reduced, as seen in Fig. 4. The neutron pairing locally increases with temperature at low values of T and high rotational frequencies. This is a clear signal of thermally assisted pairing. One of the signatures of re-entrance of the partial order is the anomalous specific heat behavior seen in Fig. 5. At $\omega = 0.5$ MeV, the specific heat exhibits a clear local dip associated with the pairing re-entrance. A pronounced low-temperature, high-frequency irregularity is also seen in the level density ρ, shown in the inset of Fig. 5.

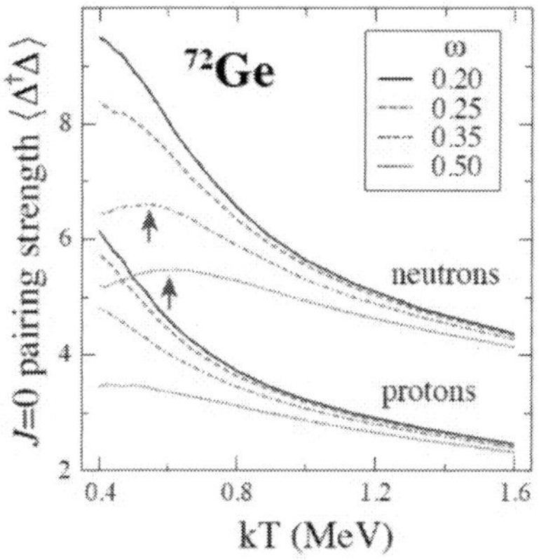

Figure 4. (color online). Neutron and proton $J = 0$ pairing strength $\langle \Delta^\dagger \Delta \rangle$ as a function of temperature for ^{72}Ge at four values of ω (in MeV). At higher rotational frequencies, neutron pairing locally increases at low temperatures reaching a maximum (marked by arrows). Taken from [10].

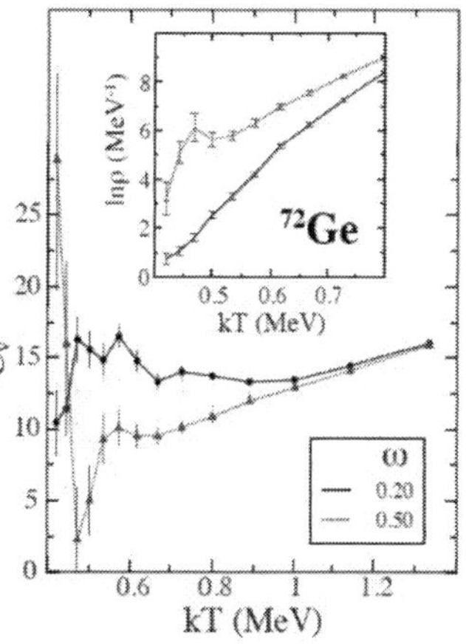

Figure 5. (color online). Specific heat calculated in SMMC for ^{72}Ge as a function of temperature at $\omega = 0.2$ and 0.5 MeV. At $\omega = 0.5$ MeV, C_v shows a local dip at low temperatures; this is a signature of pairing re-entrance. The inset shows the calculated level density. Taken from [10].

3.2. Odd-odd, $N = Z$ nuclei

We extend our studies on the effects of rotation and temperature on pairing correlations by exploring medium mass ($A > 40$) odd-odd self-conjugate $N = Z$ nuclei. The proton-neutron (pn) interaction has long been recognized to play an important role in $N = Z$ nuclei. The pn correlations can either correspond to isovector ($T = 1$) or isoscalar ($T = 0$) pairs. The pn pairing correlations are particularly important in self-conjugate odd-odd nuclei, evident in the ground state spins and isospins. The experimental ground state spins and isospins of virtually all odd-odd $N = Z$ nuclei with $A \leq 40$ (with the exception of ^{34}Cl) have ground states with $T = 0, J > 0$ and are dominated by isoscalar pairing involving pn pairs in identical orbitals. In contrasts, medium mass odd-odd $N = Z$ nuclei have $T = 1$ and $J = 0$ ground state isospins and spins (with the exception of ^{58}Cu), indicating the dominance of isovector pn pairing. The competition between isovector and isoscalar pairing is highlighted in the odd-odd $N = Z$ nuclei ^{74}Rb, where experimental results identified the $T = 0$ and $T = 1$ bands [14]. The ground state band is identified as isospin $T = 1$, arising from isovector pn correlations, with $T = 0$ states favored at higher rotational frequency, or equivalently higher excitation energy at about 1.0 MeV.

To study the effects of rotation and temperature on pn pairing correlations in odd-odd $N = Z$ nuclei, we performed SMMC calculations of ^{74}Rb and two adjacent odd-odd self-conjugate nuclei ^{70}Br and ^{78}Y. ^{78}Y has a highly deformed $J = 0$, $T = 1$ ground state with considerable quenching of the pn pairing compared to ^{74}Rb [15] and ^{70}Br is predicted to have a $J = 0$, $T = 1$ ground state [16], which has not been verified experimentally. Figure 6 shows SMMC calculations for (a) the heat capacity, C_v, (b) the $J = 0$ pairing strength, $\langle \Delta^\dagger \Delta \rangle$, and (c) the squared isospin $\langle \mathrm{T}^2 \rangle$ as a function of temperature at different cranking frequencies of $\omega = 0.0, 0.2$, and 0.5 MeV. Results show that all three nuclei exhibit similar isospin behavior. At low temperature and small frequencies $T = 1$, so that $T(T+1) = 2$ and at high frequency, $T = 0$ and $T(T+1) = 1$, shown in Fig.6(c).

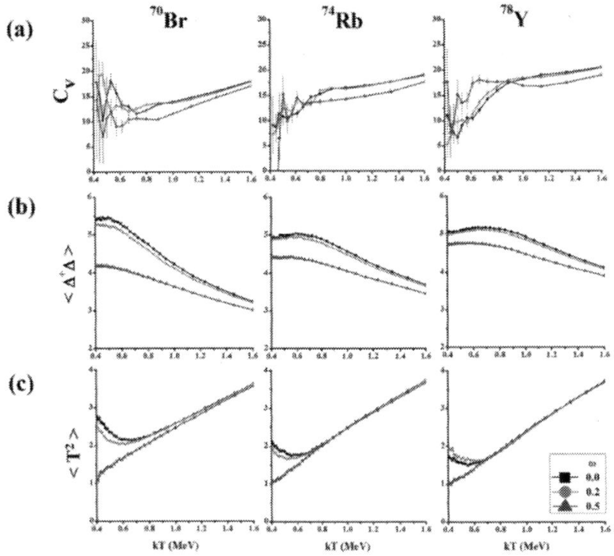

Figure 6. (a) Specific heat, (b) the $J = 0$ pairing strength, $\langle\Delta^\dagger\Delta\rangle$, and (c) $\langle T^2\rangle$ calculated in SMMC for odd-odd $N = Z$ nuclei ^{70}Br, ^{74}Rb, and ^{78}Y as a function of temperature at $\omega = 0.0, 0.2$, and 0.5 MeV.

The $J = 0$ pairing strength seen in Fig.6(b) shows contrasting behavior to that seen in ^{72}Ge, where a local increase in the pairing strength occurs at low frequency and low temperature rather than at high frequency. A clear indicator of a phase transition, such as the dip seen for ^{72}Ge, does not appear for these nuclei seen in Fig 6(a). The high frequency specific heat exhibits interesting behavior for ^{74}Rb and ^{78}Y, by crossing the low frequency specific heat and remaining unusually high. Understanding of these new behaviors requires additional study.

4. Conclusions

In summary, using the SMMC technique we explored the interplay between temperature and rotation in $fp - gds$ nuclei and its effect on pairing resulting in phase transitions. Our calculations demonstrate the presence of the partial order phenomenon associated with the reappearance of pairing at high rotational frequencies and intermediate temperatures for ^{72}Ge. Similar behavior is not currently seen for odd-odd $N = Z$ nuclei, but requires further studies to adequately understand results.

Acknowledgments

Supported by the U.S. Department of Energy under Contract No. DE-AC05-00OR22725 with UT-Battelle, LLC (Oak Ridge National Laboratory). Computational resources provided by the Oak Ridge Leadership Computing Facility.

References

[1] Cladis P E 1975 *Phys. Rev. Lett.* **35** 48–51
[2] Wei J H, Hou D and Wang X R *New J. of Phys.* **12** 053021
[3] Harrison N, Jaime M and Mydosh J A 2003 *Phys. Rev. Lett.* **90** 096402
[4] Bohr A, Mottelson B R and Pines D 1958 *Phys. Rev.* **110** 936–8
[5] Goodman A L 1981 *Nuc. Phys.* A **352** 30 – 44
[6] Dean D J, Koonin S E, Langanke K, Radha P B and Alhassid Y 1995 *Phys. Rev. Lett.* **74** 2909–12
[7] Liu S and Alhassid Y 2001 *Phys. Rev. Lett.* **87** 022501
[8] Nakada H and Alhassid Y 2008 *Phys. Rev.* C **78** 051304
[9] Kammuri T 1964 *Prog. of Theor. Phys.* **31** 595–608
[10] Dean D J, Langanke K, Nam H and Nazarewicz W 2010 *Phys. Rev. Lett.* **105** 212504
[11] Langanke K, Dean D J and Nazarewicz W 2005 *Nuc. Phys.* A **757** 360 – 72
[12] Koonin S E, Dean D J and Langanke K 1997 *Phys. Rep.* **278** 1 – 77
[13] Lang G H, Johnson C W, Koonin S E and Ormand W E 1993 *Phys. Rev.* C **48** 1518–45
[14] Rudolph D *et al.* 1996 *Phys. Rev. Lett.* **76** 376–9
[15] Uusitalo J *et al.* 1998 *Phys. Rev.* C **57** 2259–63
[16] Van Isacker P and Juillet O 2002 *Euro. Phys. J.* A **15** 233–6

International Symposium on Exotic Nuclear Structure From Nucleons (ENSFN 2012) IOP Publishing
Journal of Physics: Conference Series **445** (2013) 012030 doi:10.1088/1742-6596/445/1/012030

Exotic ideas on mixed proton-neutron symmetry and their manifestation in heavy nuclei

N Pietralla[1], C Bauer[1], J Leske[1,3], O Möller[1], T Möller[1], P von Neumann-Cosel[1], G Rainovski[2], A Scheikh-Obeid[1] and C Walz[1]

[1] Institut für Kernphysik, Technische Universität Darmstadt, 64289 Darmstadt, Germany
[2] Faculty of Physics, St. Kliment Ohridski University of Sofia, 1164 Sofia, Bulgaria

Abstract. The one-quadrupole phonon excitation of mixed symmetry, the $2^+_{1,\mathrm{ms}}$ state, is a fundamental building block of nuclear structure. This article gives a summary of our recent experimental research on this excitation mode in the $A = 90$ and $A = 130$ mass regions.

1. Introduction

The contemporary quest for the structure of exotic nuclei addresses nuclear systems with abnormal ratios of proton and neutron numbers. It, hence, focuses on the isospin dependence of nuclear structure. The isospin degree of freedom of collective nuclear structures has been studied [1] some 35 years ago in the framework of the proton-neutron version [2] of the interacting boson model [3] by Takaharu Otsuka who worked at that time with Arima and Iachello. The formulation of the IBM-2 in its F-spin limit [4, 5] has emphasized the fundamental role of collective isovector valence-shell excitations, so-called mixed-symmetry states (MSSs), for the first time [2]. After the discovery of the scissors mode [6] and the clarification of its quadrupole-collective character [7, 8] it became obvious that the isovector quadrupole excitation of the valence shell represents the building block of mixed-symmetric structures as it is particularly apparent in the framework of the Q-phonon scheme [9] for mixed-symmetry states [10, 11]. The Q-phonon scheme considers ground state correlations by only providing relative wave functions and thereby benefits from a wide applicability to good approximation [12, 13, 14].

Vibrational nuclei exhibit a one-quadrupole phonon excitation as the lowest-lying state of mixed pn symmetry, *i.e* the $2^+_{1,\mathrm{ms}}$ state. Its close relation to the 2^+_1 state is evident in the Q-phonon scheme where the wave functions of the one-quadrupole phonon excitations are in general well approximated by the expressions

$$|2^+_1\rangle \simeq Q_s\,|0^+_1\rangle = [Q_\pi + Q_\nu]\,|0^+_1\rangle \tag{1}$$

$$|2^+_{1,\mathrm{ms}}\rangle \simeq Q_m\,|0^+_1\rangle = N\left[\frac{Q_\pi}{N_\pi} - \frac{Q_\nu}{N_\nu}\right]|0^+_1\rangle \tag{2}$$

where $Q_{\pi,\nu}$ ($N_{\pi,\nu}$) denote the proton and neutron quadrupole operators (boson numbers), $N = N_\pi + N_\nu$, and $|0^+_1\rangle$ is the (in general highly correlated) ground state of a collective even-even nucleus. Despite its fundamental role in nuclear structure, the $2^+_{1,\mathrm{ms}}$ state has only recently been

[3] Present address: Helmut Fischer Gmbh, Sindelfingen, Germany

studied systematically, *e.g.*, [15, 16, 17, 18, 19]. The dominant fragments of the one-phonon $2^+_{1,\mathrm{ms}}$ state are observed at about 2 MeV excitation energy. Due to their isovector character, MSSs decay rapidly by dipole transitions and are very short lived, typically a few tens of femtoseconds. Large $M1$ matrix elements of $\approx 1\ \mu_N$ are the unique signatures for MSSs and, thus, lifetime information is needed for making safe assignments of mixed symmetry. A review article on the status of experimental information on mixed symmetry states in vibrational nuclei has been published [20]. We use the occasion of Taka Otsuka's 60th birthday to summarize our recent experimental research on one-phonon MSSs. While we have recently been asked to talk on this subject at various occasions the presentation of the present contribution is similar to what we have recently contributed to other conference proceedings.

2. Experimental Method

Projectile-Coulomb excitation has been established as a powerful method for the identification and investigation of one-phonon MSSs [17, 19]. After this approach has first been applied to the investigation of the $2^+_{1,\mathrm{ms}}$ state of ^{96}Ru at the Yale Tandem accelerator [17], we have begun a research programme on the $2^+_{1,\mathrm{ms}}$ state at Argonne National Laboratory with the nucleus ^{138}Ce as a first case study [19]. Crucial influence of sub-shell closures on mixed-symmetry structures was first observed, a phenomenon which sensitively tests the effective proton-neutron interaction in microscopic valence shell models [21, 22]. The one-phonon $2^+_{1,\mathrm{ms}}$ state of ^{136}Ce has recently been identified from similar Coulomb excitation experiments at Gammasphere [23].

A sequence of experiments on Xenon, Barium, and Cerium isotopes has been performed. The superconducting ATLAS accelerator provided the ion beams with energies corresponding to $\sim$ 85 % of the Coulomb barrier for a reaction on ^{12}C nuclei. The beam intensity amounted typically to $\sim$ 1pnA. The beam was impinging on a stationary carbon target of thickness 1 mg/cm^2. Light target ions were chosen in order to favor the one-step Coulomb excitation process over multi-step processes for ease of data evaluation. The γ-rays emitted by Coulomb-excited states of the beam nuclei were detected in the Gammasphere array which consisted of $\sim$ 100 high purity Compton suppressed Germanium detectors arranged in 16 rings. An event was defined by a γ-ray of multiplicity 1 or higher. Two corrections had to be done in order to get the total single spectra, an example of which is displayed in Fig. 1, namely the Doppler correction (recoiling velocity $\sim$ 6-8%) and the background subtraction (difference between the "in-beam" spectrum and the "off-beam" spectrum scaled to eliminate the 1461 keV ^{40}K line).

The experimental γ-ray spectra are dominated by the decays of low-spin states, such as 2^+ or 3^- states, that are predominantly populated by one-step Coulomb excitation from the ground state. For each state observed we measured the excitation cross section relative to that

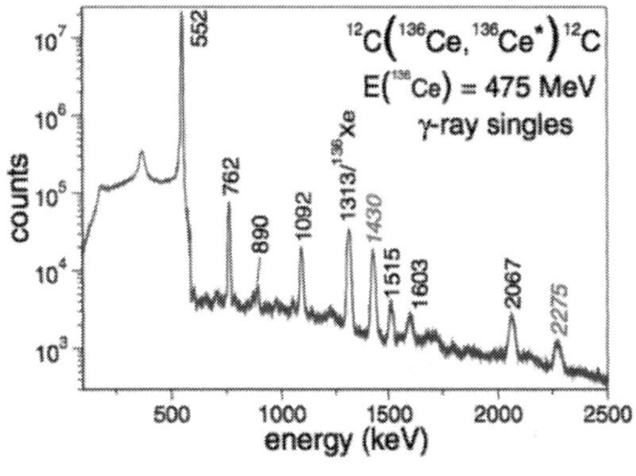

Figure 1. Background-subtracted and Doppler-corrected singles γ-ray spectrum summed over all Ge detectors of the Gammasphere array at ANL after Coulomb excitation of ^{136}Ce on a carbon target [23].

of the 2_1^+ state with an accuracy of 1 - 0.1 %. By calculating the Coulomb excitation cross sections for each excited state with the multiple-Coulomb excitation formalism and fitting them to our experimental data (normalized to the 2_1^+ state), we deduced the electromagnetic matrix elements corresponding to each transition of the excited states. The crucial multipole mixing ratios of the $2^+ \rightarrow 2_1^+$ transitions were obtained from γ-ray angular distributions if sufficient statistics have been obtained. A possible large $B(M1)$ value, signature of the MSS, is then easily derived from the data. For a further description of this method, the reader is referred to Refs. [19, 20]. This experimental technique of projectile-Coulomb excitation on a light target inside the Gammasphere array at ANL has been applied by us to 16 nuclei up to now: ^{136,138}Ce, $^{124-134}$Xe, ^{148,154}Sm, ^{194,196}Pt, ^{130,132}Ba, ^{96}Ru and ^{94}Mo. Figure 1 displays data from the projectile-Coulomb excitation reactions of a ^{136}Ce-ion beam on a carbon target.

3. Evolution of $2^+_{1,\mathrm{ms}}$ states in the $A = 130$ region

The experiments performed so far allow for a nearly complete overview on the properties of the MSS throughout the $A = 130$ region. A recent publication on the first identification of a MSS in an unstable nucleus in ^{132}Te [24] expands the experimental data on the $N = 80$ isotonic chain. A recent publication on the MSS in the nucleus ^{136}Ce [23] completes the experimental data on the stable even-even $N = 78$ isotonic chain. A set of data currently under analysis on the nucleus ^{132}Ba [25] will complete our information on the one-phonon mixed-symmetry state in the stable even-even $N = 76$ isotones. An overview on the $B(M1; 2_i^+ \rightarrow 2_1^+)$ strength distributions in the even-even nuclei in the $A = 130$ region is shown in Figure 2.

In the stable $N = 80$ isotones the excitation energy of the $2^+_{1,\mathrm{ms}}$ state increases with increasing proton number. This trend continues in the unstable nucleus ^{132}Te [24]. In the $N = 78$ isotonic chain, the energy of the MSS again increases with increasing proton number. In the neighboring $N = 76$ isotones, however, the excitation energy of the MSS decreases with increasing proton number. It is also interesting to follow the evolution of the MSS's excitation energies in the different isotopic chains. In the Ce and Ba isotopes, the excitation energy of the MSS increases with increasing neutron number, whereas in the Xe isotopes an increase in N_ν results in a decrease of $E(2^+_{1,ms})$. Apparently, the $2^+_{1,\mathrm{ms}}$ state evolves in different ways as a function of valence particle numbers. Whether or not the observed differences are related to a critical point of a nuclear shape phase transition near ^{134}Ba is unclear up to now.

From data on $E(2_1^+)$ and $E(2^+_{1,ms})$, an estimate of the proton-neutron quadrupole-quadrupole interaction V_{pn}^{QQ} according to the two-state mixing scheme in [26] has been performed on the

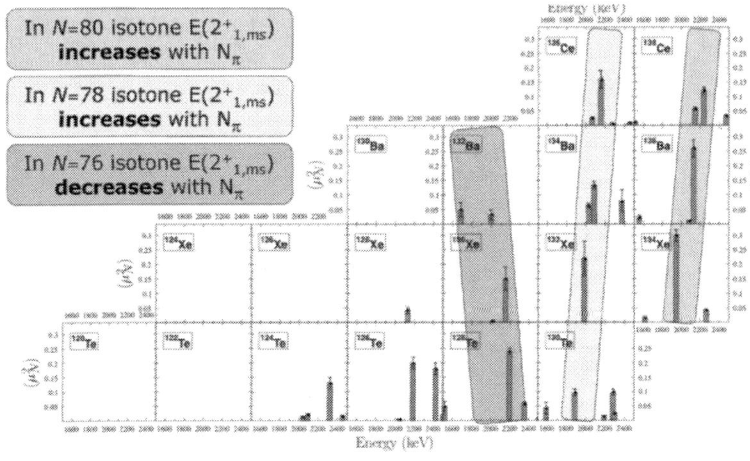

Figure 2. Overview of the $B(M1; 2_i^+ \rightarrow 2_1^+)$ strength distributions for the stable even-even nuclei in the A=130 region.

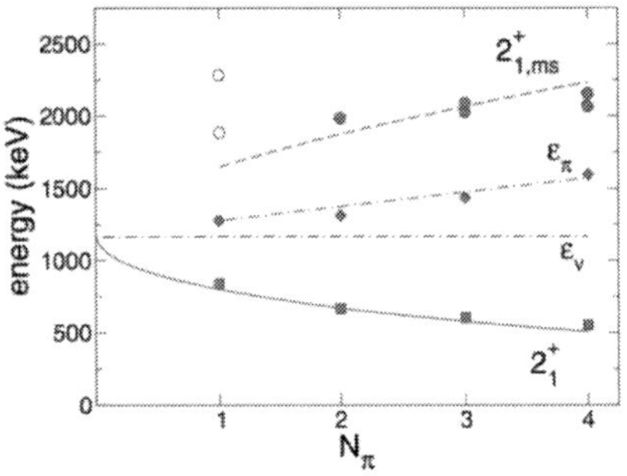

Figure 3. Simultaneous fit of the experimental energies of the 2_1^+ states (solid curve) and of the $B(M1)$-weighted average energies of the $2^+_{1,\mathrm{ms}}$ states (dashed curve) in the $N = 78$ isotones. Taken from [23].

$N = 80$ isotones [27], the Xe isotopes [28], and, just recently, on the $N = 78$ isotones [23]. The results show, that the proton-neutron quadrupole-quadrupole interaction in the $N = 78$ isotonic chain is about 14% smaller than that for the $N = 80$ isotopic chain [27] and about 6% smaller than for the Xenon isotopic chain [28]. An example of the data, taken from Ref. [23], is displayed in Fig. 3.

4. Phase of proton- and neutron-components to MSSs: The case of ^{92}Zr

We studied the formation of quadrupole collectivity in the particularly simple case of a nucleus with a low-energy structure that is dominated by one pair of valence particles each for protons and neutrons. An example is the nucleus ^{92}Zr with 2 neutrons beyond the $N = 50$ shell closure and 2 protons beyond the $Z = 38$ sub-shell closure. The lowest 2-quasiparticle (2qp) states have $\pi(1g_{9/2})^2$ and $\nu(2d_{5/2})^2$ configurations. In ^{92}Zr, the predominantly symmetric and mixed-symmetric one-phonon 2^+ states are experimentally identified as the 2_1^+ and 2_2^+ states [29, 20] with some degree of configurational isospin polarization [30].

To shed light on the microscopic origin of the effective pn-coupling strength in the valence shell we consider the quasiparticle-phonon model (QPM) [31]. The QPM wave functions are dominated by the lowest π and ν 2qp components, that show the expected in-phase and out-of-phase behavior for the 2^+_{fs} and 2^+_{ms} states. The electromagnetic properties and excitation energies are in excellent agreement with the data [32]. The magnetic moments of these states and the strong $M1$ transition between them originate almost entirely from the valence-shell configurations. However, the $B(E2)$ strengths are generated to about 80% from many components beyond the valence shell albeit their total contribution to the wave function norm is small. This observation motivates a simple three-state mixing scenario between the proton-valence shell configuration, the neutron-valence shell configuration, and the GQR for a deeper insight in the formation of the one-quadrupole phonon states with symmetric and mixed-symmetry character even on a semi-quantitative level [33]. For the nucleus ^{92}Zr with higher energy for the proton valence-shell component than the neutron valence-shell component at the $Z = 40$ sub-shell closure, this scheme inevitably requires that the neutron valence-shell component flips its phase with respect to the GQR component when going from the proton-neutron symmetric 2_1^+ state to the 2_2^+ state with predominant mixed symmetry.

Apparently, two probes with different sensitivity to protons and neutrons are needed to study this quantum interference experimentally. Electron scattering at low momentum transfer provides a measure of the charge transition radius. An (e, e') experiment was performed at the Darmstadt superconducting electron linear accelerator (S-DALINAC). An enriched (94.6 %) self-supporting ^{92}Zr target of 9.8 mg/cm^2 areal density was used. Data were taken covering a momentum transfer range between $q \sim 0.3 - 0.6$ fm^{-1} indicating no difference between the charge transition radii of the 2^+_{fs} and 2^+_{ms} states within experimental uncertainties (Fig. 4, right). Information about the neutron transition radii can be derived from the proton scattering data

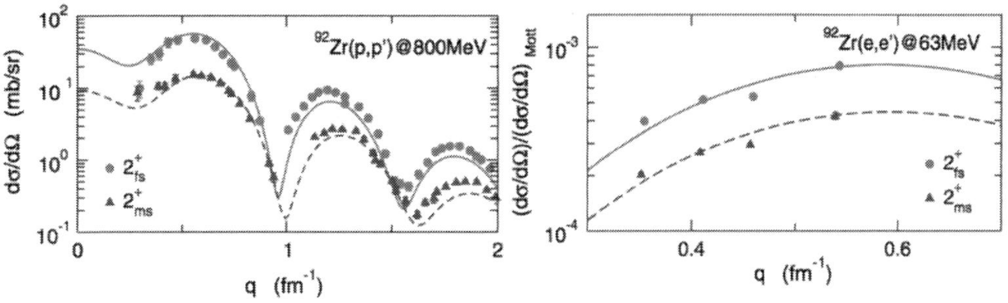

Figure 4. Form factors for the $2^+_{1,\mathrm{fs}}$ (red, solid line) and $2^+_{1,\mathrm{ms}}$ (blue, dashed line) from $^{92}Zr(p,p')$ and $^{92}Zr(e,e')$ experiments (from [33]).

of Ref. [34]. At the incident energy of 800 MeV protons interact predominantly via the isoscalar central piece of the effective projectile-nucleus interaction [35]. Clearly, the refraction pattern of the (p,p') cross section for the 2^+_{ms} state are shifted to higher q values as compared to those for the 2^+_{fs} state (Fig. 4, left) corresponding to a smaller transition radius.

Figure 5 displays the proton and neutron transition densities of the 2^+_{fs} (top) and 2^+_{ms} (bottom) states calculated in the full QPM approach. The full transition densities (solid curves) are decomposed in a collective part stemming from the GQR (dotted curves) and the predominant 2qp $\nu(2d_{5/2})^2$ neutron contributions (dashed curves).

The key point is the different radial behaviour of both parts and their relative signs. An out-of-phase coupling between the neutron valence shell contribution and the contribution from the GQR in the $2^+_{1,\mathrm{ms}}$ state leads to a destructive quantum interference that reduces the neutron transition density at large radii (due to the larger radius of the $\nu(2d_{5/2})^2$ orbital) and consequently shifts the maximum of the total neutron transition density to the interior with respect to that one for the 2^+_{fs} state, as indicated by the arrows in Fig. 5. This effect reduces the neutron transition radius of the 2^+_{ms} with respect to the 2^+_{fs} state of ^{92}Zr. In contrast, the proton transition radius remains essentially unchanged since the $\pi(1g_{9/2})^2$ part couples in-phase to the GQR contribution in both states. The combination of both data sets unambiguously demonstrates for the first time that the phase of the neutron valence-shell configurations in ^{92}Zr changes its sign between the 2^+_{fs} and the 2^+_{ms} state [33].

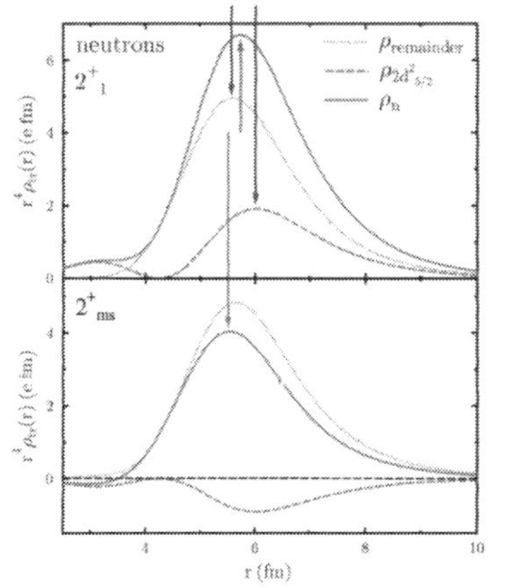

Figure 5. Neutron transition densities of the 2^+_{fs} (top) and 2^+_{ms} (bottom) states of ^{92}Zr from QPM calculations. The full transition densities (solid lines) are decomposed in parts stemming from the GQR (dotted lines) and from the main 2qp configurations (dashed lines). The arrows indicate the maxima of the corresponding full transition densities.

5. Summary

The isovector one-quadrupole phonon excitation of the valence shell, the $2^+_{1,\mathrm{ms}}$ state with F-spin $F = F_{\mathrm{max}} - 1$, has been systematically investigated in a large number of vibrational nuclei. This state is generally identified from absolute $M1$ transition strengths when the experimental sensitivity is high enough and it occurs at energies around 2 MeV featuring an $M1$ transition matrix element to the 2^+_1 state between 0.5 and 1.5 μ_N with some fragmentation. The details of its evolution as a function of particle number is not entirely understood. It may depend on the local shell structure around the Fermi level and on the evolution of quadrupole deformation.

Acknowledgments

We thank Taka Otsuka for inspiration to new experimental endeavors and for many discussions. We thank all those who have contributed to our research on various aspects of mixed-symmetry states. With respect to the data discussed in this contribution we particularly want to acknowledge the scientists and staff members at the Argonne National Laboratory, USA, at the iThemba Labs, South-Africa, and at the S-DALINAC, Germany. Support from the DFG under grants Pi393/2-2 and SFB 634, by the German-Bulgarian exchange program under Grant No. PPP 50751591 is gratefully acknowledged.

References

[1] Otsuka T 1978 *Boson model of medium-heavy nuclei* (University of Tokyo, PhD thesis)
[2] Arima A, Ohtsuka T, Iachello F and Talmi I 1977 *Phys. Lett.* B **66** 205
[3] Iachello F and Arima A 1987 *The Interacting Boson Model* (Cambridge University Press).
[4] Otsuka T, Arima A and Iachello F 1978 *Nucl. Phys.* A **309** 1
[5] Van Isacker P, Heyde K, Jolie J and Sevrin A 1986 *Ann. Phys.* (NY) **171** 253
[6] Bohle D, Richter A, Steffen W, Dieperink A, LoIudice N, Palumbo F , Scholten O 1984 *Phys. Lett.* B **137** 27
[7] Ziegler W, Rangacharyulu C, Richter A and Spieler C 1990 *Phys. Rev. Lett.* **65** 2515
[8] Pietralla N, Brentano P von, Herzberg R-D, Kneissl U, Lo Iudice N, Maser H, Pitz H H and Zilges A 1998 *Phys. Rev.* C **58** 184
[9] Otsuka T and Kim K H 1994 *Phys. Rev.* C **50** R1768
[10] Kim K-H, Otsuka T, Brentano P von, Gelberg A, Van Isacker P and Casten R F 1996 *Proc. Int. Conf. on Capture Gamma-Ray Spectroscopy and Related Topics* vol 1, ed G. Molnár, (Budapest: Springer) p 195
[11] Pietralla N *et al.* 1998 *Phys. Rev.* C **58** 796
[12] Pietralla N, Brentano P von, Casten R F, Otsuka T and Zamfir N V 1994 *Phys. Rev. Lett.* **73** 2962
[13] Pietralla N, Brentano P von, Otsuka T and Casten R F 1995 *Phys. Lett.* B **349** 1
[14] Pietralla N, Mizusaki T, Brentano P von, Jolos R V, Otsuka T and Werner V 1998 *Phys. Rev.* C **57** 150
[15] Pietralla N *et al.* 1999 *Phys. Rev. Lett.* **83** 1303
[16] Pietralla N *et al.* 2000 *Phys. Rev. Lett.* **84** 3775
[17] Pietralla N *et al.* 2001 *Phys. Rev.* C **64** 031301
[18] Pietralla N *et al.* 2003 *Phys. Rev.* C **68** 031305(R)
[19] Rainovski G, Pietralla N, Ahn T *et al.* 2006 *Phys. Rev. Lett.* **96** 122501
[20] Pietralla N, Brentano P von and Lisetskiy A F 2008 *Prog. Part. Nucl. Phys.* **60** 225
[21] LoIudice N, Stoyanov C and Tarpanov D 2008 *Phys. Rev.* C **77** 044310
[22] Bianco D, Andreozzi F, LoIudice N, Porrino A and Knapp F 2012 *Phys. Rev.* C **85** 034332
[23] Ahn T, Rainovski G, Pietralla N *et al.* 2012 *Phys. Rev.* C **86** 014303
[24] Danchev M *et al.* 2011 *Phys. Rev.* C **84** 061306
[25] Möller T *et al.*, in preparation.
[26] Heyde K and Sau J 1986 *Phys. Rev.* C **33** 1050
[27] Ahn T *et al.* 2009 *Phys. Lett.* B **679** 19
[28] Coquard L *et al.* 2010 *Phys. Rev.* C **82** 024317
[29] Werner V *et al.* 2002 *Phys. Lett.* B **550** 140
[30] Holt J D, Pietralla N, Holt J W, Kuo T T S and Rainovski G 2007 *Phys. Rev.* C **76** 034325
[31] Soloviev V 1992 *Theory of Atomic Nuclei: Quasiparticles and Phonons* (Bristol: Institute of Physics Pub.)
[32] LoIudice N and Stoyanov C 2006 *Phys. Rev.* C **73** 037305
[33] Walz C, Fujita H, Krugmann A, Neumann-Cosel P von, Pietralla N *et al.* 2011 *Phys. Rev. Lett.* **106** 062501
[34] Baker F *et al.* 1983 *Nucl. Phys.* A **393** 283
[35] Franey M A and Love W G 1985 *Phys. Rev.* C **31** 488

Spherical to deformed shape transitions in the nucleon pair shell model

S Pittel[1], Y Lei[1,2], G J Fu[1,2] and Y M Zhao[2]

[1] Bartol Research Institute and Department of Physics and Astronomy, University of Delaware, Newark, Delaware 19716, USA

[2] INPAC, Department of Physics and Shanghai Key Lab for Particle Physics and Cosmology, Shanghai Jiao Tong University, Shanghai, 200240, China

E-mail: pittel@bartol.udel.edu

Abstract. A study of the shape transition from spherical to axially deformed nuclei in the even Ce isotopes using the nucleon-pair approximation of the shell model is reported. As long as the structure of the dominant collective pairs is defined in a framework appropriate to deformed nuclei, the model is able to produce a shape transition. The resulting transition is too rapid, however, with nuclei that should be transitional being fairly well deformed. The possibility of using several pairs with each angular momentum to improve the description of the transitional region is discussed as is the possibility of using a larger single-particle space to better describe the moments of inertia in the deformed regime.

1. Introduction

A major goal of nuclear structure theory is to obtain a unified microscopic description of the variety of collective structures exhibited by atomic nuclei and the phase transitions linking them. A major advance was realized recently with the development of a unified microscopic theory of the Interacting Boson Model (IBM) [1], through the use of a procedure that maps the physics of density functionals onto the parameters of the model in the various collective domains [2].

We are interested in seeing whether these competing collective modes can be described using the Nucleon Pair Approximation (NPA) to the shell model [3, 4], a model that like the IBM is based on the introduction of collective pairs, but in which the collective pairs are not bosonized. In this work, we focus on nuclei that are known to undergo a shape transition from vibrational to rotational character, to see whether the NPA can describe such a shape transition and if so under what circumstances.

2. The Nucleon Pair Approximation

The NPA involves a truncation of the nuclear shell model in terms of collective pairs of alike nucleons. Originating in the Generalized Seniority (GS) and Broken Pair (BPA) approximations, it became a practical approach for the study of systems with a large number of pairs through the work of Chen [3], who obtained a set of recursive formulae that enabled the calculation of matrix elements for states involving many non-S pairs.

As a truncation of the shell model, the NPA inputs a single-particle space and an effective hamiltonian. It also inputs a set of collective pairs in terms of which the model space is defined.

This typically includes an $L = 0$ (S) pair and an $L = 2$ (D) pair, both for neutrons and protons, but can include additional pairs as well, either other S and D pairs or non-$S - D$ pairs.

The collective pairs have typically been chosen using the GS or BPA approach, with the S pairs obtained by a variational treatment in the GS=0 space and non-S pairs from a variational treatment in the GS=2 space. This approach is appropriate for semi-magic or near semi-magic nuclei, but not for nuclei with sufficiently large numbers of neutrons and protons where pn correlations need to be taken into account in defining the dominant pairs.

We have thus considered an alternative method for defining the dominant collective pairs for the NPA, based on the Hartree Fock Bogolyubov (HFB) approach, which dynamically incorporates the physics of the pn interaction.

For well-deformed nuclei, the dominant pairs can be obtained from an HFB treatment [5, 6]. For axially-symmetric nuclei, the HFB wave function in the canonical basis is given by

$$|\Phi >= c \exp\left(\sum_{\tau=\nu,\pi} \Gamma^{\dagger}_{\tau}\right) \mid \tilde{0}\rangle \ , \tag{1}$$

where

$$\Gamma^{\dagger}_{\tau} = d_{\tau} \sum_{\alpha m>0} \frac{v^{(\tau)}_{\alpha m}}{u^{(\tau)}_{\alpha m}} \tau^{\dagger}_{\alpha m}\tau^{\dagger}_{\alpha -m} \tag{2}$$

creates a coherent pair for particles of type τ, $|\tilde{0}\rangle$ denotes the doubly-magic core, and the v and u amplitudes are related to the eigenvalues of the density matrix in the canonical basis.

The pair creation operators $\Gamma^{\dagger}_{\tau}$ contain components with all allowed even angular momenta,

$$\Gamma^{\dagger}_{\tau} = \sum_{L} a^{(\tau)}_{L} \Gamma^{(L)^{\dagger}}_{\tau} \ , \tag{3}$$

where the square of the amplitude $a^{(\tau)}_{L}$ indicates the importance of the correlated pair of type τ and angular momentum L in the ground band intrinsic state and the operator $\Gamma^{(L)^{\dagger}}_{\tau}$ gives the structure of that correlated pair.

3. Calculations and Results

3.1. Calculational ingredients

The calculations we report here are for the even-mass Ce isotopes from ^{142}Ce through ^{148}Ce. We assume a ^{132}Sn doubly-magic core and distribute the 8 valence protons over the orbits in the $50 - 82$ shell and the $2 - 8$ valence neutrons over those in the $82 - 126$ shell.

Our hamiltonian is of the form

$$H = \sum_{\tau=\nu,\pi} \left(\sum_{j} \epsilon_{\tau j} \tau^{\dagger}_{j}\tau_{j} - \sum_{L=0,2} G_{\tau L} A^{\dagger}_{\tau L} \cdot \tilde{A}_{\tau L} \right) - \kappa Q_{\pi} \cdot Q_{\nu} \ . \tag{4}$$

It contains a single-particle energy term, multipole pairing interactions in the nn and pp channels and a separable quadrupole-quadrupole interaction in the pn channel.

The single-particle energies, extracted from the spectra of ^{133}Sb and ^{133}Sn, are listed in Table 1. The two-proton interaction parameters, extracted from the binding energy (BE) and 2^+_1 energy of ^{134}Te, are given by $G_{\pi 0} = -0.18$ MeV and $G_{\pi 2} = 0$. Those for the two-neutron interaction, extracted from the BE and 2^+_1 energy of ^{134}Sn, are given by $G_{\nu 0} = -0.13\ MeV$ and $G_{\nu 2} = -0.012\ MeV$. Finally, the strength of the pn quadrupole interaction, chosen to optimally describe the 0^+_1 and 2^+_1 states of ^{136}Te, is given by $\kappa = -0.20\ MeV$.

Table 1. Single-particle energies (in MeV) used in the calculations for the Ce isotopes described in the text, as obtained from the low-lying levels of ^{133}Sb and ^{133}Sn.

l_{j_π}	$s_{1/2}$	$d_{3/2}$	$d_{5/2}$	$g_{7/2}$	$h_{11/2}$	
ϵ_π	2.990	2.708	0.962	0.000	2.793	
l_{j_ν}	$p_{1/2}$	$p_{3/2}$	$f_{5/2}$	$f_{7/2}$	$h_{9/2}$	$i_{13/2}$
ϵ_ν	1.656	0.854	2.005	0.000	1.561	1.800

Table 2. Ratios of experimental excitation energies of the Ce isotopes. $R_{4/2}$ gives the 4_1^+ to 2_1^+ ratio, and $R_{6/2}$ the 6_1^+ to 2_1^+ ratio. The last column gives the results in the rotational limit.

	^{142}Ce	^{144}Ce	^{146}Ce	^{148}Ce	$SU(3)$
$R_{4/2}$	1.90	2.36	2.58	2.87	3.33
$R_{6/2}$	2.72	4.15	4.52	5.32	7.00

3.2. Results and discussion

We first ask whether the chain of Ce isotopes indeed exhibits a transition from vibrational to rotational character. Table 2 shows the ratios of experimental excitation energies in the even Ce isotopes. $R_{4/2}$ gives the 4_1^+ to 2_1^+ ratio, $R_{6/2}$ the 6_1^+ to 2_1^+ ratio and the last column gives the $SU(3)$ results. ^{142}Ce, with 2 valence neutrons, has a vibrational character, with a $R_{4/2}$ ratio of 1.90, while ^{148}Ce with 8 valence neutrons has a fairly rotational character with $R_{4/2} = 2.87$. The transition from vibrational to rotational is fairly smooth. We should emphasize, however, that we are only considering the properties of the ground band in this discussion.

In Fig. 1, we show the deformation energies ($E_{\text{deformed}} - E_{\text{spherical}}$) from our HFB calculations of these nuclei. As expected, the deformation energy grows with increasing neutron number, with the energy gain reaching 1.27 MeV in ^{148}Ce. This is a bit small for a well-deformed nucleus, perhaps reflecting the fact that our model space does not include all the orbits that could contribute meaningfully. Nevertheless, the results suggest that the HFB calculations are producing the observed shape transition in the ground band.

Next we turn to our NPA results. In all calculations we included the lowest S pair, the lowest D pair and the lowest G pair (with $J^\pi = 4^+$), for neutrons and protons, allowing however only a single G pair for each type of particle. Fig. 2 compares several sets of NPA results with the experimental data. The spectra denoted BPA make use of the broken pair prescription for the S, D and G pairs. The spectra denoted HFB make use of the axially-symmetric HFB prescription. The spectra denoted COM include the S and D pairs from the two prescriptions in the same calculation, with their nonorthogonality taken into account. We explain later why we have included the latter results.

There are several points to note. First, the traditional broken pair (BPA) results are unable to describe the vibrational to rotational shape transition as we proceed through the chain. The HFB results, in contrast, give rise to a clear shape transition. While ^{142}Ce does not exhibit a rotational pattern, adding just two neutrons leads to a rotational spectrum, which becomes even more pronounced with the addition of further neutrons.

We can get a clearer understanding as to why it is important to use dynamically-determined collective pairs by looking at the ground state binding energies that emerged relative to ^{132}Sn. These results are shown in Fig. 3. When we use the BPA prescription for the collective pairs of the model we obtain much less binding than with the HFB pairs, which include crucial pn

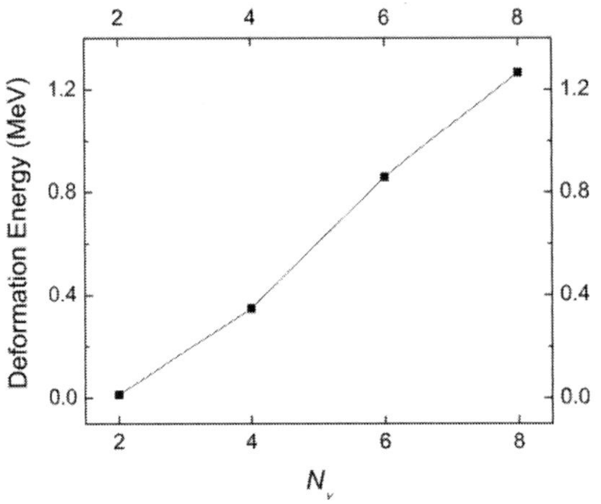

Figure 1. Deformation energies ($E_{\rm deformed} - E_{\rm spherical}$) from axially-symmetric HFB calculations for the even-mass Ce isotopes as a function of the number of valence neutrons N_ν

correlation effects in their structure.

While the method with HFB pairs does produce a shape transition from vibrational to rotational nuclei, it produces it too rapidly. As is evident from Fig. 2, it yields a well-deformed spectrum for ^{144}Ce, which from Table 2 should be transitional. It is not surprising that we are unable to adequately describe transitional nuclei, since an axially-symmetric HFB treatment should not be very meaningful for such nuclei as it is unstable against zero-point fluctuations. A possible thought is to include more than one pair for each angular momentum in this regime. As a first step in this direction, we have carried out calculations in which we include both the HFB pairs and the BPA pairs, properly orthogonalized. Those are the results denoted COM in Fig. 2. The inclusion of a second S and D pair for each type of nucleon improves the description substantially, giving a better reproduction of the properties of *all* nuclei considered. However, the shape transition it produces is still too rapid. Nevertheless, we believe that including more than one pair with each angular momentum is a possible approach for future consideration, as it can be readily included in the analysis, but further thought as to which pairs to include is needed.

It is also worth noting that even in the most deformed nucleus considered, ^{148}Ce, the moment of inertia from our best NPA calculation, that denoted COM, is still too small and the spectrum is thus too spread-out. Interestingly the same result emerged in the first efforts to microscopically derive the IBM in deformed nuclei through a mapping the of fermionic density functional onto a corresponding one in the boson space. Recently it was shown that it is necessary to add a further term to the IBM hamiltonian of the form $\alpha J(J+1)$ to better reproduce the moment of inertia of the rotational systems that were studied [7]. The NPA seems to also have need for an additional contribution to the rotational moment of inertia, with perhaps a similar origin. It is not clear at this time, however, how to incorporate it in the NPA formalism. Minimally, we should probably first increase the size of our model space to include other orbitals thought to be important in this region, as this might already improve the moments of inertia.

4. Summary and closing remarks

In this paper, we have addressed the issue of whether the NPA is able to describe the transition from vibrational to rotational nuclei in real nuclear systems. We focused on the even-mass Ce

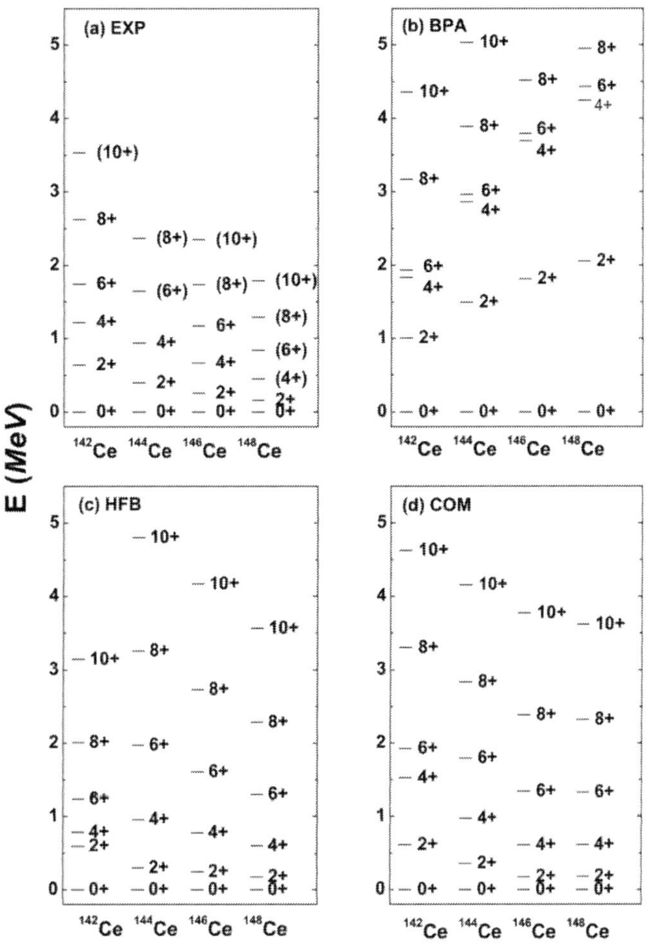

Figure 2. Comparison of the spectra calculated using the NPA for the even *Ce* isotopes (panels b-d) with the corresponding experimental spectra (panel a, denoted EXP). Panel (b) uses a model space based on BPA pairs, panel (c) uses a model space based on HFB pairs, and panel (d) (denoted COM) uses a direct sum of those used in panels (b) and (c).

isotopes, where such a shape transition seems to occur experimentally in the ground band. We see that when we use the traditional collective pairs derived from the broken pair approach, the model cannot describe the shape transition. In contrast, a description that uses collective pairs from the HFB approximation, which includes crucial pn correlation effects, is able to produce the observed shape transition. The shape transition produced is a bit too sharp, however, reflecting the fact that the correlated pairs produced in the HFB calculations are not adequate for the transitional region. Furthermore, the moments of inertia it produces are too small.

The difficulty in obtaining appropriate collective pairs to use in the NPA in the various nuclear regimes is reminiscent of the difficulty that arose in early efforts to microscopically derive the

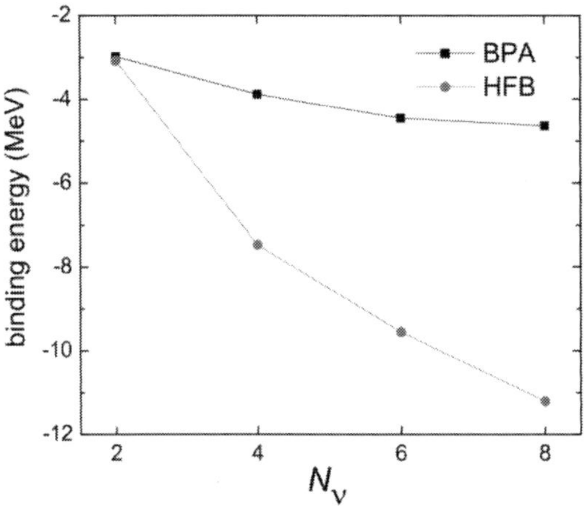

Figure 3. Comparison of the binding energies arising from the NPA calculations for the *Ce* isotopes described in the text, based on the use of BPA collective pairs and HFB collective pairs.

IBM for these different regimes. It is only recently with the advent of methods that map the fermionic collective surface onto an associated bosonic collective surface that it has been possible to obtain a unified microscopic derivation of the IBM. Since there is no analogous procedure for the NPA, we have suggested the possibility of using several collective pairs in those regimes in which a single HFB pair for each angular momentum is not sufficient. Further work to identify how to optimally obtain those pairs in transitional regimes is still needed, as is further thought on how to increase the moments of inertia that emerge within the NPA method for traditional shell-model spaces.

Acknowledgments

The work of S. P.,Y. L. and G. F. was supported in part by the National Science Foundation under Grant No. PHY-0854873, and that of Y. M. Z., Y. L., and G. J. F. by the National Science Foundation of China under Grants No. 10975096 and 11145005, and by the Science & Technology Committee of Shanghai city under Grant No. 11DZ2260700.

References

[1] Iachello F and Arima 1987 *The Interacting Boson Model* (Cambridge: Cambridge University Press)
[2] Nomura K, Shimizu N and Otsuka T 2008 *Phys. Rev. Lett.* **101** 142501
[3] Chen JQ 1997 *Nucl. Phys.* A **626** 686
[4] Zhao YM, Yoshinaga N, Yamaji S, Chen JQ and Arima A 2000 *Phys. Rev.* C **62** 014304
[5] Pittel S and Dukelsky J 1983 *Phys. Lett.* **128B** 9
[6] Maglione E and Vitturi A 1983 *Prog. Part. Nucl. Phys.* **9** 87
[7] Nomura K, Otsuka T, Shimizu N and Guo L 2011 *Phys. Rev.* C **83** 041302

Band structure of neutron rich Se and Ge isotopes

N Yoshinaga[1] **and K Higashiyama**[2]

[1]Department of Physics, Saitama University, Saitama City 338-8570, JAPAN
[2]Department of Physics, Chiba Institute of Technology, Narashino, Chiba 275-0023, JAPAN

E-mail: yoshinaga@phy.saitama-u.ac.jp

Abstract. The band structure of the neutron-rich Se and Ge isotopes has been studied in terms of the full-fledged shell model. The monopole and quadrupole pairing plus quadrupole-quadrupole interaction is employed as an effective interaction. The model reproduces well the energy levels of high-spin states as well as the low-lying states. In order to investigate the structure of the high-spin states and low-lying collective states, the energy spectra in the shell model are compared with those in the quantum-number-projected generator coordinate method. It is shown that the triaxial components play essential roles in describing the γ bands.

1. Introduction

The intriguing properties of the even-even Se and Ge isotopes in the mass region $A \sim 80$ have been investigated in a number of previous experimental and theoretical studies [1]. These isotopes belong to a typical transitional region. The structure of their low-lying states can be attributed to the interplay of rotational and vibrational collective motions. For high-spin states, γ-ray spectroscopy of the near-yrast states in the $N = 44$ and 46 isotones of Se (^{80,82}Se) was carried out for deep-inelastic reactions [2]. Recently, full-fledged shell-model calculations have been performed on the even-even and odd-mass nuclei in this mass region using the phenomenological monopole and quadrupole pairing plus quadrupole-quadrupole ($P+QQ$) interaction [3]. One year later large scale shell model calculations were performed to determine effective interactions in the fpg- shell (JUN45) [4]. In this mile-stone paper Otsuka, Honma and others did a lot of efforts to obtain effective interactions in the whole range of the fpg-shell. Their shell model describes very well the expected properties of the fpg-shell nuclei such as the triaxiality, isomer states, the oblate ground-state deformation and the shape coexistence.

2. Shell model framework

In the shell model calculation we take Z=28 and N=50 as magic cores. Then, Se isotopes have 6 valence protons, and Ge isotopes, 4 valence protons. Here neutrons are treated as holes. We use the $P+QQ$ interaction:

$$\hat{H} = \sum_{jm\tau=\nu,\pi} \varepsilon_{j\tau} c^{\dagger}_{jm\tau} c_{jm\tau} - \sum_{\tau=\nu,\pi} \left[G_{0\tau} \hat{P}^{\dagger(0)}_{\tau} \hat{P}^{(0)}_{\tau} + G_{2\tau} \hat{P}^{\dagger(2)}_{\tau} \tilde{\hat{P}}^{(2)}_{\tau} + \kappa_{\tau} : \hat{Q}_{\tau} \cdot \hat{Q}_{\tau} : \right] - \kappa_{\nu\pi} \hat{Q}_{\nu} \cdot \hat{Q}_{\pi}, \quad (1)$$

where the strengths of the parameters are the same as in the previous shell model calculations [3]. As shown in figures 1 and 2 our calculation reproduces quite well the energy levels for the Se and Ge isotopes. We show $B(E2)$ transition rates in figure 3. The reduction of $B(E2)$ occurs

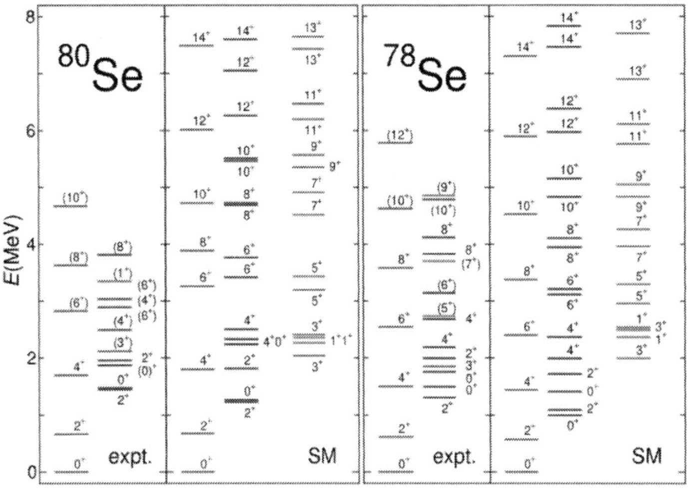

Figure 1. Comparison between the experimental spectra (expt.) and shell model results (SM) for ^{80,78}Se isotopes.

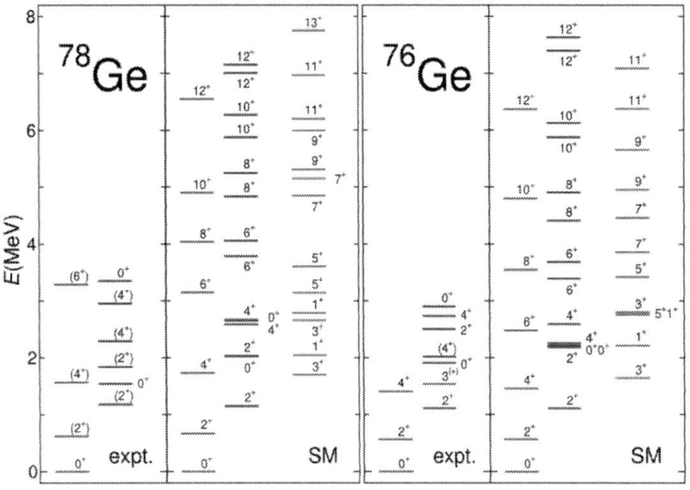

Figure 2. Same figure as in figure 1, but for ^{78,76}Ge isotopes.

from spin 8 to 6 for N=46 isotones (^{80}Se and ^{78}Ge), but no abrupt change is seen for N=44 isotones (^{78}Se and ^{76}Ge).

Two questions arise here. (i) Why the 8_1^+ states seem to be irregular and $B(E2)$ transitions become small from 8_1^+ to 6_1^+ states such as in N=46 for Se and Ge isotopes? (ii) Is the triaxiality important for a description of γ bands? If so, is the system either γ-unstable or γ-rigid?

To answer the first question, we have compared the results of the shell model (SM) calculations with the pair truncated shell model (PTSM) calculations [5, 6, 7]. In this model the collective S, D, G nucleon pairs, and the non-collective $(0g_{9/2})^2$ pairs (H pairs) are assumed to be the building blocks. The S and D pairs play essential roles in describing the low-lying states, while the pair of $0g_{9/2}$ neutrons is indispensable for a description of high-spin states. We have found that for nuclei with N=46 a sudden alignment of two $0g_{9/2}$ neutrons occurs, but only a smooth change in structure occurs for N=44. All the details are given in [3]. The ground state energies

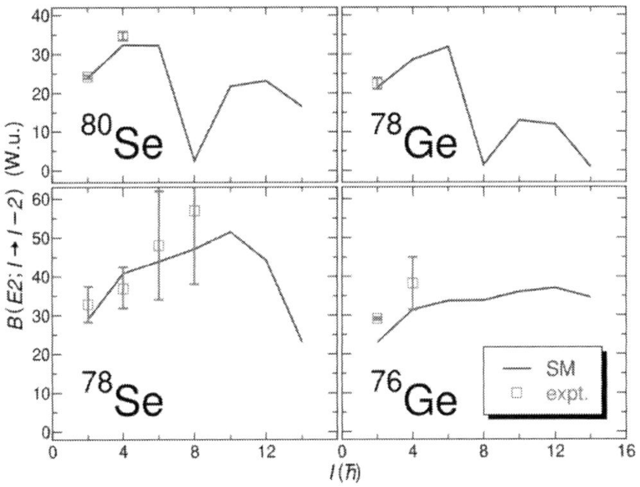

Figure 3. Comparison of the calculated $B(E2)$ values (solid lines) for the yrast states of ^{80,78}Se and ^{78,76}Ge with experimental data (open squares).

for ^{78}Ge are found to be $E_{\rm SM} = -11.0529$ MeV and $E_{\rm PTSM} = -10.6827$ MeV, respectively, for the SM and the PTSM calculations.

3. Quantum-number-projected generator coordinate method

To answer the second question we apply the quantum-number-projected generator coordinate method (GCM) under the same interaction as used in previous SM studies [3].

In the present scheme, the spins of the neutron and proton systems (I_ν and I_π) are projected out separately, and the total spin of each state is constructed by angular momentum coupling [8, 9]. To generate functions for the GCM, we employ the intrinsic Nilsson states for the neutron or proton system. The intrinsic Nilsson states are constructed by the following procedure. First we consider the intrinsic Nilsson hamiltonian for either neutron or proton space:

$$\hat{h}_{\rm Nil} = \sum_{jm} \varepsilon_j c^\dagger_{jm} c_{jm} - \hbar\omega\beta \left[\cos\gamma \hat{Q}_0 - \frac{\sin\gamma}{\sqrt{2}} \left(\hat{Q}_2 + \hat{Q}_{-2} \right) \right] \tag{2}$$

with

$$\hat{Q}_\mu = \sum_{jm,j'm'} \langle jm | \frac{r^2}{b^2} Y^{(2)}_\mu | j'm' \rangle c^\dagger_{jm} c_{j'm'}. \tag{3}$$

Here β and γ indicate axial and triaxial quadrupole deformations, respectively. Then N-particle Nilsson intrinsic states are written as

$$|\Phi(\beta,\gamma)\rangle = b^\dagger_1 b^\dagger_2 \cdots b^\dagger_{N-1} b^\dagger_N | - \rangle = \prod_{i=1}^{N} b^\dagger_i | - \rangle. \tag{4}$$

The intrinsic deformed state $|\nu\rangle = b^\dagger_\nu | - \rangle$ is related with the spherical basis state $|i\rangle = c^\dagger_i | - \rangle$ as

$$b^\dagger_\nu = \sum_i F_{i\nu} c^\dagger_i, \tag{5}$$

by diagonalizing the Nilsson hamiltonian $\hat{h}_{\rm Nil}$. Then the GCM wavefunction for the ρth state of a spin I_τ in neutron or proton space ($\tau = \nu$ or π) is given by

$$|\Psi^{(\tau)}_{I_\tau M_\tau \rho}\rangle = \sum_k \sum_{K_\tau = -I_\tau}^{I_\tau} \mathcal{F}^{I_\tau k}_{K_\tau \rho} \hat{P}^{I_\tau}_{M_\tau K_\tau} |\Phi_\tau(\beta_k, \gamma_k)\rangle, \tag{6}$$

where $\hat{P}^{I_\tau}_{M_\tau K_\tau}$ is the spin projection operator, $\mathcal{F}^{I_\tau k}_{K_\tau \rho}$, the weight function to be determined by the GCM, and k stands for a representative point of deformation (β, γ). Then, the many-body wavefunction for an even-even nucleus can be written as [8, 9]

$$|\Psi_{IM}(I_\nu \rho, I_\pi \sigma)\rangle = [|\Psi^{(\nu)}_{I_\nu \rho}\rangle \otimes |\Psi^{(\pi)}_{I_\pi \sigma}\rangle]^{(I)}_M, \tag{7}$$

where I is the total spin and M, its projection.

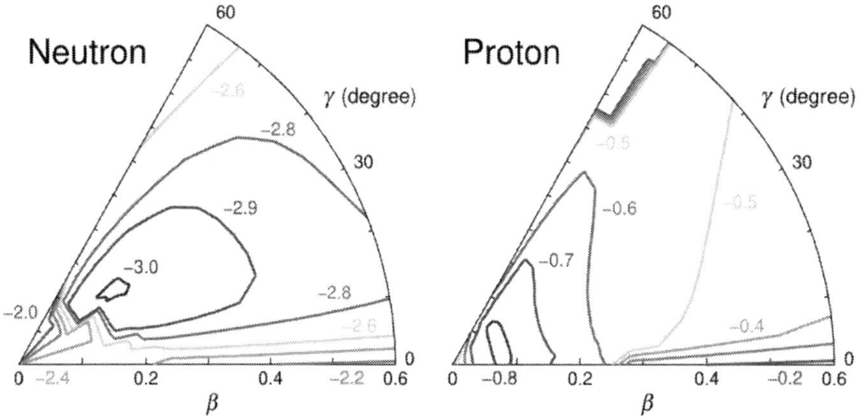

Figure 4. The contour plot of the energy gain calculated for neutrons (left panel) and protons (right panel) in ^{78}Ge. The contour line separation is 0.1 MeV.

In figure 4, the contour plots of the potential energy surface (PES) in the β-γ plane for Nilsson states using equation (4) are shown for neutrons and protons of ^{78}Ge. For neutron space, triaxiality appears, but it is not clear for proton space. In the GCM calculations we take $\beta = 0.1, 0.2, \cdots, 0.6$ and $\gamma = 10°, 20°, 30°, 40°$, and $50°$. In figure 5, the theoretical energy spectra of the GCM are compared with the SM results and the experimental data for ^{78}Ge. It is found that the energy levels obtained by the SM is well reproduced by the GCM. The ground state energy obtained by the GCM for the ^{78}Ge is $E_{\rm GCM} = -10.5059$ MeV.

Now let us improve our GCM calculations by including the pairing correlation. We introduce the nucleon pair creation operator as

$$\Lambda^\dagger(\beta, \gamma) = \sum_{\alpha > 0} f_\alpha(\beta, \gamma) b^\dagger_\alpha b^\dagger_{\bar{\alpha}}, \tag{8}$$

where $\bar{\alpha}$ indicates the time reversal state of α. Then the intrinsic N-particle states for neutrons or protons are written as

$$|\Phi(\beta, \gamma)\rangle = [\Lambda^\dagger(\beta, \gamma)]^{N/2}|-\rangle. \tag{9}$$

The structure coefficients $f's$ are determined by variation for a definite deformation (β, γ):

$$\delta \frac{\langle \Phi(\beta, \gamma)|\hat{H}|\Phi(\beta, \gamma)\rangle}{\langle \Phi(\beta, \gamma)|\Phi(\beta, \gamma)\rangle} = 0. \tag{10}$$

In figure 6, the contour plots of the PES are shown for neutron space and proton space, respectively. Now they are very smooth functions of β and γ, and they have no definite minimum. The pairing interaction washes away the deformations.

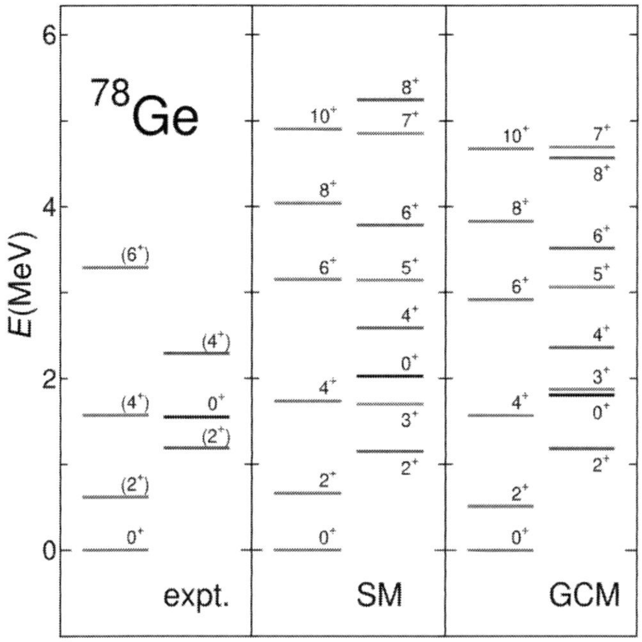

Figure 5. Comparison of experimental energy spectrum (expt.) with that in the SM(SM) and that obtained by the GCM (GCM) for ^{78}Ge.

In the GCM calculations we take for axial deformations (31 points), β = 0.00, 0.04, 0.08, $\cdots$, 0.60 and γ = 0°, 60°. For triaxial deformations we take (20 points) β = 0.10, 0.20, 0.30, γ = 10°, 20°, 30°, 40°, 50°, and (β, γ) = (0.40, 20°), (0.40, 40°), (0.50, 10°), (0.50, 30°), (0.50, 50°). In figure 7 energy spectra are compared for experimental results, SM, GCM (triaxial) and GCM (axial), respectively. We compare between the SM and the GCM results (2 cases) for ^{78}Ge. The ground state energies for the ^{78}Ge are $E_{\rm SM} = -11.0530$ MeV, $E_{\rm GCM(triaxial)} = -11.0048$ MeV, and $E_{\rm GCM(Axial)} = -10.9643$ MeV, respectively for the SM, the GCM with triaxial deformations and the GCM only with axial deformations. A large improvement is seen for the triaxial deformation for the ground state energies. The triaxiality is important in this region, especially in the description of the γ band. The potential energy surface is shallow in the direction and it shows the γ-unstable nature or IBM O(6) like picture. The generator coordinate model calculations show the importance of triaxiality in describing the γ-band, but the PES is shallow, indicating the γ-unstable interpretation.

4. Summary

In this paper SM results are analyzed in terms of the PTSM and the GCM. The PTSM analysis shows the alignment of neutron in the $g_{9/2}$ orbital at the yrast spin of 8^+. The GCM calculations show the importance of triaxiality in describing the γ-band, but the PES is shallow, indicating the γ-unstable interpretation.

Acknowledgments

We would like to celebrate the 60th birthday of Professor Otsuka. One of the authors (N Y) is grateful for his tireless education and long-term collaboration. This work was supported by Grant-in-Aid for Scientific Research (C) (24540251) from Japan Society for the Promotion of Science, and by Grant-in-Aid for Young Scientists (B) (22740171) from The Ministry of Education, Culture, Sports, Science and Technology of Japan.

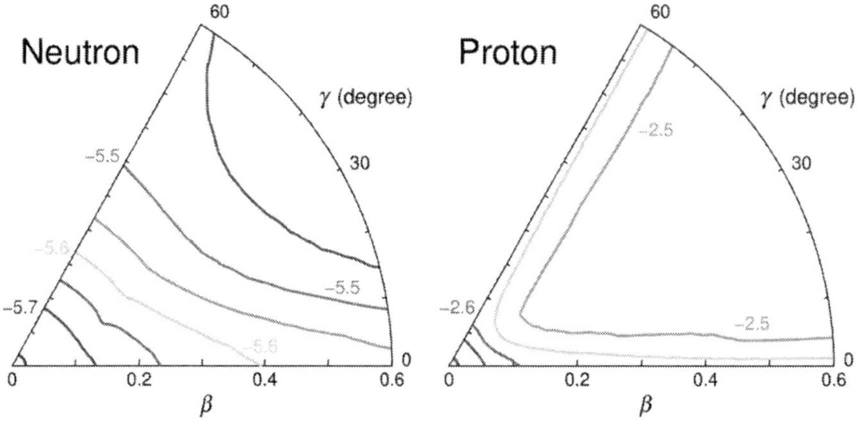

Figure 6. Same as figure 4, but for those states with the pairing correlation included. The contour line separation is 0.05 MeV

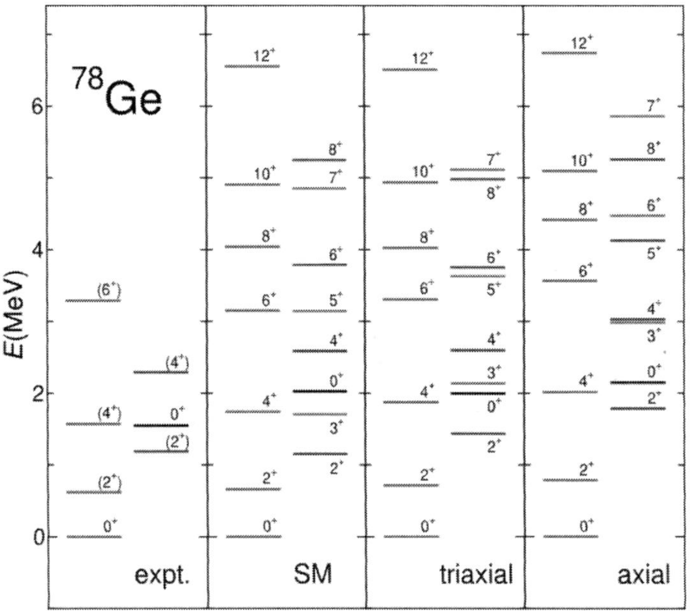

Figure 7. Energy levels of the yrast and quasi-γ bands in ^{78}Ge. From the left, energy levels in the experiment (expt.), the SM (SM), the triaxial GCM (triaxial), and the axial GCM (axial) are displayed.

References

[1] Speidel K H *et al* 1998 *Phys. Rev.* C **57** 2181
[2] Jones G A *et al* 2007 *Phys. Rev.* C **76** 054317
[3] Yoshinaga N Higashiyama K and Regan P H 2008 *Phys. Rev.* C **78** 044320
[4] Honma M Otsuka T Mizusaki T and Hjorth-Jensen M 2009 *Phys. Rev.* C **80** 064323
[5] Higashiyama K Yoshinaga N and Tanabe K 2003 *Phys. Rev.* C **67** 044305
[6] Yoshinaga N and Higashiyama K 2004 *Phys. Rev.* C **69** 054309
[7] Higashiyama K and Yoshinaga N 2011 *Phys. Rev.* C **83** 034321
[8] Enami K Tanabe K and Yoshinaga N 2002 *Phys. Rev.* C **65** 064308
[9] Enami K Higashiyama K Tanabe K and Yoshinaga N 2002 *Phys. Rev.* C **66** 047301

International Symposium on Exotic Nuclear Structure From Nucleons (ENSFN 2012) IOP Publishing
Journal of Physics: Conference Series **445** (2013) 012033 doi:10.1088/1742-6596/445/1/012033

Deformed halo nuclei probed by breakup reactions

Takashi Nakamura

Department of Physics, Tokyo Institute of Technology, 2-12-1 O-Okayama, Meguro, Tokyo 152-8551, Japan

E-mail: nakamura@phys.titech.ac.jp

Abstract. Breakup reactions play important roles in elucidating the structures near the drip lines, such as nuclear halo. The recent experimental results using the Coulomb and nuclear breakup reactions for the neutron-drip-line nuclei at the new-generation RI beam facility, RIBF at RIKEN, are presented. Focuses are put on the results on the newly found halo nucleus ^{31}Ne, which is intriguing also in that this nucleus is in the island-of-inversion and thus could be strongly deformed. The results on other Ne/Mg/Si neutron rich isotopes ranging from N=20 towards N=28 are also briefly reported. The first breakup experiments using SAMURAI facility at RIBF and future perspectives are also presented.

1. Introduction

Structural Evolution of atomic nuclei towards the neutron and proton drip lines in the nuclear chart is one of the most important and intriguing issues in current nuclear physics. In particular, the location of the neutron drip line is the key to understand the stability of atomic nuclei around the bound limit of neutron-proton quantum system. In spite of such importance, the neutron drip line has been determined only up to the oxygen isotopes. The drip-line nuclei exhibit peculiar feature of nuclear structures as in neutron halo nuclei found in the vicinity of the neutron drip line. However, the neutron halo nuclei so-far known have been limited to the light system, and the heaviest neutron halo nucleus experimentally known had long been ^{19}C [1]. The $1n$ halo nucleus ^{31}Ne [2] and the $2n$ halo nucleus ^{22}C [3], heavier than ^{19}C, have been identified only very recently. In what region and how the heavier halo nuclei can be formed are key questions in discussing the nuclear structure near the drip line.

It has been established that the shell is evolving from the stability line towards the neutron drip line. One of highlights of such phenomena is the island of inversion for the neutron-rich nuclei near $N \sim 20$ [4], where the ground state becomes dominated by the intruder configuration. The island contains the strongly deformed ^{32}Mg [5] and neighboring Ne/Na/Mg [6, 7, 8] isotopes. Theoretically, the shell model also shows such an island [4, 9, 10]. Note also that the loosely-bound feature enhances the shell evolution. The interplay between the halo formation and the shell evolution could thus be studied at the neutron-rich edge the island of inversion.

The neutron-rich nuclei with neutron number $20 < N < 28$ in the island of inversion can indeed provide us with some insights about the shell evolution and the halo phenomena. The halo is formed on the condition that the nucleus has a very small $1n$ or $2n$ separation energy ($S_{\rm n}, S_{\rm 2n} < 1$ MeV), and that the least bound neutron has low or no angular momentum, s or p. The latter condition is not met for nuclei with $20 < N < 28$ if we adopt the conventional shell order since the least bound neutron in $0f_{7/2}$ moves with a large centrifugal barrier. On the other hand, Utsuno and Otsuka et al. predicted the effective single particle energy (SPE) between $1p_{3/2}$ and $0f_{7/2}$ becomes small or inverted for neutron-rich nuclei for $Z \sim 10$ [10], which may make the p-wave configuration dominant.

As was pointed out by Hamamoto [11], nuclei with neutrons in the nearly degenerate $f_{7/2}$ and $p_{3/2}$ shells, which couple strongly to each other by quadrupole-quadrupole interaction, can be quadrupole deformed. This could be one cause of the island of inversion. The one-particle motion in a deformed potential (Nilsson picture) also provides insights about such nuclei. The weakly-bound neutron has also an interesting behavior. Namely, the lower ℓ configuration becomes dominant when $S_{\rm n}$ approaches zero [12, 13, 14]. For instance, the Nilsson orbit of $[330]1/2^-$ is occupied by $f_{7/2}$ neutron if $S_{\rm n}$ is deeper than about 3 MeV, while $p_{3/2}$ becomes dominant below this energy.

In general, in heavier nuclei, least-bound neutron(s) may tend to occupy orbitals with higher angular momentum, which hinders the halo formation. On the other hand, if the shell evolution occurs, the deformation is developed, and the dominance of s or p orbital becomes significant, then the halo structure may be abundant along the neutron drip line from light to heavy nuclei. Two neutron halo in heavier nuclei is also intriguing since the two halo neutrons may be strongly correlated in space, called dineutron, as predicted by Migdal [15]. Hence, the understanding of the formation of halo in heavier drip-line nuclei is very important in current nuclear physics.

This paper reports the recent experimental study of the neutron rich B/C/Ne/Mg/Si isotopes by Coulomb and nuclear breakup reactions RIBF around 240 MeV/nucleon, in order to investigate the halo structure and single particle configuration of exotic nuclei. Main focus here is ^{31}Ne, which has N=21. The part of the result is in Refs.[2, 16]

2. Breakup reactions of exotic nuclei

One of the important experimental probes for nuclei far from the stability is breakup reactions at intermediate/high energies. Because the Fermi level of drip-line nuclei lies close to the continuum, weakly bound levels can be easily excited to the continuum, which corresponds to breakup reactions. The breakup reactions thus play important roles since the cross section is significantly large for such nuclei, and is sensitive to the wave function of the least-bound nucleon(s). The breakup reaction shows very different characteristics, depending on the target [17, 18]. A heavy target such as Pb induces primarily Coulomb breakup, while a light target such as C induces nuclear breakup.

The Coulomb breakup can extract the $E1$ response of the nucleus since it is essentially the photo absorption process. The Coulomb breakup cross section is written as,

$$\frac{d\sigma(E1)}{dE_{\rm x}} = \frac{16\pi^3}{9\hbar c} N_{E1}(E_{\rm x}) \frac{dB(E1)}{dE_{\rm x}}, \tag{1}$$

where $N_{E1}(E_{\rm x})$ denotes the number of the virtual photons for the electric dipole transition with the excitation energy $E_{\rm x}$ [19]. Such simplicity that the cross section can be factorized allows us to extract the $E1$ strength distribution.

One of the important and unique features of halo nuclei is the strong enhancement of the electric dipole strength ($B(E1)$) just above the neutron decay threshold, which we call soft $E1$ excitation. This excitation is contrast to the $E1$ response of normal nuclei near the stability, whose $E1$ strength is concentrated almost fully on the Giant Dipole Resonance (GDR) at high excitation energies of about 20 MeV. Recently, another dipole excitation at lower excitation energies, called pygmy dipole resonance (PDR) has been observed for ^{132}Sn [20]. It should be noted that the soft $E1$ excitation and PDR has different characteristics in that the former is non-collective excitation, while the latter is considered to be collective. The PDR peaks at around $E_{\rm x} \sim$ 8-10 MeV, higher than the soft $E1$ excitation but lower than the GDR peak.

The mechanism of the soft $E1$ excitation has been understood well for the $1n$-halo nuclei as the revelation of the direct breakup. In direct breakup mechanism, the $E1$ response reflects the halo spatial distribution [21, 22, 23]. There, the matrix element is simply written as

$$\frac{dB(E1)}{dE_{\rm rel}} = \mid \langle \Phi_f(\vec{r},\vec{q}) \mid e_{\rm eff}^{E1}\hat{T}(E1) \mid \Phi_i(\vec{r})\rangle \mid^2, \tag{2}$$

where $\Phi_i(\vec{r})$ and $\Phi_f(\vec{r},\vec{q})$ show the wave function of the ground state (halo state), and the final state in the continuum of the neutron relative to the core, respectively. $\vec{r}$ is the relative coordinate of the neutron relative to the c.m. of the core. $\Phi_f(\vec{r},\vec{q})$ is also a function of the relative momentum $q = \sqrt{2\mu E_{\rm rel}}/\hbar$. The electric dipole operator is $\hat{T}(E1)(= rY^{(1)}(\Omega))$. This formula shows that the $B(E1)$ spectrum is approximately the Fourier-Bessel transform of the radial wave function of the ground state. This suggests that the large amplitude in the radial wave function at large r as in the halo state causes the large $E1$ amplitude at small q, or small $E_{\rm rel}$(relative energy) in the $B(E1)$ spectrum. Note that the $E1$ operator involving r causes further amplification [24]. This characteristics makes the soft $E1$ excitation a very sensitive tool to probe the halo wave function as applied successfully to ^{11}Be [21, 22, 23], ^{15}C [25, 26], and ^{19}C [1].

The soft $E1$ excitation of two neutron halo nuclei is much more complicated, but has intriguing aspects. The current understanding for two-neutron halo cases is also non-resonant origin, but the initial two-neutron correlation and the final state interaction affect largely the $E1$ spectra. It should be noted that for two-neutron halo nuclei, integrated sum of the low-energy $E1$ strength reflects the spatial correlation [27, 28]. In Ref.[28], the dineutron correlation, which is a strong spatial nn correlation and has a mixture of different parity configurations, for the halo neutrons in ^{11}Li is suggested, which is responsible for the strong $E1$ strength of this nucleus.

Nuclear breakup of an unstable nucleus with a light target at intermediate/high energies offers a useful spectroscopic tool as well. The momentum distribution of the core fragment after $1n$ removal can be used to extract the orbital angular momentum of the removed neutron, as demonstrated recently even for the unbound final state [29]. When a measurement is combined with that of the γ ray emitted from the core fragment, the core configuration of the projectile with the spectroscopic factor C^2S can also be extracted. For instance, when we study the $1n$ removal of ^{31}Ne into ^{30}Ne, then the momentum distribution of ^{30}Ne can probe the angular momentum of the neutron removed. When the γ ray emitted from the ^{30}Ne core fragment is measured in coincidence, one can determine the ratio of ^{30}Ne ground state configuration to the excited one. A combined analysis of Coulomb and nuclear breakup further enhances this capability due to different sensitivities to the spatial distribution of the wave function around the surface of exotic nuclei.

3. Inclusive Coulomb and nuclear breakup of neutron rich C/Ne/Mg/Si isotopes

In the Coulomb breakup, the $E1$ energy spectrum characterizes the soft $E1$ excitation, as mentioned above. A kinematically complete measurement of the breakup reaction, where the four-momentum vectors of the outgoing particles are measured in coincidence, is required to extract the excitation energy on event-by-event basis. The experiment is rather complicated where neutron(s) and charged fragments at relativistic energies should be measured in coincidence. Although such a measurement is possible within current experimental techniques, an alternative inclusive measurement, where only the integrated $-1n(-2n)$ cross section is measured, has often advantage over the kinematically complete one, in that the experiment is simpler and no measurement for the neutron is required. Such a measurement is in particular important at the earlier stage of the new facility, as in that of the new-generation RI beam facility, RIBF, at RIKEN.

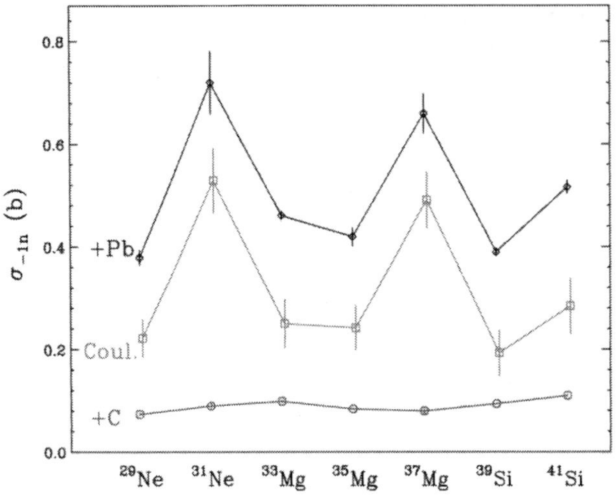

Figure 1. $1n$ removal cross sections of ^{29,31}Ne, ^{33,35,37}Mg, and ^{39,41}Si on C (blue circles) and Pb (black diamonds) at 220-240 MeV/nucleon. The Coulomb breakup cross sections on Pb (red squares) are estimated by subtracting the nuclear breakup contribution estimated from the ones with the carbon target. The enhancement of the Coulomb breakup cross section is clearly seen for ^{31}Ne and ^{37}Mg.

The inclusive measurement, although it provides less information about the structure, has signature on the soft $E1$ excitation. The inclusive Coulomb breakup cross section corresponds to measure the integrated cross section as shown in

$$\sigma(E1) = \int_{S_{\rm n}}^{\infty} \frac{16\pi^3}{9\hbar c} N_{E1}(E_{\rm x}) \frac{dB(E1)}{dE_{\rm x}} dE_{\rm x}. \tag{3}$$

The expression shows that the cross section is basically the integration of $dB(E1)/dE_{\rm x}$ weighted by the photon number $N_{E1}(E_{\rm x})$ over the excitation energy. As the photon spectrum falls exponentially with $E_{\rm x}$ [2, 17], $\sigma(E1)$ becomes significant only when the $B(E1)$ is concentrated at low excitation energies as in the case of soft $E1$ excitation for halo nuclei. In fact, it can be estimated that for ^{31}Ne the cross section of about 0.5 b is expected when it is a halo nucleus that causes a soft $E1$ excitation, while the GDR contributes only about 0.06 b [17]. It should be noted that for a $1n$ plus core system, the upper limit of the integration could be limited to $E_{\rm x} \sim S_{2\rm n}$ since above that the nucleus decays into $2n$+fragment when the measurement requires only $-1n$ channel. With that the GDR contribution will become negligible. Hence the soft $E1$ excitation can even be more easily distinguished.

The inclusive breakup cross sections ($1n$ and/or $2n$ removal cross sections) of ^{19,20,22}C, ^{29,31}Ne, and ^{33,35,37}Mg at 220-240 MeV/nucleon have been measured at ZDS (Zero Degree Spectrometer) at the next-generation RI beam facility, RIBF, at RIKEN [2, 16]. The $1n$ removal cross sections for Ne, Mg, and Si isotopes are shown in Fig. 1.

As shown, the cross section for ^{31}Ne and ^{37}Mg among others are significantly larger ($\sim$ 0.5 b), expected for the halo nucleus. As detailed in Ref. [2], the nucleus ^{31}Ne has been identified as a new $1n$ halo nucleus, evidenced by this enhanced Coulomb breakup cross section.

We can further investigate the structure of these nuclei by using the combinatorial analysis of nuclear and Coulomb breakup. The preliminary analysis combined with the γ ray coincidence shows the rather small ground-state ^{30}Ne core configuration ($\sim$ 30%) with small $S_{\rm n}$ of about

100 keV with J^{π}=3/2$^-$. The result is well compatible with the large scale shell model, which predicts such ground state spin-parity. The shell model also shows the large configuration mixing of ^{30}Ne(0_1^+) $\otimes$ $\nu 1p_{3/2}$, ^{30}Ne(2_1^+) $\otimes$ $\nu 1p_{3/2}$, and ^{30}Ne(2_1^+) $\otimes$ $\nu 0f_{7/2}$, which is also consistent with the preliminary result of the momentum distribution of ^{30}Ne.

Such a configuration mixing can be regarded as the development of the deformation. Hamamoto [14] describes the valence neutron of ^{31}Ne lies in the Nilsson orbit [330]1/2$^-$ or [321]3/2$^-$. More recently, the ground state properties of ^{31}Ne were also discussed using the particle-rotor model, which takes into account the rotational excitation of the ^{30}Ne core [30]. Since the mean field potential is quadrupole deformed, which causes the configuration mixing involving p-wave in ^{31}Ne, we may call such a system "deformed halo nucleus". Our preliminary analysis also finds that the structure of the deformed core plus p-wave halo neutron also takes place for ^{37}Mg. Such a halo system could be common in the neutron-drip part of the island of inversion region.

4. Breakup reaction experiments at SAMURAI facility

As mentioned, ideally, the breakup reaction should be measured in complete kinematics. With that, in the Coulomb breakup one can map $B(E1)$ as a function of the excitation energy with the aid of Eq. (1). Note that the $B(E1)$ distribution contains information of the ground state wave function, such as C^2S (spectroscopic factor), ℓ (orbital angular momentum), and $S_{\rm n}$ (separation energy) of the halo state. The kinematically complete measurement is also a suitable tool to study the dineutron correlation.

The SAMURAI (**S**uperconducting **A**nalyser for **MU**lti-particles from **RA**dio **I**sotope Beams) has been constructed to realize such kinematically complete measurement with high efficiency and with the intense RI beam available at RIBF. SAMURAI comprises the superconducting magnet with a maximum magnetic field of 3.1 Tesla ($B_{max}L$=7.1 Tm), beam detectors, heavy ion detectors, and the neutron detector array NEBULA (**NE**utron-detection system for **B**reakup of **U**nstable nuclei with **L**arge **A**cceptance).

The SAMURAI facility has successfully commissioned in March, 2012, followed by the day-one campaign experiment in May, 2012, using the ^{48}Ca beam at 345 MeV/nucleon, which includes the kinematically complete measurement of Coulomb breakup of two neutron halo nuclei ^{19}B and ^{22}C. We had the beam intensity of ~100 counts per second (cps) for ^{19}B and of ~ 15 cps for ^{22}C, about 3-4 orders of magnitude larger than those obtained in the previous facility (RIPS). These nuclei are considered to be important in understanding the dineutron correlation of the $2n$ halo systems. The ^{22}C could be the last s-wave dominant two neutron halo system that the current, and near-future-planned experimental facilities can reach. In SAMURAI, we also plan to study the deformed halo nuclei in kinematically complete measurements in the near future. We expect that the mechanism of the revelation of such system will be clarified soon.

5. Summary

The importance and usefulness of the breakup reactions at intermediate/high incident energies in the new-generation RI beam facility has been demonstrated. Inclusive breakup measurements at the new-generation RI beam facility, RIBF at RIKEN, were briefly presented, where evidence for the $1n$ halo structure in ^{31}Ne is shown. The large configuration mixing suggested from the combined analysis of Coulomb and nuclear breakup reactions and the large scale shell model calculation indicates that the ^{31}Ne is a deformed halo nucleus, where p-wave neutron is orbiting the deformed core potential. Such a unique system is also suggested for ^{37}Mg.

The new facility SAMURAI at RIBF, which aims at conducting the kinematically complete measurements of breakup reactions of very neutron-rich nuclei, has commissioned in 2012. The first series of physics experiments, one of which is the kinematical complete measurement of the Coulomb breakup of ^{19}B and ^{22}C at about 240 MeV/nucleon, have been successfully performed.

This demonstrates the powerfulness of the facility, so that we expect that such breakup reactions are applied to more exotic and heavier nuclei along the drip lines in the near future, including the deformed halo nuclei near N=20 and N=28.

Acknowledgments

The author would like to thank the collaborators of the ^{31}Ne experiment [2]. The author also would like to thank the SAMURAI day-one campaign collaboration. The present work was supported in part by a Grant-in-Aid for Scientific Research (B) (No. 22340053), and that on Innovative Areas (No. 24115005) from the Ministry of Education, Culture, Sports, Science and Technology (MEXT, Japan).

References

[1] Nakamura T *et al.* 1999 *Phys. Rev. Lett.* **83** 1112
[2] Nakamura T *et al.* 2009 *Phys. Rev. Lett.* **103** 262501
[3] Tanaka K *et al.* 2010 *Phys. Rev. Lett.* **104** 062701
[4] Warburton E K, Becker J A and Brown B A 1990 *Phys. Rev.* C **41** 1147
[5] Motobayashi T *et al.* 1995 *Phys. Lett.* B **364** 9
[6] Thibault C *et al.* 1975 *Phys. Rev.* C **12** 644
[7] Yanagisawa Y *et al.* 2003 *Phys. Lett.* B **566** 84
[8] Doornenbal P *et al.* 2009 *Phys. Rev. Lett.* **103** 032501
[9] Caurier E, Nowacki F, Poves A and Retamosa J 1998 *Phys. Rev.* C **58** 2033
[10] Utsuno Y, Otsuka T, Mizusaki T and Honma M 1999 *Phys. Rev.* C **60** 054315
[11] Hamamoto I 2012 *Phys. Rev.* C **85** 064329
[12] Misu T, Nazarewicz W and Åberg S 1997 *Nucl. Phys.* A **614** 44
[13] Hamamoto I 2004 *Phys. Rev.* C **69** 041306(R)
[14] Hamamoto I 2010 *Phys. Rev.* C **81** 021304(R)
[15] Migdal A B 1972 *Yad. Fiz.* **16** 427; English translation 1973 *Sov. J. Nucl. Phys.* **16** 238
[16] Kobayashi N *et al.* 2012 *Phys. Rev.* C **86** 054604
[17] Nakamura T and Kondo Y 2012 *Clusters in Nuclei* vol 2 (*Lecture Notes in Physics* vol 848), ed C Beck (Berlin: Springer) pp 67-119
[18] Aumann T and Nakamura T 2013 *Phys. Scr.* **T152** 014012
[19] Bertulani C and Baur G 1988 *Phys. Rep.* **163** 299
[20] Adrich P *et al.* 2005 *Phys. Rev. Lett.* **95** 132501
[21] Nakamura T *et al.* 1994 *Phys. Lett.* B **331** 296
[22] Palit R *et al.* 2003 *Phys. Rev.* C **68** 034318
[23] Fukuda N *et al.* 2004 *Phys. Rev.* C **70** 054606
[24] Otsuka T *et al.* 1994 *Phys. Rev.* C **49** R2289
[25] Datta Pramanik U *et al.* 2003 *Phys. Lett.* B **551** 63; Datta Pramanik U *et al.* 2002 *Prog. Theor. Phys. Suppl.* **146** 427
[26] Nakamura T *et al.* 2009 *Phys. Rev.* C **79** 035805
[27] Aumann T *et al.* 1999 *Phys. Rev.* C **59** 1252
[28] Nakamura T *et al.* 2006 *Phys. Rev. Lett.* **96** 252502
[29] Kondo Y *et al.* 2010 *Phys. Lett.* B **690** 245
[30] Urata Y, Hagino K and Sagawa H 2011 *Phys. Rev.* C **83** 041303(R)

Matter radii of ^{22,23}O

R Kanungo

Astronomy and Physics Department, Saint Mary's University, Halifax, Canada

E-mail: ritu@triumf.ca, rkanungo@ap.smu.ca

Abstract. The anomaly in the interaction cross section of ^{23}O is discussed through a new experiment performed at the FRS, GSI. The matter radius of ^{23}O is derived through a Glauber model analysis. Results using *ab initio* coupled cluster theory as well as densities from mean field models are presented.

1. Introduction

The structure of neutron-rich nuclei are an intriguing subject of investigation as we learn about unexpected deviations from our well-established model of nuclei. The presence of neutron-halo and skin in neutron-rich nuclei have opened a wealth of new information [1]. These exotic forms have been discovered through the measurement of matter radii and charge radii. Since the discovery of the neutron halo in ^{11}Li, the matter radii of neutron-rich nuclear isotopes have been quite effectively derived from measurement of interaction cross sections [2]. A nuclear halo, is manifested by an abrupt large increase in the cross section and hence the matter radius of a nucleus compared to its preceding isotopic neighbour. The radius therefore deviates greatly from the well known $r_0A^{1/3}$ rule. Such a structure has been pictured a one or two valence neutrons that are very weakly bound, extend to large distances from the neighbouring isotope, core, thereby forming a low-density neutron halo around the core.

One and two neutron halos have been identified in the light nuclei from helium to carbon. Extending further, a clear confirmation on halos in the isotopic chains of nitrogen and oxygen are not firmly established. The first measurement of interaction cross section for these isotopes reported quite an astounding observation of enormously large radii for the N=15 isotopes, namely, ^{22}N and ^{23}O [3]. In a conventional shell model arrangement of the nuclear orbitals the odd neutron in these nuclei should occupy the $2s_{1/2}$ orbital. Such a configuration can lead to the largest possible size in the framework of the adopted core+neutron halo model. The valence neutron in ^{23}O however is not very weakly bound. This leads to a calculated cross section within the $^{22}O_{gs}$(core) + neutron ($2s_{1/2}$) being much smaller than the experimental data. It was therefore a mystery as to what new additional effect might lead to such an enhancement. Subsequently it was proposed that the large interaction cross section, and matter radius, arises due to an enlarged size of ^{22}O as a core than the bare ^{22}O [4].

On the other hand, one-neutron removal reaction from ^{23}O, measuring the momentum distribution of the fragment ^{22}O, was described by ^{23}O having the structure dominated by ^{22}O in the ground state and valence neutron with spectroscopic factor of 0.97±0.19 [5] in the $2s_{1/2}$ orbital. The Coulomb dissociation measurement arrived at a similar conclusion with the $2s_{1/2}$ spectroscopic factor of 0.78±0.13 [6].

The momentum distribution measured form one-neutron knockout from the neighbouring drip-line nucleus ^{24}O was explained with solely neutrons being knocked out of the $2s_{1/2}$ orbital with a spectroscopic factor of 1.74±0.19 [7]. Theoretical interpretation of the removal cross section from a more *ab initio* standpoint using coupled cluster theory [8] also predicted a spectroscopic factor consistent with Ref.[6]. Therefore it appears that the two valence neutrons in ^{24}O occupy the $2s_{1/2}$ orbital and each of them couple to ^{23}O in its ground state, which means the ground state spin of ^{23}O is $1/2^+$. This is therefore not favourable of a picture with ^{23}O ground state having a strong component with ^{22}O core in an excited state, which was thought to be one of the reasons for its possible enlargement.

2. Experiment

The situation was therefore inconclusive with the discrepancy between the interaction cross section measurement and the neutron removal reactions regarding the configuration of ^{23}O. A re-investigation of the interaction cross section of ^{23}O was therefore important and was undertaken at the fragment separator FRS at GSI. The details of the experimental setup is described in Ref.[9]. The layout of the experimental setup is shown schematically in Fig.1.

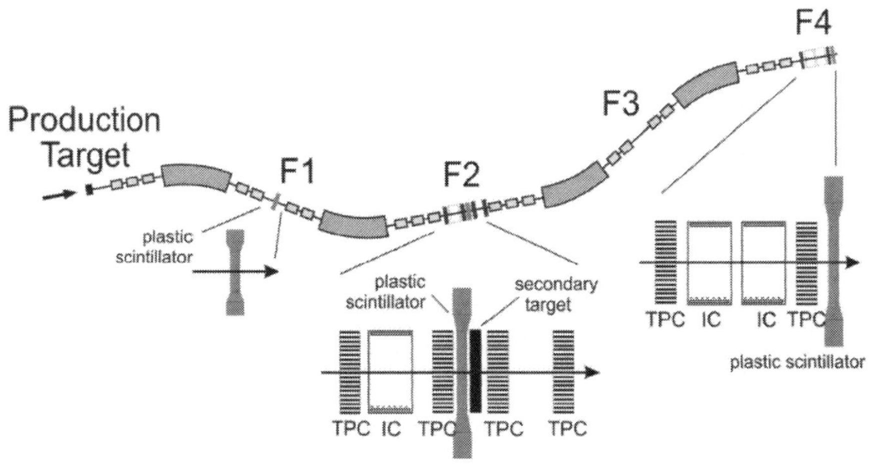

Fig.1 Experimental setup at the FRS, GSI. IC represents multi sampling ionization chamber and TPC represents, time projection chambers.

The measurement was done using the method of transmission, where ^{23}O before and after the reaction target needs to be identified and counted event by event. This was done by using the Bρ-ΔE-TOF technique. The identification before the target was done using the section F1 to F2 of the FRS. Two plastic scintillators located at these foci were used to measure the time of flight (TOF). Time Projection Chambers were used as beam tracking detectors to determine the beam profile on the target. The position measurement combined with the magnetic field information provided the magnetic rigidity (Bρ). The TOF and Bρ information determines the mass/charge (A/Q) of the incident beam species. A multisampling ionization chamber was used for the charge (Q = Z for fully stripped nuclei) identification. The beam energy at the reaction target was ~ 900*A* MeV. The carbon reaction target of 4 g/cm^2 was located at the dispersive second focus, F2. Two tracking detectors located after the reaction target provided the beam profile of the outgoing beam particles before they passed through the second stage of the FRS spectrometer. This allowed an assessment of any transmission losses. The unreacted ^{23}O nuclei after the target were transported from F2 to the final focus F4 using the second half of the FRS. The identification at F4 followed the same method as used before the target.

The open symbols in Fig.2 show the interaction cross section from Ref.[3]. The interaction cross section for $^{22,23}O$ measured in this new experiment is shown by the filled circles. They are found to be smaller than the earlier measurements.

3. Results and discussion

The root mean square matter radii of $^{22,23}O$ were derived through a Glauber model analysis of the data including the higher order terms that are missing in the optical limit approximation [10]. The calculations were done using two different parametric forms of the density distributions. The root mean square matter radius derived with a Fermi type density resulted in 2.75±0.15 fm for ^{22}O and 2.95±0.23 for ^{23}O [9]. A conservative approach was taken in this determination by searching for all combinations of half-radius and diffuseness of the Fermi density profile that reproduced the experimental cross section. The results using the harmonic oscillator density distribution were consistent with those derived with the Fermi densities giving values of 2.75±0.07 fm for ^{22}O and 2.97±0.11 fm for ^{23}O [9]. This shows that the root mean square radii extracted from the interaction cross sections are largely independent of the choice of the form the density distribution. It maybe mentioned here that one can place further restrictions on the radii by measuring the cross sections with different targets and/or beam energies.

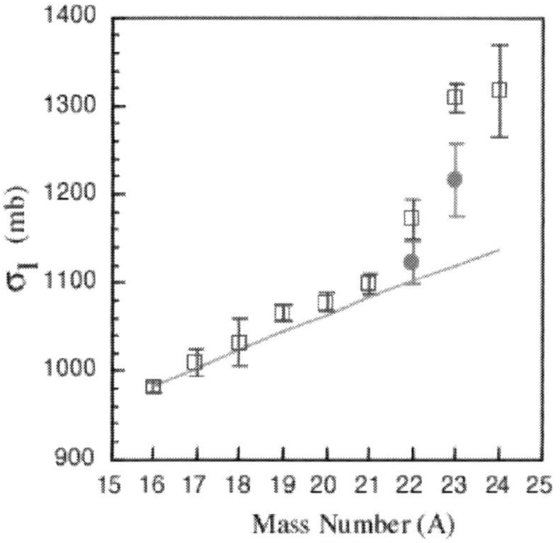

Fig.2 The interaction cross section of $^{A}C+^{12}C$. The open squares are from Ref.[3] and filled circles are from Ref.[8].

In order to investigate what information on ^{23}O structure we can obtain from the present data, few body Glauber model calculations were performed which describes ^{23}O in a ^{22}O(core) + neutron model. The resulting cross section with two different possibilities of the valence neutrons being in the $2s_{1/2}$ or $1d_{5/2}$ orbital is shown in Fig.3. It is seen that within the lower limit of the experimental uncertainty, the data can be consistently described with the valence neutron occupying the $2s_{1/2}$ orbital. Such a description would be consistent with the observation from the knockout reaction. A potential model calculation of the oxygen isotopic chain [11] also is in agreement with the data within the uncertainty. We note however, that the central value of the interaction cross section is higher than the results of a few-body Glauber model or the potential model.

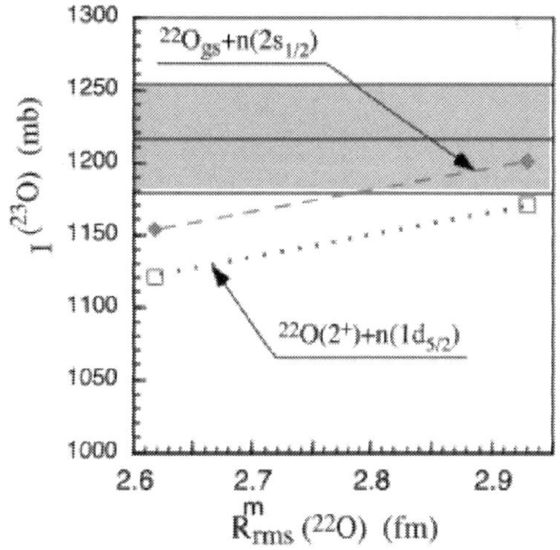

Fig.3 The interaction cross section of $^{23}O+^{12}C$ at 900*A* MeV shown by the horizontal shaded area. The lines are few-body Glauber model calculations with the configurations as labeled in the figure.

Coupled cluster calculations were also performed

by Hagen et al., which described the ground state of ^{23}O as a super- position of 1h and *1p2h* excitations on top of the correlated ground state of ^{24}O [12]. The first set of calculations were performed using the low-momentum version of the nucleon-nucleon interaction from chiral effective field theory. Fig. 4 shows the results with density distributions from coupled cluster theory with different momentum cutoff parameters λ [13] that were used in the Glauber model analysis. The cross sections with λ = 4.0 and 3.8 fm^{-1} are in good agreement with the data. The variation of the cross section magnitude with λ probably shows the contributions of the neglected three-nucleon forces. A later coupled cluster calculation was performed including the effects of the three-nucleon forces and of the continuum [14]. The three-nucleon forces were approximated as density-dependent nucleon- nucleon forces. The results of such a calculation are consistent with the matter radii derived from the present experiment as shown in Fig.3 of Ref.[14]. It is noted however that this calculation predicts ^{22,23}O slightly more overbound compared to experimental findings from mass measurements. The effect of overbinding might relate to calculated radii being therefore slightly smaller. Calculations with other *ab initio* approaches in the future may provide us with further information.

Figure 4 also shows the Glauber model calculations using density distributions from mean field potentials. Both the calculations agree with the present experimental data for ^{23}O but are higher for ^{22}O. In general, we notice that the evolution pattern of matter radii from ^{21}O to ^{22}O and from ^{23}O to ^{24}O shows an increasing trend in the mean field calculations while the coupled cluster calculations show an almost constant flat trend. It will be interesting to explore with further precise measurements of the cross sections the odd-even pairing effect, as observed from the coupled cluster results. In this context, it is noteworthy to mention a recent investigation of odd-even staggering from the interaction cross sections of $^{22-24}$O isotopes that was studied by Hagino and Sagawa [17]. The present data for ^{23}O was found to lead to a consistent description of the staggering parameter compared to theoretical systematics. Ref.[17] employed the optical limit approximation of the Glauber model. The density of ^{24}O was constructed with Hartree-Fock-Bogoliubov method using a Woods-Saxon mean field potential together with a density dependent zero-range pairing interaction. The density distributions of $^{22-23}$O did not take into account the pairing interaction. It was shown that the staggering parameter depends very weakly on energy based on which it is discussed that it could be a useful way to study pairing correlations in neutron-rich nuclei.

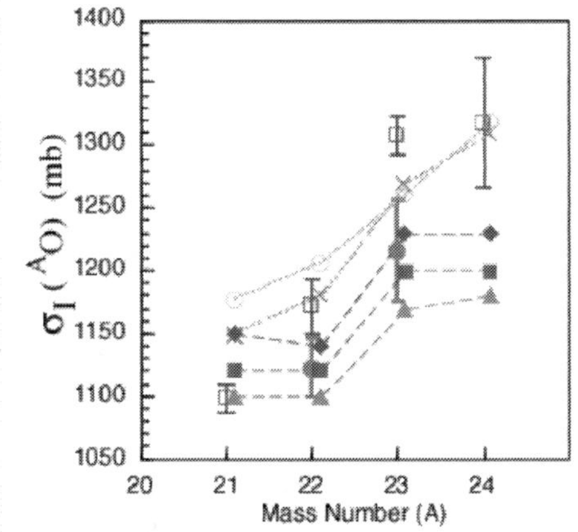

Fig.4 The interaction cross section of oxygen isotopes from Ref.[3] open squares and Ref.[9] filled circles. Results with densities from mean field potentials are shown by cross marks (Ref.[15]) and open circles (Ref.[16]). The filled diamond/squares/triangles are coupled cluster calculations with λ = 4.0/3.8/3.6 fm^{-1}.

It will be interesting to look into the matter radii predictions of monte carlo shell model calculations such as by Otsuka and collaborators in understanding the evolution of structure of the oxygen isotopes.

References

[1] Tanihata I, Savajols H and Kanungo R 2013 *Prog. Part. Nucl. Phys.* **68** 215
[2] Ozawa A, Suzuki T and Tanihata I 2001 *Nucl. Phys.* A **693** 32
[3] Ozawa A et al. 2000 *Phys. Rev. Lett.* **84** 5493

[4] Kanungo R, Ozawa A and Tanihata I 2001 *Phys. Lett.* B **512** 261
[5] Cortina-Gil D et al. 2004 *Phys. Rev. Lett.* **93** 062501
[6] Nociforo C et al. 2005 *Phys. Lett.* B **605** 79
[7] Kanungo R et al. 2009 *Phys. Rev. Lett.* **102** 152501
[8] Jensen Ø et al. 2011 *Phys. Rev.* C **83** 012305R
[9] Kanungo R et al. 2011 *Phys. Rev.* C **84** 061304R
[10] Horiuchi W et al. 2007 *Phys. Rev.* C **75** 044607
[11] Fortune H T and Sherr R 2013 *Eur. Phys. J.* A **49** 26
[12] Hagen G, Papenbrock T, Dean D J and Hjorth-Jensen M 2010 *Phys. Rev.* C **82** 034330
[13] Bogner S K, Furnstahl R J and Perry R J 2007 *Phys. Rev.* C **75** 061001
[14] Hagen G et al. 2012 *Phys. Rev. Lett.* **108** 242501
[15] Abu-Ibrahim B et al. 2009 *J. Phys. Soc. Jpn.* **78** 044201
[16] Brown B A et al. 2001 *Phys. Rev.* C **65** 014612
[17] Hagino K and Sagawa H 2012 *Phys. Rev.* C **85** 037604

Ab initio calculations for Be-isotopes with JISP16

P Maris

Department of Physics and Astronomy, Iowa State University, Ames, IA 50011, USA

E-mail: pmaris@iastate.edu

Abstract. We present recent results from no-core configuration interaction calculations for ^{8}Be, ^{10}Be, and ^{12}Be using the phenomenological two-body interaction JISP16. We calculate the binding energies of the ground state and the excitation energies of the low-lying positive-parity states. We discuss the contributions from the proton and neutron intrinsic spin and orbital motion to the total spin for several states, and use this to identify states which may be dominated by α-cluster configurations. In addition, we also calculate other observables such as dipole and quadrupole moments, as well as transition rates for select E2 transitions.

1. No-Core Configuration Interaction approach

Configuration Interaction (CI) methods have been used in recent years to make increasingly accurate large scale ab initio calculations in nuclear structure, see e.g. Refs. [1, 2] and references therein. In this method, the many-body Schrödinger equation

$$H\,\Psi_i(\vec{r}_1,\ldots,\vec{r}_A) \;=\; E_i\,\Psi_i(\vec{r}_1,\ldots,\vec{r}_A) \tag{1}$$

becomes a large sparse matrix problem with eigenvalues E_i. Improved algorithms to construct this matrix and to determine its lowest eigenstates, as well as efficient use of increasing computational resources are critical for these successes [3, 4, 5, 6, 7, 8].

In the No-Core CI (NCCI) approach [1, 2, 9] the wavefunction Ψ of a nucleus consisting of A nucleons (protons and neutrons) is expanded in an A-body basis of Slater determinants Φ_k of single-particle wavefunctions $\phi_{nljm}(\vec{r})$

$$\Psi(\vec{r_1},\ldots,\vec{r_A}) = \sum c_k \Phi_k(\vec{r_1},\ldots,\vec{r_A})\,, \tag{2}$$

with $\Phi_k(\vec{r_1},\ldots,\vec{r_A}) = \mathcal{A}[\phi_{n_1l_1j_1m_1}(\vec{r}_1)\,\phi_{n_2l_2j_2m_2}(\vec{r}_2)\ldots\phi_{n_Al_Aj_Am_A}(\vec{r}_A)]$ and $\mathcal{A}$ the antisymmetrization operation. Conventionally, one uses a harmonic oscillator (HO) basis for the single-particle wavefunctions, but it is straightforward to extend this approach to a more general single-particle basis [10]. The single-particle wavefunctions are labelled by the quantum numbers n, l, j, and m, where n and l are the radial and orbital HO quantum numbers, with $N_i = 2n_i + l_i$ the number of HO quanta; j is the total single-particle spin, and m its projection along the z-axis. The many-body basis states Φ_k have well-defined total spin-projection, which is simply the sum of m_i of the single-particle states, $M = \sum m_i$ (hence the name M-scheme), but they do not have a well-defined total spin J. Some of the benefits of this scheme is that it is straightforward to implement, and that in two runs (one for positive and one for negative parity), we get the complete low-lying spectrum, including the ground state, even if the spin of the ground state is not known a priori.

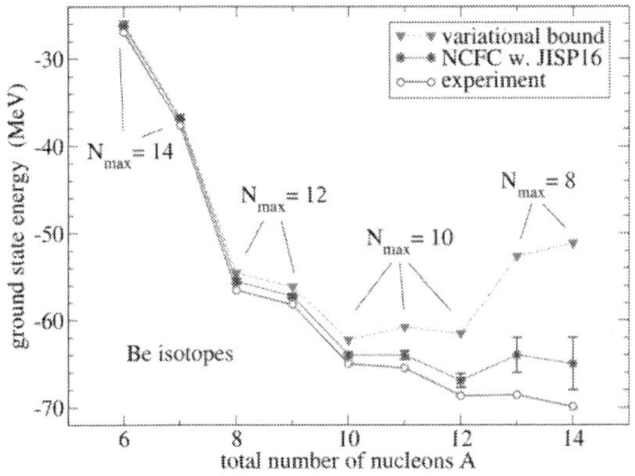

Figure 1. Ground state energy of Be-isotopes: results with JISP16 compared to data [17].

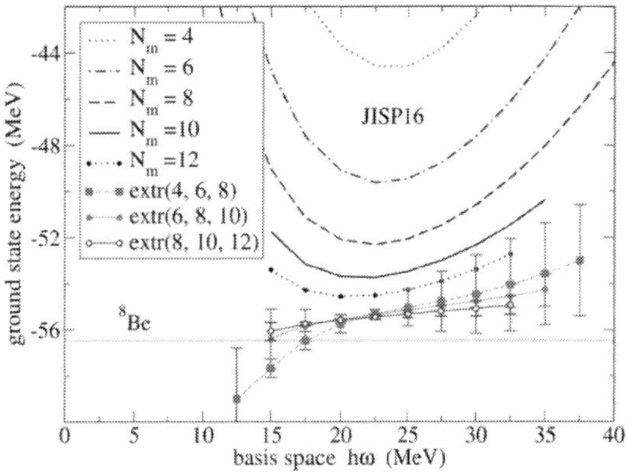

Figure 2. Ground state energy of ^{8}Be with JISP16 as function of the basis space $\hbar\omega$.

In a complete basis, this method would give exact results for a given potential. However, practical calculations can only be done in a finite-dimensional truncation of a complete basis. Different truncation schemes have different convergence rates: Full Configuration Interaction (FCI) employs a widely-know truncation in which all many-body basis states are retained that can be constructed from a finite set of single-particle states. However, this truncation has a rather slow convergence rate (as function of the dimensionality of the basis space) in typical NCCI calculations [11, 12]. The so-called $N_{\rm max}$ truncation, which is a truncation on the total number of HO quanta in the many-body basis: $\sum N_i \leq N_0 + N_{\rm max}$, is generally much more efficient for NCCI calculations. Here, N_i is the number of quanta of each single-particle state in the many body basis state; N_0 is the minimal number of quanta for that nucleus; and $N_{\rm max}$ is the truncation parameter. Furthermore, for HO single-particle states, this truncation leads to an exact factorization of the center-of-mass wavefunction and the relative wavefunction [1, 2, 10, 13].

2. Recent results for Be isotopes with JISP16

Here we present new results for the Beryllium isotopes using JISP16 [14], which is a phenomenological nonlocal NN-interaction written as a finite matrix in a HO basis. It is constructed to reproduce the available NN scattering data using the J-matrix inverse scattering approach. In addition, phase-equivalent transformations have been used to modify its off-shell properties in order to achieve a good description of selected states in light nuclei [14]. It gives a good description of most narrow states in light nuclei up to about $A = 12$ [15, 16] without additional three-nucleon forces. For our calculations we use the code MFDn [4, 5, 6, 7] which has been demonstrated to scale to over 200,000 cores, and we consider basis spaces with dimensions up to 3.3 billion basis states, and nearly 4 trillion nonzero matrix elements.

Because of the variational principle, any result for the ground state energy in a finite basis forms a strict upper bound for the exact ground state energy. Indeed, our calculated energies in finite bases are all above the experimental ground state energies, see Fig. 1. With an exponential extrapolation [15] we obtain No-Core Full Configuration (NCFC) results for the ground state energies (with numerical error bars due to the extrapolation uncertainties) that are in general much closer to the experimental data than these variational upper bounds. It turns out that with JISP16 all Be-isotopes are underbound [3]: the light isotopes, ^{6}Be and ^{7}Be, by a fraction of an MeV, and the neutron-rich isotopes ^{13}Be and ^{14}Be by about 5 MeV. Nevertheless, the overall pattern of the ground state energies is in reasonable agreement with the data [17].

2.1. Results for ^{8}Be

In Fig. 2 we show the ground state energy of ^{8}Be as function of the basis parameter $\hbar\omega$ for a range of different $N_{\rm max}$ values, as well as the extrapolation [15] to the infinite basis using results at three consecutive $N_{\rm max}$ values and fixed $\hbar\omega$, as indicated in the legend. Near the minimum of the ground state energy the convergence appears to be reasonable, with an estimated error (due to the extrapolation) of about 0.3 MeV in the ground state energy. Note that both our calculated and the experimental ground state energies are above the threshold for two α-particles: we are dealing with a (narrow) resonance, rather than a true bound state.

JISP16 does give the correct ground state energy for an α-particle [14, 15]. Hence we know that in the infinite basis space, the energy of the lowest state should be that of two α-particles. On the other hand, our extrapolated ground state energy is about 1 MeV above the threshold for two α-particles. We appear to have 'pseudo-convergence' to a resonance which is above the two-α threshold. This could explain why our extrapolated ground state energy still has a significant and systematic dependence on the HO parameter $\hbar\omega$. Nevertheless, the point proton rms radius trends towards a value of about 2.3 fm, but with a large dependence on $\hbar\omega$, see Fig. 3. Again, this appears to be 'pseudo-convergence', since we know that in the infinite basis space, the lowest state is that of two α-particles, and hence the radius should go to infinity.

Despite the fact that the ground state is well above the threshold for two α-particles, the spectrum of the positive parity states looks in reasonable agreement with the data [18], see Fig. 4. The excitation energies of the two lowest excited states, the 2^+ and 4^+ states, are very well converged, and agree with the experimental values. Although neither the quadrupole moments nor the B(E2)'s are converged in these basis spaces, the ratio of Q_{4^+}/Q_{2^+}, as well as the ratio's B(E2;$2^+ \to 0^+$)$/Q^2_{2^+}$ and B(E2;$4^+ \to 2^+$)$/Q^2_{2^+}$ are in qualitative agreement with a rotational model [19, 20]. Furthermore, more than 95% of their total spins comes from the proton and neutron orbital motion, whereas the proton and neutron intrinsic spin contributes less than 5%. This is all consistent with a picture of these two states being rotational states of the ground state, and their internal structure being dominated by two α-cluster configurations.

Next we have several pairs of states with the same spin: a pair of 2^+, 1^+, and 3^+ states, with the lower having approximately isospin $T = 1$, and the higher one having approximately $T = 0$. And finally a 4^+, $T = 0$ appears at about 20 MeV, with a similar excitation energy as the higher of the two 3^+ states. All of these states are in reasonable agreement with the data, but not as well converged as the lowest 2+ and 4^+ states. None of these states appear to have an α-cluster

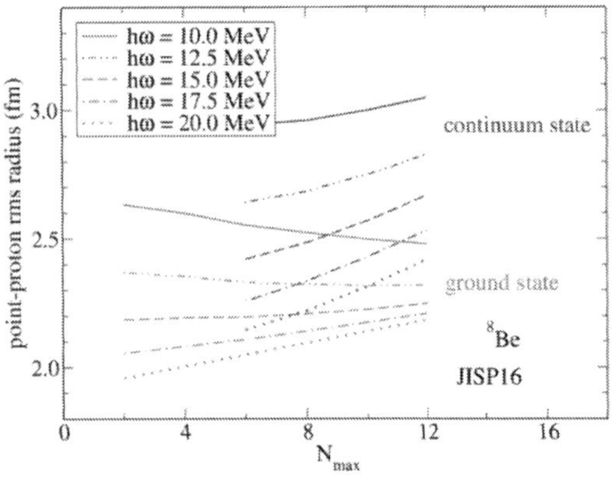

Figure 3. Point-proton rms radius for the lowest 0^+ states of ^{8}Be as function of $N_{\rm max}$.

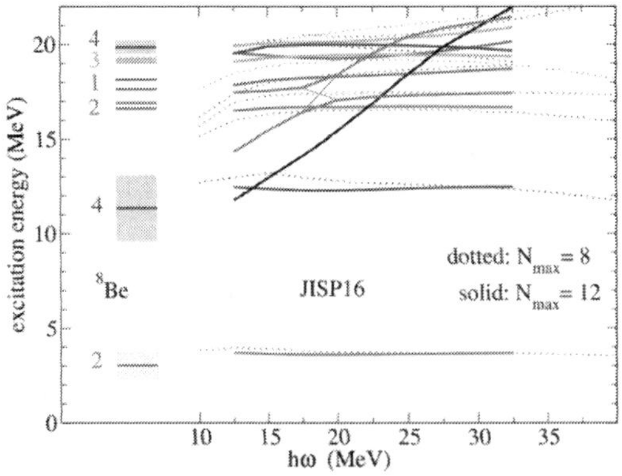

Figure 4. Positive parity spectrum of ^{8}Be with JISP16 up to 20 MeV as function of $\hbar\omega$.

structure: the proton and neutron intrinsic spins contribute at least 20% to the total spin for each of these states.

In addition, we see in Fig. 4 two states that are likely 'continuum states': a 0^+ and a 2^+ state (both isospin 0) which appear in the largest model space calculations, $N_{\rm max} = 10$ and 12. These states have a very strong, almost linear, $\hbar\omega$ dependence. The 0^+ is clearly identifiable, since there are no other 0^+ states in the low-lying spectrum, but the 2^+ 'continuum state' interferes with the isospin 0 resonance at about 17 MeV. The point proton rms radius of the 0^+ state increases rapidly with $N_{\rm max}$ and shows no signs of convergence, see Fig. 3. Most likely these two states are continuum states, consisting of two unbound α-particles.

2.2. Results for ^{10}Be

In contrast to ^{8}Be, ^{10}Be is a true bound state, well below the threshold of ^{4}He + ^{6}He, as indicated by the dashed line in the left panel of Fig. 5. After the extrapolation to the infinite basis space, we find a binding energy of 64.0(3) MeV for the 0^+ ground state, compared to the experimental value of 64.98 MeV; that is, JISP16 underbinds this state by about 1 MeV.

The first excited state is a 2^+ state, its excitation energy is rapidly converging in our calculations, and in good agreement with the data [18]. The second 2^+ state is not very well converged, but after the extrapolation of the binding energy to the infinite basis its excitation energy is in agreement with the data, given the extrapolation uncertainties. However, the first excited 0^+ is missing from our low-lying spectrum. The excitation energy of the lowest excited 0^+ state is above 10 MeV, even after extrapolating the binding energies. As of yet it is unclear why we do not find a low-lying 0^+; it could be an indication that something is missing in the phenomenological NN interaction, or the structure of that state could be such that it cannot be well represented in HO basis. Note that also in ^{12}C the lowest excited 0^+ is above 10 MeV with JISP16 [15], whereas experimentally the lowest 0^+ state is around 7.7 MeV (the Hoyle state). It is quite possible that the absence of a low-lying 0^+ in ^{12}C with this interaction and in these basis spaces is related to the absence of a low-lying 0^+ in ^{10}Be.

The magnetic moment of the lowest 2^+ state is reasonably well converged, $\mu = 1.7(1)\ \mu_N$, but that of the second 2^+ is not at all converged — though interestingly, the *sum* of the magnetic moments of the second and third 2^+ states seems to be converged at $\mu(2^+_2) + \mu(2^+_3) \approx 5.5\ \mu_N$, while $\mu(2^+_2)$ ranges from 1.5 to 3 μ_N for $\hbar\omega$ between 20 and 30 MeV at $N_{\rm max} = 8$. This is mainly due to variations in the contributions from the proton intrinsic spin: over this range in $\hbar\omega$ the proton intrinsic spin contributes between 10% and 30% to the total spin of these two states.

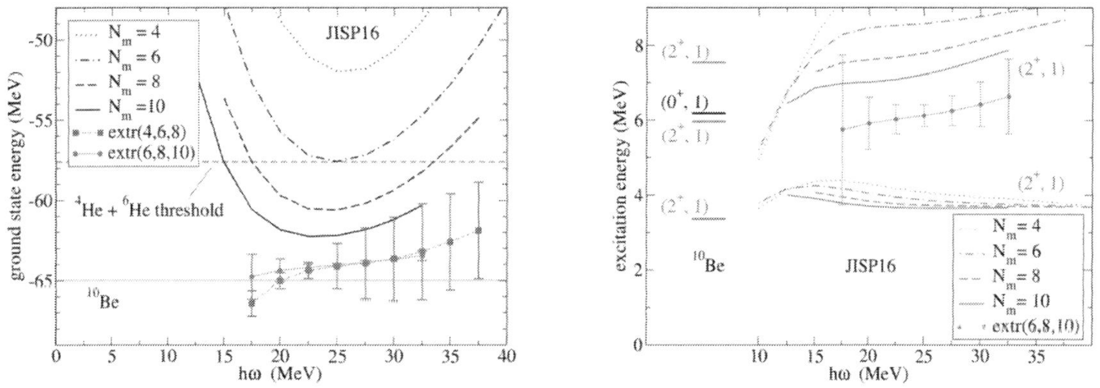

Figure 5. Ground state energy of ^{10}Be (left), and low-lying positive parity spectrum (right).

This suggests that neither the second nor the third 2^+ state is an α-cluster state. Calculations in larger basis spaces are needed to resolve this. On the other hand, the lowest 2^+ state has a spin decomposition that is reasonably well converged, with more than 70% of the spin coming from the proton orbital motion and more than 20% from the neutron orbital motion, whereas the proton and neutron intrinsic spins contribute each less than 5%. Thus it is conceivable that the first 2^+ state is dominated by two α-cluster configurations plus two neutrons.

Neither the quadrupole moments of the 2^+ states nor the B(E2)'s from the 2^+ states to the ground states are converged, but it is clear from our caclulations that the quadrupole moment of the first 2^+ state is negative, around -4 to $-6\,e\,\mathrm{fm}^2$, while that of the second 2^+ state is positive, around $+4$ to $+6\,e\,\mathrm{fm}^2$. This is in qualitative agreement with GFMC calculations using AV18 plus IL7 [22]. Furthermore, the B(E2) from the first 2^+ to the ground state is strong, around 6 to $8\,e^2\,\mathrm{fm}^4$ in our calculations, whereas that of the second 2^+ is weak, of the order of $0.1\,e^2\,\mathrm{fm}^4$ in our calculations. This is in reasonable agreement with recent experiments [21]: $\mathrm{B(E2;}2^+_1 \to 0^+) = 9.2(3)\ e^2\,\mathrm{fm}^4$ and $\mathrm{B(E2;}2^+_2 \to 0^+) = 0.11(2)\ e^2\,\mathrm{fm}^4$.

2.3. Results for ^{12}Be

Finally, ^{12}Be is an unstable nucleus with a lifetime of 21 ms. It is a semi-magic nucleus: it has eight neutrons, which closes the p-shell. In Fig. 6 we show the ground state energy of ^{12}Be as function of the basis space parameter $\hbar\omega$ for a range of different N_{max} values, as well as the extrapolation (at fixed $\hbar\omega$) to the infinite basis space. We find a binding energy of 66.9(6) MeV for the 0^+ ground state, compared to the experimental value of 68.65 MeV. That is, our calculations suggests that JISP16 underbinds this nucleus by almost 2 MeV.

The first excited state is a 2^+ state, in agreement with experiment (see the right panel of Fig. 6), though the excitation energy appears to be somewhat too large: about 3.1(4) MeV, compared to 2.1 MeV experimentally [23, 24]. It is interesting to note that for this 2^+ state the neutron orbital motion contributes less than 10% to the total spin, and their intrinsic spin less than 2% to the total spin. Most of the spin is carried by the proton orbital motion, with a significant fraction coming from the proton intrinsic spin. This indicates that this state is not dominated by α-clustering. Furthermore, this 2^+ state has a large B(E2) to the ground state, of the order of 6 to 10 $e^2\,\mathrm{fm}^4$, in agreement with the experimental value of 8.0(3.0) $e^2\,\mathrm{fm}^4$ [24].

The next excited state in the calculated spectrum is a 0^+, but this state is not very well converged: its excitation energy drops significantly as N_{max} increases. An exponential extrapolation of the binding energy of this state suggests that it has an excitation energy of about

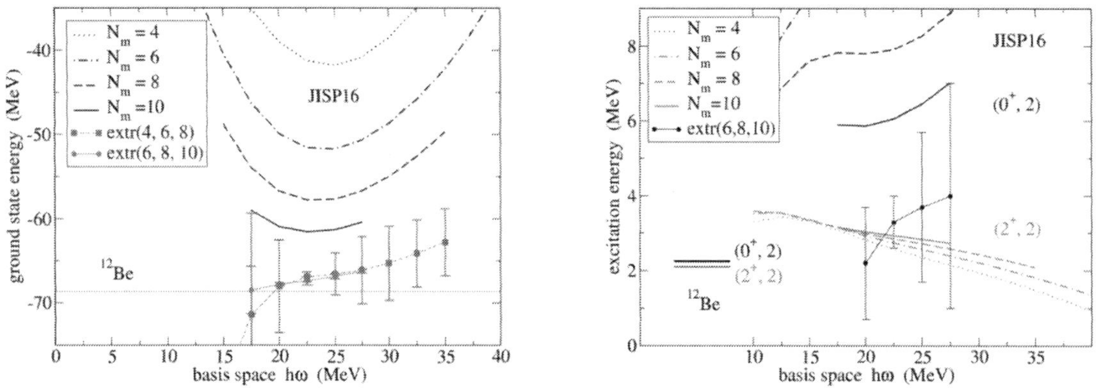

Figure 6. Ground state energy of ^{12}Be (left), and low-lying positive parity spectrum (right).

3 MeV in the limit of an infinite basis space, but with a rather large extrapolation uncertainty. Although its (extrapolated) excitation energy agrees with data, its B(E2) strength to the first excited 2^+ state is more than an order of magnitude too small compared to experiment: we find a B(E2) strength that is less than 0.2 $e^2\,\mathrm{fm}^4$, compared to the experimental value of 7.0(6) $e^2\,\mathrm{fm}^4$ [23]. Of course, as the energy of this 0^+ state gets closer to that of the ground state with increasing N_{max}, there may be more mixing between these two states, which could strongly influence their B(E2) transition strengths; in addition there are also higher excited 2^+ states that are not yet converged, with decreasing excitation energies as N_{max} increases, which may or may not mix with the first excited 2^+ state.

Acknowledgments
I would like to thank M. Caprio, T. Abe, N. Shimizu, T. Otsuka, Y. Utsuno, T. Yoshida, and J. Vary for useful and stimulating discussions. This work was supported in part by U.S. Department of Energy Grants DESC0008485 (SciDAC-3/NUCLEI), DE-FG02-87ER40371, by the U.S. NSF Grant 0904782, and through JUSTIPEN under grant number DE-FG02-06ER41407. Computational resources were provided by the National Energy Research Scientific Computing Center (NERSC), which is supported by the DOE which is supported by the DOE Office of Science under Contract DE-AC02-05CH11231, and by an INCITE award, Nuclear Structure and Nuclear Reactions, from the DOE Office of Advanced Scientific Computing. This research used resources of the Oak Ridge Leadership Computing Facility at ORNL, which is supported by the DOE Office of Science under Contract DE-AC05-00OR22725.

References

[1] Navrátil P, Quaglioni S, Stetcu I and Barret B R 2009 *J. Phys. G: Nucl. Part.* **36** 083101
[2] Barrett B R, Navrátil P and Vary J P 2013 *Prog. Part. Nucl. Phys.* **69** 131
[3] Vary J P, Maris P, Ng E, Yang C and Sosonkina M 2009 *J. Phys.: Conf. Ser.* **180** 012083
[4] Sternberg P, Ng E G, Yang C, Maris P, Vary J P, Sosonkina M and Le H V 2008 *Proc. of the 2008 ACM/IEEE conf. on Supercomputing* SC '08 (Piscataway, NJ: IEEE Press) 15
[5] Maris P, Sosonkina M, Vary J P, Ng E G and Yang C 2010 *Procedia Computer Science* **1** 97
[6] Aktulga H M, Yang C, Ng E G, Maris P and Vary J P 2012 *Lecture Notes in Computer Science* **7484** 830
[7] Aktulga H M, Yang C, Ng E G, Maris P and Vary J P 2012 *Improving the Scalability of a Symmetric Iterative Eigensolver for Multi-core Platforms* submitted to *Concurrency and Computation: Practice and Experience*
[8] Shimizu N, Abe T, Tsunoda Y, Utsuno Y, Yoshida T, Mizusaki T, Honma M and Otsuka T 2012 *Prog. Theor. Exp. Phys.* **2012** 01A205
[9] Navrátil P, Vary J P and Barrett B R 2000 *Phys. Rev. Lett.* **84** 5728
[10] Caprio M, Maris P and Vary J 2012 *Phys. Rev.* C **86** 034312
[11] Abe T, Maris P, Otsuka T, Shimizu N, Utsuno Y and Vary J P 2012 *Phys. Rev.* C **86** 054301
[12] Maris P 2012 *J. Phys.: Conf. Ser.* **402** 012031
[13] Elliott J P and Skyrme T H R 1955 *Proc. R. Soc. London* A **232** 561
[14] Shirokov A M, Vary J P, Mazur A I and Weber T A 2007 *Phys. Lett.* B **644** 33
[15] Maris P, Vary J P and Shirokov A M 2009 *Phys. Rev.* C **79** 014308
[16] Cockrell R, Maris P and Vary J P 2012 *Phys. Rev.* C **86** 034325
[17] Audi G, Wapstra A H and Thibault C 2003 *Nucl. Phys.* A **729** 337
[18] Tilley D R *et al.* 2004 *Nucl. Phys.* A **745** 155
[19] Caprio M A, Maris P and Vary J P 2013 *Phys. Lett.* B **19** 179
[20] Maris P, Caprio M A and Vary J P 2013 *in preparation*
[21] McCutchan E A *et al.* 2009 *Phys. Rev. Lett.* **103** 192501
[22] McCutchan E A *et al.* 2012 *Phys. Rev.* C **86** 014312
[23] Shimoura S *et al.* 2007 *Phys. Lett.* B **654** 87
[24] Imai N *et al.* 2009 *Phys. Lett.* B **673** 179

Exotic cluster structures in the mean-field theory

M A Klatt[1], T Ichikawa[2], K Iida[3], N Itagaki[2], J A Maruhn[4], K Matsuyanagi[2,5], K Mecke[1], S Ohkubo[6,7], P-G Reinhard[1] and B Schuetrumpf[4]

[1] Institut fuer Theoretische Physik, Universitaet Erlangen-Nuernberg, 91058 Erlangen, Germany
[2] Yukawa Institute for Theoretical Physics, Kyoto University, Kyoto 606-8502, Japan
[3] Department of Natural Science, Kochi University, 2-5-1 Akebono-cho, Kochi 780-8520, Japan
[4] Institut fuer Theoretische Physik, Universitaet Frankfurt, 60438 Frankfurt, Germany
[5] RIKEN Nishina Center, Wako 351-0198, Japan
[6] Research Center for Nuclear Physics, Osaka University, Ibaraki, Osaka 567-0047, Japan
[7] University of Kochi, Eikokuji-cho, Kochi 780-8515, Japan

Email: maruhn@th.physik.uni-frankfurt.de

Abstract. Investigations of exotic cluster-like phenomena in the framework of the Skyrme-Hartree-Fock approach are reported. The occurrence of highly excited isomeric states is discussed in connection with the question of their stability in static and time-dependent Hartree Fock (TDHF) calculations. We find rotational stabilization of a 4α chain structure in ^{16}O occurring for a limited range of angular momenta. A toroidal configuration of ^{40}Ca was also stabilized by rotation and provides a very interesting example of rotation about a symmetry axis with a strictly quantized total angular momentum. Finally we look at the formation of nuclear pasta phases in a time-dependent approach and their classification.

1. Complex isomeric states in Hartree-Fock calculations

In an early paper on cluster structure of light nuclei [1] it was already found that in Hartree-Fock calculations exotic configurations can appear that appear to be as stable as the ground state but upon closer examination are found to in reality be unstable against some type of collective deformation.

Fig. 1 illustrates how this works in practice: a chain configuration of three α-particles remains stable for several thousand static iterations in a Hartree-Fock code but then suddenly switches over to the ground state of ^{12}C via a bending deformation described by the Q_{31} multipole accompanied by a drastic change in quadrupole deformation β. It is especially interesting in this context that the typical convergence criteria like the relative change in energy per iteration ΔE and uncertainty in the single-particle energies Δh are at the same level for the chain state as for the ground state.

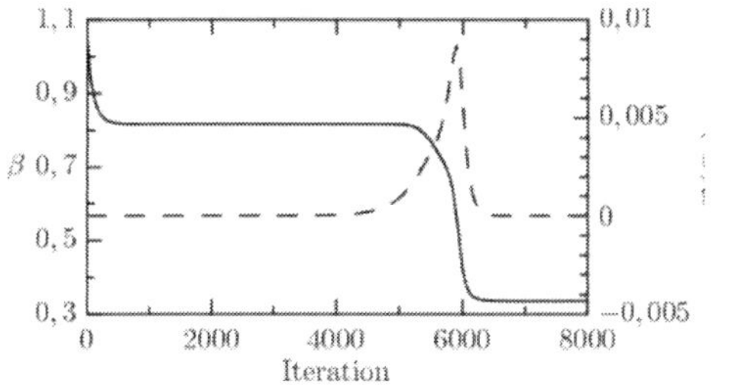

Figure 1. Quadrupole (full curve) and bending deformation (dashed) as functions of iteration number.

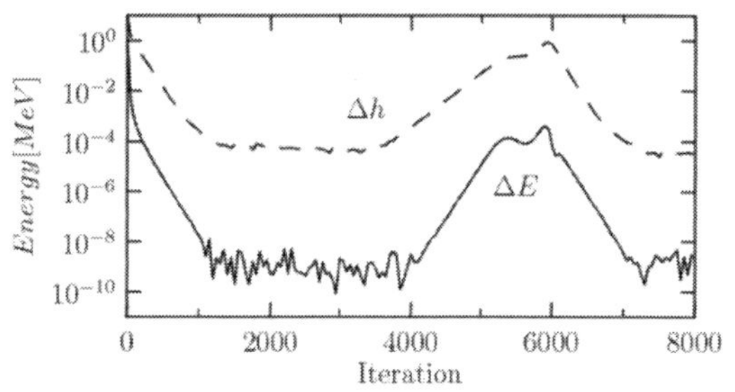

Figure 2. Convergence criteria as functions of iteration number

The reason for this behavior is not fully understood yet. It may depend on the numerical algorithm to what extent symmetries – like in this case axial symmetry – are preserved. Starting the calculation with a three α-particles not arranged on a line led to slow increase of bending and then transition into the compact shape. On the other hand, a dynamic distortion showed that instability is not instantaneous but only happens after several oscillations: it appears that just any excitation breaking axial symmetry is not enough but a more specific bending mode has to be present which comes in owing to the nonlinear coupling between these modes.

2. Stabilization of the 4α-chain state

In a recent paper [2] (discussed also in [3]) we investigated the possible stabilization of such α-chains by rotation. For that purpose static Hartree-Fock calculations with a cranking constraint were performed. The initial configuration consisted of three or four α-particles not completely collinear to avoid spurious persistence of axial symmetry and see whether the particles would be pulled onto the axis by the centrifugal force. The initial configuration is illustrated in Fig. 3(a).

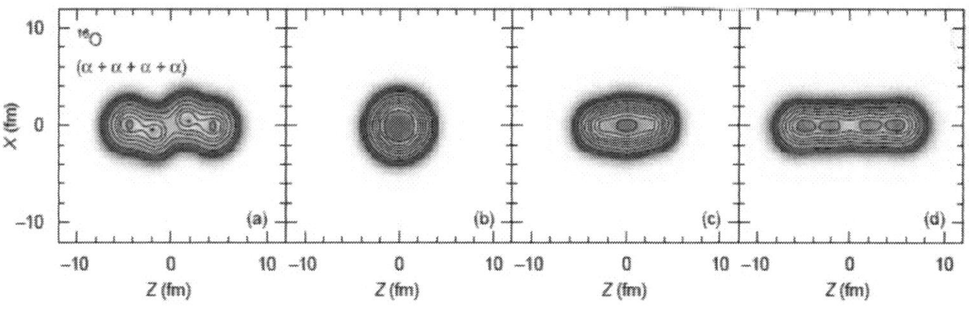

Figure 3. Initial state and different final configurations for the rotating 4α-chain.

For the ^{12}C chain state no stabilization was observed: for slow rotation the chain folded back into the compact configuration, while for fast rotation fission occurred.

Depending on the angular frequency used in the cranking constraint, for the ^{16}O case various final configurations are achieved. Fig. 4 shows which of the configurations shown in Fig. 3 is achieved at different cranking ω by plotting the rotational energy factor as a function of iteration. The most interesting of these is (d), which corresponds to the configuration being stretched into the chain by rotation. The lowest curve in the figure corresponds to fission.

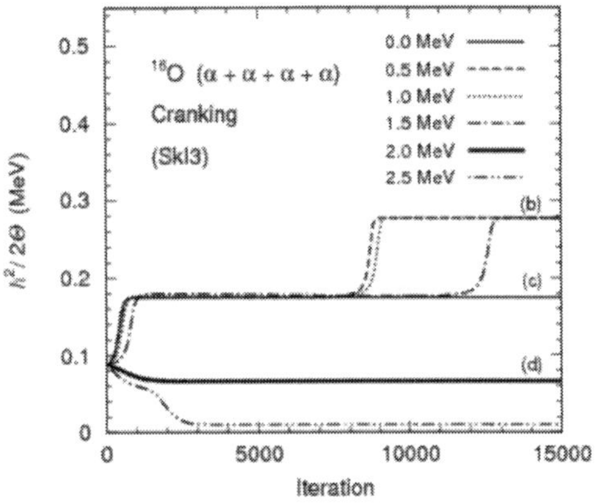

Figure 4. Convergence to different final deformations depending on ω.

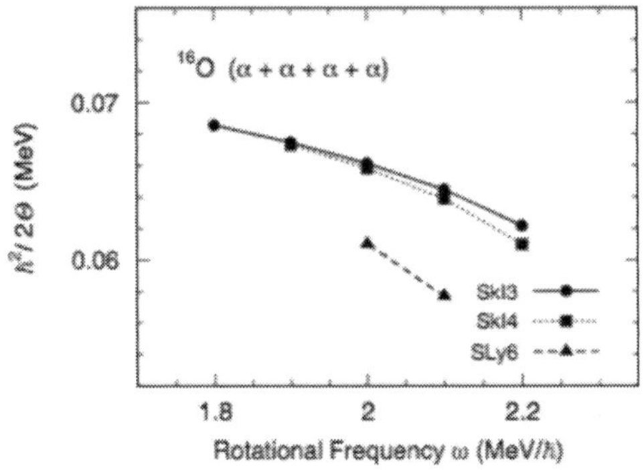

Figure 5. Rotational energy factor vs. ω for three different Skyrme forces.

The moment of inertia was calculated in two ways: from the angular momentum divided by ω and from the rigid-body expression. Both yielded essentially the same values for the chain configuration, which is not surprising considering the large deformation.

The final result for the chain configuration is shown in Fig. 5. It clearly shows that both the range of stable frequencies and the size of the deformation differ significantly between the SkI3/4 and Sly6 cases. Centrifugal stretching is also noticeable.

Time-dependent calculations were also performed for a ^{8}Be+^{8}Be collision and showed that for suitable initial orientations of the nuclei the chain state could appear as a long-lived (> 1000 fm/c) intermediate state.

3. Torus stabilization by rotation

A more exotic scenario is that of a nuclear torus as originally suggested by C. Y. Wong [4], their collective rotation and possible stabilization effects was also studied in [5,6], where, however, rigid-body rotation was assumed. For an ideally axisymmetric torus such a rotation about its symmetry axis is forbidden in quantum mechanics. There is an alternative in such cases via the alignment of the single-particle orbital and spin angular momenta producing a collective rotation as already suggested by Bohr and Mottelson [7] and realized in *K*-isomers. In contrast to rigid-body rotation this leads to a strongly quantized total angular momentum of the nucleus.

In a first investigation [8] we examined the nucleus ^{40}Ca in a rotating toroidal configuration. This was achieved by initializing a cranked static Hartree-Fock calculation with a ring of 10 α-particles. It was found that for a certain range of cranking frequencies ω a torus configuration with azimuthal symmetry was achieved. The quantal angular momentum, however, was constant over a wide range of values of ω. Fig. 6 illustrates the general behavior for different ω: for slow rotation. The values of J appearing are 20, 60, and 100, but only for J=60 the configuration remains stable for a very large number of iterations. The values occurring here can be understood quite simply from the radially displaced harmonic oscillator model [9]. In practice the wave functions can be approximated quite well as Gaussians in the distance from the central circle of the torus, and by azimuthal factors like $\exp(im\phi)$. The two spin orientations are almost degenerate. The total angular momentum is then generated by an asymmetric occupation of 10 *m*-values with four nucleons for each state. Occupying m from -4 to +5 yields a total projection of 20, for -3 to +6 we get 60, and for -2 to +7 finally a value of 100 results. This explains the large jump in projection values observed. It is made a bit smoother once spin is included, but the basic features are apparent in this way.

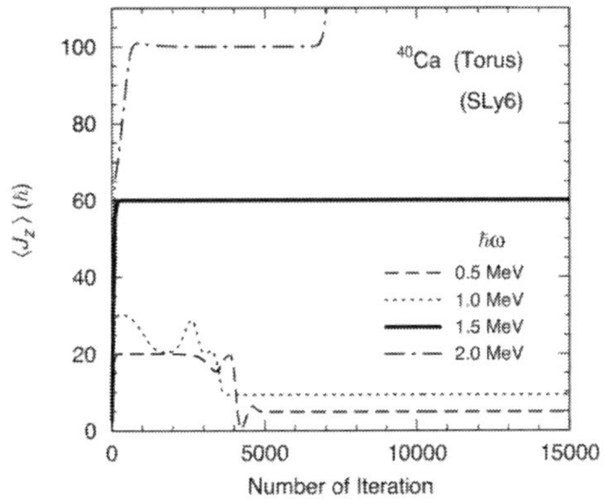

Figure 6. Convergence of angular momentum for the torus for different constraint ω.

The true stability of the configuration at J=60 is not easy to establish in static calculations, although starting with a distorted ring-like configuration showed convergence back to azimuthal symmetry for some range of distortion magnitudes. Therefore also a dynamic calculation was done to check how resilient the torus is with respect to a time-dependent sausage-like perturbation. In practice a positive time-dependent external potential was applied to act as a disrupting mechanism. It was of Gaussian shape with σ=1.8 fm and temporally also had a Gaussian profile with the duration of 20 fm/c. For a peak value of 40 MeV there was still no break-up of the torus but an initial disturbance was quickly repaired. For larger values, however, the ring broke up. The sequence of shapes shown in Fig. 7 illustrates this and also shows the collective rotation: the point of disruption clearly moves in a counter-clockwise direction.

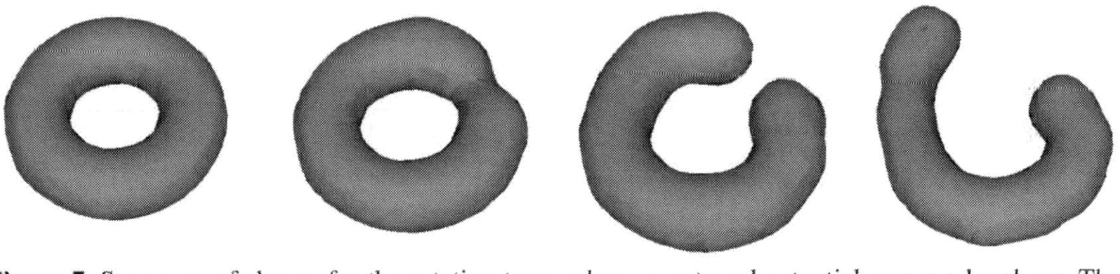

Figure 7. Sequence of shapes for the rotating torus when an external potential causes a break-up. The external potential acts at the right-hand side of the torus.

If the state with the torus configuration is formed at J = 60, the macroscopic circulating current strongly violates the time-reversal symmetry in the intrinsic state. It is interesting to investigate how this fascinating new state can be observed in experiments. The state with the torus configuration at J = 60 would have an extremely large magnetic moment ($\mu = 30\mu_N$). This would lead to a procession motion under an external magnetic field.

A question then arises how such a "femtoscale magnet" rotates spontaneously. When the spherical symmetry is broken, the collective rotation emerges spontaneously, in principle, to restore the broken symmetry. However, it is unclear whether such a state built with significant amount of circulating current can rotate about the perpendicular direction to the symmetry axis or not. If not, such a state would be an anomalous one, which has not yet been recognized in the experiments. Even if the state can rotate, the rotational band built on such a state would show interesting behaviors for their M1/E2 transition strengths and moment of inertia.

Journal of Physics: Conference Series **445** (2013) 012036 doi:10.1088/1742-6596/445/1/012036

4. Time-dependent formation of nuclear pasta

The "pasta" phases of nuclear matter have been investigated intensively in various models, including Skyrme-force Hartree-Fock [10]. We have performed investigations in a time-dependent scenario to investigate the transition from some unordered initial state into the pasta phases depending on the density and temperature assumed initially [11].

For this purpose we set up a random initial state in a Cartesian periodic box. This state contained a number of α-particles at randomly selected positions (but rejecting positions with the particles too close) plus a background gas of neutrons in plane-wave states to produce a proton fraction of 1/3. Both the α-particles and the neutrons were given initial velocities from a thermal distribution. The typical grid consisted of 16^3 grid points with a spacing of 1 fm.

The calculated time dependence then showed a relatively rapid formation of one of the familiar pasta structures, which of course still contain time-dependence in the sense of vibrations remaining in the structures. This usually happens within less than 1000 fm/c. Since only a relatively small number of calculations could be performed, large fluctuations are present which, e. g., sometimes lead to different pasta structures for similar initial conditions.

While the average density is simply given by the number of particles in the computational grid, the temperature is given only indirectly. Since TDHF preserves the total energy, not the temperature, the latter has to be calculated for the final pasta configuration, which is not identical to the artificial temperature used to assign the initial velocities. In practice we estimated it by calculating the excitation energy relatively to the HF ground state and using the Fermi-gas formula. The resulting phase structure is shown in Fig. 8, while a graphical impression of the pasta structures attained is given in Fig. 9.

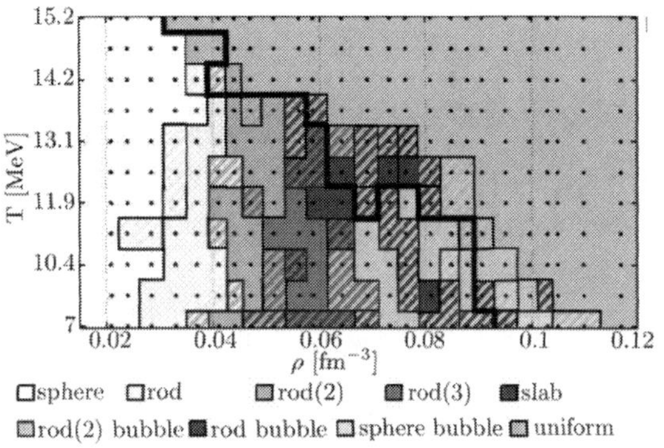

Figure 8. Phase diagram of the pasta structures reached in dependence on density and temperature. Taken from [11].

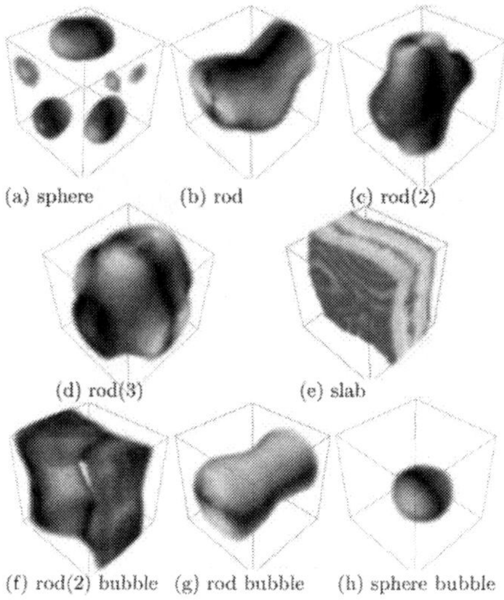

Figure 9. Overview of the pasta structures reached in the simulations. Blue color indicates the lowest and red the highest densities.

Since the geometric structure was quite complicated in many cases, we used the method of Minkowski scalars [12] to distinguish the various cases. This was already applied to similar situations in [13]. For this purpose the density threshold is selected to separate the system into "empty" and "filled regions" with clearly defined surfaces dividing them. Using these surfaces the geometric invariants can be calculated; we found the integral mean curvature and Euler characteristic most useful. The results are quite stable for a broad range of threshold densities. The main results are summarized in the Table 1, which shows that indeed the classification can be done practically completely using just these two quantities.

Table 1. Classification of pasta configurations using the integral mean curvature and Euler characteristic. The letter "b" denotes the "bubble" configuration with high and low density regions exchanged.

Shape	Sphere	Rod	Rod(2)	Rod(3)	Slab	Rod(2) b	Rod b	Sphere b
W2	>0	>0	>0	- to +	≈0	<0	<0	<0
W3	>0	=0	<0	<0	=0	<0	=0	>0

Acknowledgments
This work was undertaken as part by the Yukawa International Project for Quark-Hadron Sciences (YIPQS), and was partly supported by the GCOE program "The Next Generation of Physics, Spun from Universality and Emergence" from MEXT of Japan. J.A.M. was supported by the Frankfurt Center for Scientific Computing and by the BMBF under contracts 06FY9086 and 06ER9063. One of the authors (JAM) would like to thank the Japan Society for the Promotion of Science (JSPS) for an invitation fellowship for research in Japan.

References

[1] Maruhn J, Kimura M, Schramm S, Reinhard P G, Horiuchi H and Tohsaki A 2006 α-cluster structure and exotic states in a self-consistent model for light nuclei *Phys. Rev.* C **74** 044311
[2] Ichikawa T, Maruhn J, Itagaki N and Ohkubo S 2011 Linear Chain Structure of Four-α Clusters in ^{16}O *Phys. Rev. Lett.* **107** 112501
[3] D'Agostino M 2011 Rod-Shaped Nucleus *Phys. Rev. Focus* **28** 10
[4] Wong C Y 1972 Toroidal nuclei *Physics Letters* B **41** 446–50
[5] Wong C-Y 1984 Rotating toroidal nuclei in heavy-ion reactions *Phys. Rev.* C **30** 1949–52
[6] Royer G, Haddad F and Jouault B 1996 Rotating bubble and toroidal nuclei and fragmentation *Nuclear Physics* A **605** 403–16
[7] Bohr A and Mottelson B R 1981 The structure of angular momentum in rapidly rotating nuclei *Nuclear Physics* A **354** 303–16
[8] Ichikawa T, Maruhn J A, Itagaki N, Matsuyanagi K, Reinhard P G and Ohkubo S 2012 Existence of exotic torus configuration in high-spin excited states of ^{40}Ca *Preprint* arXiv:1207.6250 [nucl-th]
[9] Wong C Y 1973 Toroidal and spherical bubble nuclei *Annals of Physics* **77** 279–353
[10] Newton W G and Stone J R 2009 Modeling nuclear "pasta" and the transition to uniform nuclear matter with the 3D Skyrme-Hartree-Fock method at finite temperature: Core-collapse supernovae *Phys. Rev.* C **79** 055801
[11] Schuetrumpf B, Klatt M A, Iida K, Maruhn J, Mecke K and Reinhard P-G 2012 Time-Dependent Hartree-Fock Approach to Nuclear Pasta at Finite Temperature *Preprint* arXiv:1210.8334 [nucl-th]
[12] Mecke K and Stoyan D 2000 *Statistical Physics and Spatial Statistics - The Art of Analyzing and Modeling Spatial Structures and Pattern Formation* vol 554 (Springer)
[13] Watanabe G, Sato K, Yasuoka K and Ebisuzaki T 2004 Phases of hot nuclear matter at subnuclear densities *Phys. Rev.* C **69** 055805

Structures of ground and excited states in C isotopes

Y Kanada-En'yo[1], F Kobayashi[1] and T Suhara[2]
[1]Department of Physics, Kyoto University, Kyoto 606-8502, Japan
[2]Institute of Physics, University of Tsukuba, Tsukuba 305-8571, Japan

Abstract. Structures of ground and excited states of C isotopes are theoretically investigated with the method of antisymmetrized molecular dynamics. For ^{14}C, it is suggested that a linear-chain 3α structure can be stabilized by the excess neutrons and may construct an rotational band near the ^{10}Be+α threshold energy. In ^{11}C and its mirror nucleus ^{11}B, a cluster gas state and a linear-like cluster state are suggested in excited states. Possible assignment of the cluster states in ^{11}B with the experimental energy levels are discussed. Deformations of proton and neutron densities of the ground bands in neutron-rich C are also discussed.

1. Introduction

Cluster aspect of unstable nuclei is one of the hot topics in physics of unstable nuclei. For instance, cluster structures of Be isotopes have been intensively investigated in theoretical and experimental works[1]. It was revealed that a variety of cluster states exist in neutron-rich Be owing to a 2α core and surrounding excess neutrons. In addition to such two-center cluster structures suggested in Be, three-center cluster structures in C isotopes have been attracting a great interest. Rich cluster phenomena are expected in neutron-rich C due to the 3α core formation and valence neutrons.

Carbon nuclei are the fascinating nuclei that show coexistence of two kinds of features, shell-model and cluster aspects. The typical example is ^{12}C which shows the coexistence of the $p_{3/2}$ sub-shell and the 3α cluster features. The ground state is considered to be admixture of the closed $p_{3/2}$ configuration and the 3α core structure. In the excited states near and above the 3α threshold energy, spatially well-developed 3α cluster structures are seen. In particular, a 3α cluster gas state suggested in the ^{12}C(0_2^+) is one of the recent hot topics and it is often discussed in relation with alpha condensation [2]. Moreover, searching for a linear-chain structure of 3 α clusters has been a long standing problem.

Another interesting problem is the structure change of the ground band in C isotopes with increase of the neutron number. In the systematic study of ground states of C isotopes, it was found that neutron structure rapidly changes depending on the neutron number [3]. In contrast to the drastic change of the neutron part, proton structure is rather stable in C because of the shell effect at $Z = 6$. As a result, a neutron-skin structure may develop in C near the neutron drip line. Another interesting feature in the ground bands of unstable C nuclei concerns deformation of proton and neutron densities. The decoupling, i.e., the difference between proton and neutron shapes was suggested in ^{10}C, ^{16}C, and ^{18}C. The recently measured $E2$ transition strengths for the $2_1^+ \to 0_1^+$ in ^{10}C[4], ^{16}C[5, 6], ^{18}C[6, 7], and ^{20}C[8] are useful information to clarify deformation properties of the ground bands in C isotopes. For instance, the quenching of $E2$ transition strengths in ^{10}C and ^{16}C can be described by the decoupling of proton and neutron shapes.

Those phenomena tell us that cluster is one of the essential features in unstable nuclei as well as stable nuclei. Various kinds of cluster structure emerge depending on the excitation energy and also depending on the number and kind of core clusters as well as on the number of excess neutrons. In low-lying states, clusters are tightly bound in general and cluster features are characterized by the cluster correlation or the cluster formation. On the other hand, in highly excited states, one may often see remarkable cluster structures where clusters develop spatially. In the case of neutron-rich nuclei, excess neutrons play an important role in cluster structures.

In this paper, structures of ground and excited states of C isotopes are theoretically investigated with the method of antisymmetrized molecular dynamics. Three-center cluster structures in excited states of ^{14}C and ^{11}C are studied. Ground state structure of C isotopes is also investigated. The quenching and enhancement of $E2$ transition strength are discussed in relation with proton and neutron shape difference.

2. Formulation

In the framework of antisymmetrized molecular dynamics (AMD) [9], an A-nucleon wave function is given by a Slater determinant of Gaussian wave packets;

$$\Phi_{\rm AMD}(\mathbf{Z}) = \frac{1}{\sqrt{A!}}\mathcal{A}\{\varphi_1,\varphi_2,...,\varphi_A\}, \tag{1}$$

where the ith single-particle wave function is written as,

$$\varphi_i = \phi_{\mathbf{X}_i}\chi_i\tau_i, \tag{2}$$

$$\phi_{\mathbf{X}_i}(\mathbf{r}_j) \propto \exp\{-\nu(\mathbf{r}_j - \frac{\mathbf{X}_i}{\sqrt{\nu}})^2\}, \tag{3}$$

$$\chi_i = (\frac{1}{2}+\xi_i)\chi_\uparrow + (\frac{1}{2}-\xi_i)\chi_\downarrow. \tag{4}$$

Here $\phi_{\mathbf{X}_i}$ and χ_i are spatial and spin functions, and τ_i is the isospin function which is fixed to be up(proton) or down(neutron). We take an optimized value of the width parameter ν for each nucleus. Accordingly, an AMD wave function is expressed by a set of variational parameters, $\mathbf{Z} \equiv \{\mathbf{X}_1, \mathbf{X}_2, \cdots, \mathbf{X}_A, \xi_1, \xi_2, \cdots, \xi_A\}$, which indicate Gaussian centers and spin orientations for all nucleons. Those parameters are determined by the energy variation. The AMD wave function is quite similar to a wave function of fermionic molecular dynamics (FMD)[10, 11] where more generalized wave functions are used.

To study nuclear structure the energy variation and the spin and parity projections are performed using AMD wave functions. In the simple version of AMD, the variation is performed after the parity projection and then the spin projection is done after the variation. The variation after spin-parity projections is performed in the AMD+VAP calculation. For ^{14}C and ^{11}B, the AMD+GCM method with the β-γ constraint [12] is also applied. More details of the present calculations are described in Refs. [13, 14, 15, 16]. General formulation of the AMD method is explained in Refs. [3, 17].

In the AMD wave function, all single nucleons are treated independently with localized Gaussians. Although any clusters are not assumed *a priori*, multi-cluster structures can be described by grouping of single-nucleon Gaussian wave packets in the spatial configuration. On the other hand, if all Gaussian centers gather around a certain position, the AMD wave function becomes equivalent to a harmonic oscillator shell-model wave function around the position due to the effect of antisymmetrization. Thus, the AMD model space can describe both cluster and mean-field features with assembling and disassembling of Gaussian wave packets. If a system favors a specific cluster channel, such a cluster structure will be automatically obtained in the energy variation.

The effective nuclear interactions used in the present calculations are the Volkov force and the MV1 force [18, 19] with the spin-orbit force supplemented by Coulomb force.

3. Cluster structures of excited states in C

In the case of ^{12}C, we see a variety of cluster structures based on 3α cluster structures. The ground state of ^{12}C is considered to be the compact 3α cluster state with a mixing of the closed $p_{3/2}$-shell configuration. The second 0+ state is the dilute cluster gas state of three α where α clusters are weakly interacting like a gas as proposed by Tohsaki *et al.* [2] and other groups. Another interesting problem is whether the linear-chain 3α structure is stable or not. According to AMD and FMD calculations[20, 21], it was suggested that the straight-line linear chain structure is not stable in ^{12}C but only a chain-like, an open triangle 3α configuration may appear in a higher 0^+ state a few MeV above the 0^+_2 state. This suggestion is supported by the recent experimental observation of 0^+ states around 10 MeV excitation energy by Itoh *et al.* In ^{11}C and its mirror nucleus ^{11}B, similar cluster structures consisting of two α and ^{3}He(triton) clusters were suggested [14, 16]. That is, a compact triangle structure, a dilute cluster gas-like state, and a linear-chain structure are theoretically suggested. In particular, the developed $2\alpha + t$ cluster structure of the ^{11}B($3/2^-_3$) shows analogy with the 3α structure of ^{12}C(0^+_2). In the experimental side, candidates of rotational band members for the excited band on the top of the $3/2^-_3$ state in ^{11}B were suggested by Yamaguchi *et al.* [22].

As mentioned before, the straight-line chain structure of three α clusters may not be stable in ^{12}C. In neutron-rich C isotopes, we will see further rich cluster phenomena in excited states because of three-center cluster structure with excess neutrons. Let us consider structure of ^{14}C. By adding two neutrons to the ^{12}C system, we can expect developed cluster structures consisting of 3 α clusters with two excess neutrons in excited states of ^{14}C. Ground and excited states of ^{14}C are calculated with a method of β-γ constraint AMD in combination with the GCM[15]. A variety of developed 3α-cluster core structures are suggested in excited states of ^{14}C. It was found that the 3α linear-chain structure can be stabilized by additional neutrons in ^{14}C. As a result, a $K^\pi = 0^+$ rotational band was suggested to be constructed from the elongate shape with the linear-chain structure in the energy region above the ^{10}Be+α threshold energy.

4. Deformation and $B(E2)$ of the ground bands in C

The structures of Be, B, and C isotopes were investigated with the AMD method [3, 13]. The results suggest that the ground state structure rapidly changes as the neutron number increases in each isotope chain. In Be isotopes, the well-developed 2α-cluster structure is known in ^{8}Be. With the increase of the neutron number the cluster structure weakens in ^{10}Be and enhances again in ^{12}Be. In contrast to the rapid change of the cluster structure in Be, cluster core structure is rather stable in neutron-rich C against the neutron number increasing. Namely, 3α-cluster structure in ^{12}C weakens in ^{14}C. With further increase of the neutron number, proton structure is almost unchanged in spite of the rapid change of the neutron structure. Consequently, the neutron skin structure may enhance in neutron-rich C near the drip line. Moreover, the shape difference between proton and neutron densities were suggested in ^{16}C having prolate neutron and oblate proton shapes.

Let us discuss more details of the ^{16}C structure. ^{16}C consists of six protons and ten neutrons. Six proton systems tend to deform oblately, while ten neutrons favor a prolate deformation. Usually, proton and neutron deformations are expected to be consistent with each other because of the proton-neutron interaction. However, the results suggest that ^{16}C is a unique nuclei where proton and neutron structures can decouple to each other. As a result of the decoupling, the neutron excitation is dominant in the 2^+_1 state. Since the system is deformed, the excitation energy of the 2_1^+ state should be small. Nevertheless, proton contribution in the 2^+ excitation is expected to be minor, and hence, the $E2$ transition strength for $2^+ \to 0^+$ may be suppressed. This suggestion is supported by the experimental measurements of the $E2$ strength and inelastic scattering cross sections [6, 13, 23, 24]. In contrast to the decoupling proton and neutron deformations in ^{16}C, ^{20}C has an oblate neutron deformation which is consistent with the oblate proton deformation. It means that ^{16}C is the proton-neutron decoupling system while ^{20}C is the consistent system. Such the structure change may affect the $E2$ transition strength.

Recently, $E2$ strengths for ^{18}C and ^{20}C as well as that for ^{16}C were experimentally observed by life time measurements [5, 6, 7, 8]. They are useful data to clarify the ground band properties especially concerning proton structure. Figure 1 shows the $B(E2)$ values in C isotopes. It is experimentally known that the 2^+ excitation energies in ^{16}C, ^{18}C, and ^{20}C are relatively small compared with those in lighter C isotopes. It suggests that these are deformed nuclei. The experimental $B(E2)$ value is small for ^{16}C indicating that the proton contribution is small in the spin 2 rotation. On the other hand, the recently measured $B(E2)$ value for ^{20}C is larger than that for ^{16}C. The AMD and AMD+VAP calculations reproduce the systematics of $E2$ strengths in C isotopes. The quenching of $B(E2)$ for ^{16}C is described by the minor proton contribution because of the decoupling of proton and neutron shapes, while the enhancement of $B(E2)$ for ^{20}C can be understood by the consistency of proton and neutron deformations in ^{20}C.

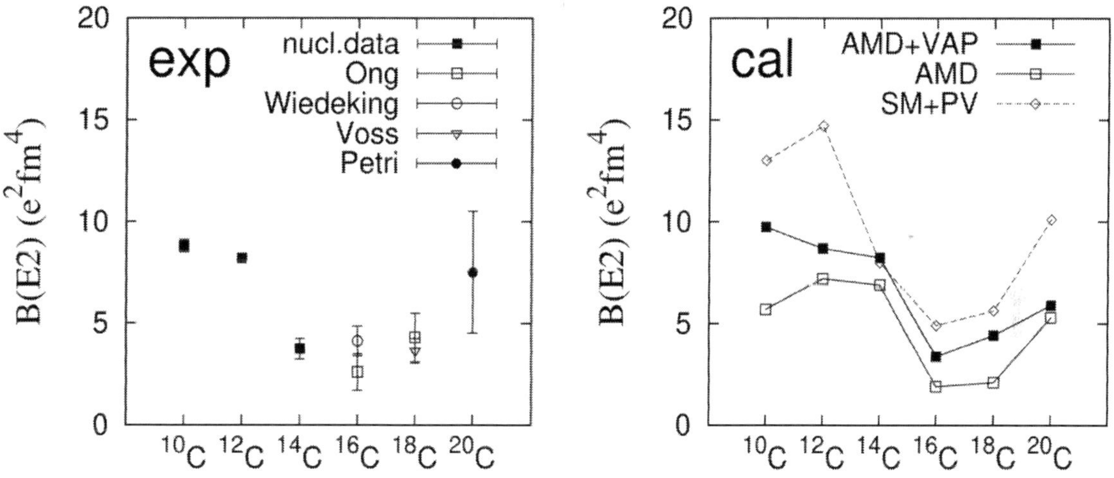

Figure 1. $E2$ transition strengths of C isotopes. The experimental data of $E2$ transition strengths are those in the compilation[25] and those measured by Ong *et al.*[6], Wiedeking *et al.*[5], Petri *et al.*[8], Voss *et al.*[7], McCutchan *et al.*[4]. The theoretical values are results calculated with AMD+VAP using MV1-case1(m = 0.62)+LS(u_{ls} = 2600 MeV) force and those with AMD using MV1-case3(m = 0.576)+LS(u_{ls} = 900 MeV) force. The shell model calculations are taken from Ref. [26].

5. Summary

Structures of ground and excited states of C isotopes were discussed based on theoretical calculations with the antisymmetrized molecular dynamics method. Possible cluster states were suggested in excited states of ^{14}C, ^{11}C and ^{11}B. In the study of ground states of C isotopes, the structure change with the increase of neutron number was suggested. The quenching and the enhancement of $E2$ transition strengths in ^{16}C and ^{20}C were discussed in relation with decoupling and coupling between proton and neutron deformations, respectively.

Acknowledgments

Discussions during YIPQS workshops held at YITP are helpful to advance this work. The computational calculations of this work were performed by using the supercomputers at YITP and those at KEK. This work was supported by Grant-in-Aid for Scientific Research from Japan Society for the Promotion of Science (JSPS). It was also supported by the Grant-in-Aid for the Global COE Program "The Next Generation of Physics, Spun from Universality and Emergence" from the Ministry of Education, Culture, Sports, Science and Technology (MEXT) of Japan.

References

[1] Oertzen W, Freer M and Kanada-En'yo Y 2006 *Phys. Rep.* **432** 43
[2] Tohsaki A, Horiuchi H, Schuck P and Röpke G 2011 *Phys. Rev. Lett.* **87** 192501
[3] Kanada-En'yo Y and Horiuchi H 2011 *Prog. Theor. Phys. Suppl.* **142** 205
[4] McCutchan E A *et al.* 2012 *Phys. Rev.* C **86** 014312
[5] Wiedeking M *et al.* 2008 *Phys. Rev. Lett.* **100** 152501
[6] Ong H J *et al.* 2008 *Phys. Rev.* C **78** 014308
[7] Voss P *et al.* 2012 *Phys. Rev.* C **86** 011303
[8] Petri M *et al.* 2011 *Phys. Rev. Lett.* **107** 102501
[9] Kanada-En'yo Y, Horiuchi H and Ono A 1995 *Phys. Rev.* C **52** 628 ; Kanada-En'yo Y and Horiuchi H 1995 *Phys. Rev.* C **52** 647
[10] Feldmeier H 1990 *Nucl. Phys.* A **515** 147
[11] Feldmeier H, Bieler K and Schnack J 1995 *Nucl. Phys.* A **586** 493
[12] Suhara T and Kanada-En'yo Y 2010 *Prog. Theor. Phys.* **123** 303
[13] Kanada-En'yo Y 2005 *Phys. Rev.* C **71** 014310
[14] Kanada-En'yo Y 2007 *Phys. Rev.* C **75** 024302
[15] Suhara T and Kanada-En'yo Y 2010 *Phys. Rev.* C **82** 044301
[16] Suhara T and Kanada-En'yo Y 2012 *Phys. Rev.* C **85** 054320
[17] Kanada-En'yo Y, Kimura M and Horiuchi H 2003 *Comptes rendus Physique* **4** 497
[18] Volkov A B 1965 *Nucl. Phys.* **74** 33
[19] Ando T, Ikeda K and Tohsaki A 1980 *Prog. Theory. Phys.* **64** 1608
[20] Kanada-En'yo Y 2007 *Prog. Theor. Phys.* **117** 655 [Erratum-ibid. 2009 **121** 895]
[21] Chernykh M, Feldmeier H, Neff T, von Neumann-Cosel P and Richter A 2007 *Phys. Rev. Lett.* **98** 032501
[22] Yamaguchi H *et al.* 2011 *Phys. Rev.* C **83** 034306
[23] Takashina M, Kanada-En'yo Y and Sakuragi Y 2005 *Phys. Rev.* C **71** 054602
[24] Elekes Z *et al.* 2004 *Phys. Lett.* B **586** 34
[25] Tilley D R *et al.* 2004 *Nucl. Phys.* A **745** 155
[26] Sagawa H, Zhou X R, Zhang X Z and Suzuki T 2004 *Phys. Rev.* C **70** 054316

Journal of Physics: Conference Series **445** (2013) 012038 doi:10.1088/1742-6596/445/1/012038

Density profiles in the intrinsic frame of light nuclei obtained from Monte Carlo shell model calculation

T Yoshida [1], **N Shimizu**[1], **T Abe**[2] **and T Otsuka**[1,2,3]

[1]Center for Nuclear Study, University of Tokyo, 7-3-1 Hongo, Bunkyo, Tokyo 113-0033, Japan
[2]Department of Physics, University of Tokyo, 7-3-1 Hongo, Bunkyo, Tokyo 113-0033, Japan
[3]National Superconducting Cyclotron Laboratory, Michigan State University, East Lansing, Michigan, USA

E-mail: `yoshida@cns.s.s-tokyo.ac.jp`

Abstract. The wave functions of the ground states for ^{8,10}Be which are obtained from the Monte Carlo shell model (MCSM) are investigated. A method to define an intrinsic state in the MCSM is discussed. The appearance of two-α-cluster structure in ^{8}Be and the property of the valence neutrons in ^{10}Be are discussed.

1. Introduction

In the large region of the nuclear chart, the shell model picture is essential to understand the nuclear magic numbers related to the single-particle motion of a nucleon. The conventional shell model calculation enables us to take all relevant degrees of freedom of nucleon motion into account when the particles are set to be in the valence shell on top of the core nucleus. For the lightest part of the nuclear chart (mass number lower than $A = 12$), shell model calculations without assuming an inert core can be performed. On the other hand, α cluster configuration exists in light nuclei, which has an important role to determine the transition strength at the α-breakup-threshold-energy region. The appearance of α cluster structure has been studied for a few decades [1].

Recently, *ab initio* calculations have been started for the lightest part. One example is the Green's function Monte Carlo (GFMC) [2] approach. In its analysis of $A = 8$ nuclei, the two-α-cluster shape emerges in *a priori* from the density distribution. In order to estimate the transition properties, it is important to confirm the existence of the cluster configuration by calculating the density distribution of light nuclei from shell-model side. However, it is difficult to discuss α-cluster configuration in view of intrinsic shape utilizing shell model calculations. This is because the single-particle orbit of the shell model is expanded around the one center of the harmonic oscillator potential.

Recently, the size of the model space which is tractable in the shell model calculation has been increased due to the development of parallel computers and methodology of shell model calculations. Therefore, it is natural to expect that α-cluster configurations are included in the single-particle orbit of the shell model if the model space is large enough. In fact, the deformation and cluster-like shape for Li isotopes have been investigated by using no core full configuration (NCFC) approach [3].

Another approach is the Monte Carlo shell model (MCSM) [4] method which has been applied to the various nuclei by using the recent parallel computer with sophisticated energy-minimum search where the conjugate gradient method and energy-variance extrapolation [5, 6] are combined. The MCSM is expected to be applicable toward the large region of the nuclear chart both with and without the inert core (no-core MCSM). In the proceedings, we show the density profiles of wave function obtained in the no-core-MCSM calculation. For this purpose, we define the intrinsic state from the no-core-MCSM wave function in the following section. We focus on the ground state of ^{8,10}Be which are expected to have the two-α-cluster configuration. In ^{10}Be, the density for the valence neutrons is discussed.

2. Formulation

In this section, we show how to obtain the intrinsic structure for the no-core-MCSM wave function. In order to investigate the cluster-like state, it is important to extract the intrinsic state from the wave function whose angular momentum is a good quantum number in the no-core MCSM.

In the no-core-MCSM calculation, the wave function is projected to the state which has a good angular momentum quantum, J. We use a Slater determinant, $|\phi_n\rangle$, to describe a basis state which is defined as

$$|\phi_n\rangle = \prod_{\alpha=1}^{N_p} \sum_{i=1}^{N_{sp}} c_i^\dagger D_{i\alpha}^{(n)} |-\rangle, \tag{1}$$

where $|-\rangle$ is a vacuum state and the $c_i^\dagger$'s are creation operators on the single-particle orbit, i. The resultant wave function is given by a superposition of the angular-momentum-projected Slater determinants,

$$|\Psi\rangle = P_{MK}^{J^\pi}|\Phi\rangle, \quad |\Phi\rangle = \sum_{n=1}^{N_b} f_n|\phi_n\rangle. \tag{2}$$

In this study, we use the values, $\pi = +$, $J = 0$, $M = 0$ and K-quantum number, $K = 0$. The coefficients f_n's are determined by diagonalizing the Hamiltonian matrix, $\langle\phi_i|H|\phi_j\rangle$. In the no-core-MCSM procedure, we tune the value of $D^{(n)}$ for each Slater determinant by the conjugate gradient method in order to minimize the expectation value of the Hamiltonian with J-projected wave functions. Here, we use the JISP16 NN interaction which reproduces NN scattering data and properties of deuteron and other light nuclei [7].

In this J-projected wave function, the cluster-like structure is difficult to be visualized. To analyse the wave function in detail, we go back to the wave function before the angular momentum projection, $|\Phi\rangle$, which is the linear combination of the unprojected basis states. However, this state is not considered as an intrinsic state because the principal axis of a basis state, $|\phi_i\rangle$, is not necessarily in the same direction with that of other basis states. As a result, the intrinsic structure such as cluster configurations may be hidden.

In order to extract the wave function of the intrinsic state, we rotate each basis state so that its diagonalized Q-moment is aligned along the z-axis, following the concept of the GFMC [2]. The intrinsic wave function $|\Phi^{\rm intr}\rangle$ is defined as

$$|\Phi^{\rm intr}\rangle \equiv \sum_n f_n R(\Omega_n)|\phi_n\rangle = \sum_n f_n|\phi_n^R\rangle, \tag{3}$$

with the transformed basis state $|\phi_n^R\rangle$. Here, $R(\Omega_n)$ is the rotation operator with an Euler angle Ω_n, which is determined by the condition, $\langle\phi_n^R|Q_{i,j}|\phi_n^R\rangle = \delta_{i,j}q_i$, where $Q_{i,j(=x,y,z)}$ are the operators of the Q-moment. We note that this state, $|\Phi^{\rm intr}\rangle$, exactly has the same energy after the angular momentum projection. After that, it is probable that these aligned basis states have large overlap with each other and make a distinct principal axis toward the z-axis when the condition $q_z > q_y > q_x$ is set. The intrinsic density is obtained by the expectation value of the one-body-density operator,

$$\rho^{\text{intr}}(r) = \langle \Phi^{\text{intr}} | \sum_i \delta(r - r_i) | \Phi^{\text{intr}} \rangle, \tag{4}$$

where r_i denotes the position of the i-th nucleon.

3. Result

Figure 1 shows the proton densities for the 0^+ ground state of ^{8}Be. Here, the densities of the wave function Φ (before alignment) are shown in the left panel. The intrinsic densities of the wave function Φ^{intr} are shown in the right panel. As shown in the results of $N_b = 1$, a clear deformation and its neck structure to be called dumbbell shape appear. We can see that as the number of the Slater determinants (N_b) increases, the densities before the alignment (the left panel) becomes vaguer because of the mixture of different directions of principal axes of the basis states. On the other hand, the intrinsic densities (the right panel) have clearer dumbbell-like structure for each N_b. In addition, the density distribution of the intrinsic state is almost unchanged with respect to N_b. This result indicates the appearance of cluster structure. Since the number of particles for opposite sides of the principal axis are almost the same, this state can be considered as two α clusters. This result is similar to that of the GFMC [2].

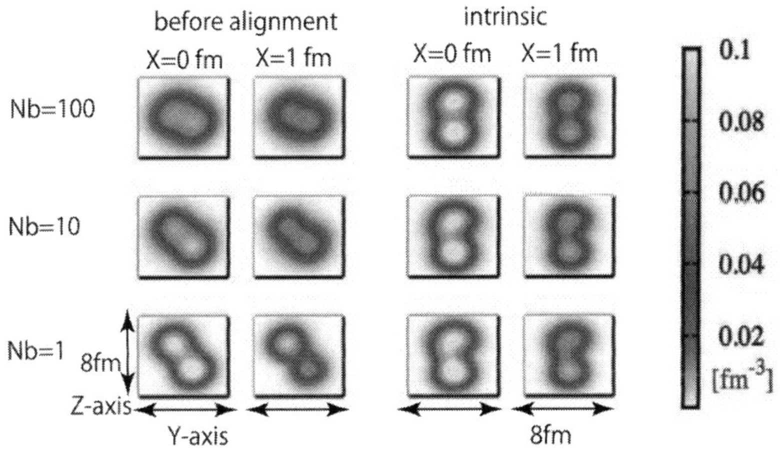

Figure 1. ^{8}Be proton densities before alignment of principal axes of unprojected basis states, $|\Phi\rangle$ (left panel) and of intrinsic state, $|\Phi^{\text{intr}}\rangle$ (right panel) for various N_b and slices along yz plane. The numbers of basis states are $N_b = 1, 10$ and 100 for the lower, middle and upper figures, respectively. Slices along yz plane are $x = 0$ fm plane (left) or $x = 1$ fm plane (right) for each panel. The size of each box is 8 fm $\times$ 8 fm. The model space and harmonic oscillator energy are N_{shell}=4 and $\hbar\omega$=20 MeV, respectively.

We perform the same analysis for the ground state of ^{10}Be. We depict the densities as shown in Fig. 2. The valence-neutron density is estimated by subtracting the proton density from the neutron density. We can see that the valence neutrons are located mainly around the node of ^{8}Be-like core. This picture is consistent with the behavior of π-orbit in the molecular orbital model of ^{10}Be [8].

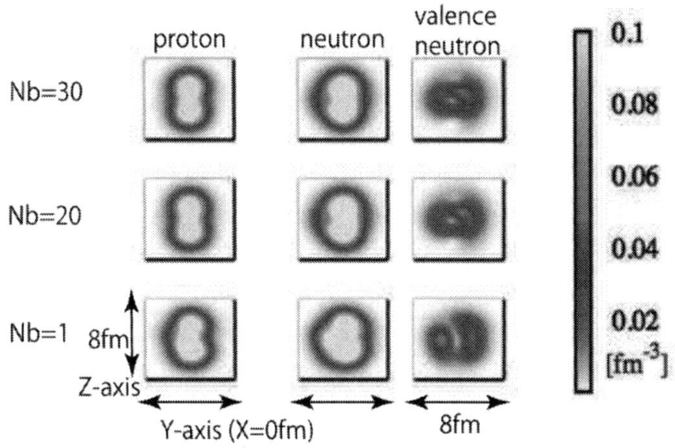

Figure 2. The densities of the 0^+ ground state for ^{10}Be are shown. The number of basis states is N_b = 1, 20 and 30 (lower, middle and upper, respectively). The model space and harmonic oscillator energy are N_{shell}=4 and $\hbar\omega$=25 MeV, respectively. The intrinsic densities for proton, neutron and valence neutrons are shown (left, middle and right, respectively).

In summary, we analyse the density distribution of the 0^+ ground state of ^{8,10}Be. We use the method by which one can extract the intrinsic structure from the wave function in the no-core MCSM. We show that the neck shape indicating the two α clusters emerges within a relatively small number of the harmonic oscillator quanta. We notice that the shape which appears in the no-core MCSM has the similar property in the GFMC in terms of the density for the 0^+ ground state of ^{8}Be. In the analysis for the 0^+ ground state of ^{10}Be, we see the two-α-cluster shape and π-orbit-like configuration of the valence neutrons. This corresponds to the result of the α cluster model. However, it is necessary to clarify whether the present analysis is valid with larger model spaces, with Coulomb interaction and without contamination of spurious center of mass motion as a future study.

Acknowledgements
This work has been supported by the SPIRE Field 5 from MEXT, and the CNS-RIKEN joint project for large-scale nuclear structure calculations. The numerical calculations were performed on the T2K Open Supercomputers at the University of Tokyo and Tsukuba University.

References
[1] Fujiwara Y, Horiuchi H, Ikeda K, Kamimura M, Katō K, Suzuki Y and Uegaki E 1980 *Prog. Theor. Phys. Suppl.* **68** 29
[2] Wiringa R B, Pieper S C, Carlson J and Pandharipande V R 2000 *Phys. Rev.* C **62** 014001
[3] Cockrell C, Vary J P and Maris P 2012 *Phys. Rev.* C **86** 034325
[4] Abe T, Maris P, Otsuka T, Shimizu N, Utsuno Y and Vary J P 2012 *Phys. Rev.* C **86** 054301
[5] Otsuka T, Honma M, Mizusaki T, Shimizu N and Utsuno Y 2001 *Prog. Part. Nucl. Phys.* **47** 319
[6] Shimizu N, Abe T, Tsunoda Y, Utsuno Y, Yoshida T, Mizusaki T, Honma M and Otsuka T 2012 *Prog. Theor. Exp. Phys.* **2012** 01A205
[7] Shirokov A M, Vary J P, Mazur A I, Zaytsev S A and Weber T A 2005 *Phys. Lett.* B **621** 96
[8] Itagaki N and Okabe S 2000 *Phys. Rev.* C **61** 044306

Institute of Physics
Dirac House, Temple Back
Bristol BS1 6BE UK

CURRAN ASSOCIATES INC
proceedings.com